2012

THE YEAR OF THE MAYAN PROPHECY

Daniel Pinchbeck

PIATKUS

PIATKUS

First published in Great Britain in 2007 by Piatkus Books
This paperback edition published in 2008 by Piatkus Books
First published in the US in 2006 by Jeremy P. Tarcher, an imprint of the
Penguin Group (USA) Inc.

A CIP catalogue record for this book
is available from the British Library

ISBN 978 0 7499 2853 7

The author gratefully acknowledges permission to quote from:
Excerpt from 'The Hollow Men' in *Collected Poems* (1909–1962) by T.S. Eliot,
© 1936 by Harcourt Inc., and renewed 1964 by T.S. Eliot, reprinted by permission of
the publisher.
'The Second Elegy', © 1982 by Stephen Mitchell, from *The Selected Poetry of Rainer Maria Rilke* by Rainer Maria Rilke, translated by Stephen Mitchell. Used by permission.
Verse by Saul Williams. Used by permission of the author.

Text design by Gretchen Achilles

Data manipulation by Phoenix Photosetting, Chatham, Kent
www.phoenixphotosetting.co.uk
Printed and bound by
Mackays of Chatham, Chatham, Kent

Piatkus Books
An imprint of
Little, Brown Book Group
100 Victoria Embankment
London EC4Y 0DY

An Hachette Livre UK Company

www.piatkus.co.uk

Love never faileth: but whether there be prophecies, they shall fail; whether there be tongues, they shall cease; whether there be knowledge, it shall vanish away. For we know in part, and we prophesy in part. But when that which is perfect is come, then that which is in part shall be done away.

1 CORINTHIANS 13:8

For my daughter

INTRODUCTION

with uninhibited fingers for the unfathomable

Our civilization is on a path of ever-increasing acceleration, but what are we rushing toward?

Anxieties are multiplying. The environment is disintegrating. The heat is rising as the ozone layer thins. Jihad faces off against McWorld in senseless wars and televised atrocities. Populations are displaced as cities disappear beneath toxic flood tides. Rogue nations stockpile nuclear arsenals. Presidents assure us they will "make the pie higher," while increasing inequity and rubber-stamping torture. Military analysts prepare for resource wars fought over water and grain; indigenous prophecies point to an imminent polar reversal that will wipe our hard drives clean. Technology advances according to the exponential curve described by Moore's Law, its unforeseen consequences following Murphy's Law. New Age cults scan the skies for extraterrestrial apparitions. Old Age religions anticipate Armageddon. The linear progress and "end of history" once promised by liberal theorists seem a hallucinatory figment of another reality. Activists and radicals, horrified by the scorched-earth effects of globalization, insist "another world is possible." But few people have any idea of what that world would be.

This book advances a radical theory: that human consciousness is rapidly transitioning to a new state, a new intensity of awareness that will manifest itself as a different understanding, a transformed realization, of time and space and self. By this thesis, the transition is already under way—though largely subliminally—and will become increasingly evident as we approach the year 2012. According to the sacred calendar of the Mayan and Toltec civilizations of Mesoamerica, this date signifies the end of a "Great Cycle" of more than five thousand years, the conclusion of one world age and the beginning of the next.

Traditionally, the completion of the Great Cycle was associated with the return of the Mesoamerican deity Quetzalcoatl, "Sovereign Plumed Serpent," depicted in sculptures and temple friezes as a fusion of bird and snake, representing the union of spirit and matter. Mexican archaeologist Enrique Florescano writes: "Quetzalcoatl is the god who hands down civilization, reveals time, and discerns the movement of the stars and human destiny." The hypothesis I propose is that the completion of the Great Cycle and the return of Quetzalcoatl are archetypes, and their underlying meaning points toward a shift in the nature of the psyche. If this theory is correct, the transformation of our consciousness will lead to the rapid creation, development, and dissemination of new institutions and social structures, corresponding to our new level of mind. From the limits of our current chaotic and uneasy circumstances, this process may well resemble an advance toward a harmonic, perhaps even utopian, situation on the Earth.

Such a radical proposition may seem absurdly far-fetched and beyond rational analysis, more fable than fact—certainly, it does not seem to match the current direction of world affairs. However, it is my view that this transition can be approached sensibly, considered in a way that does not insult our reasoning faculties. I will do my best to unpack and explain what I mean by this transformation of consciousness, giving the reader guidelines for understanding the process, as I envision it. I will argue that this quantum jump into a new context has been carefully prepared by our history, when it is viewed from a certain perspective. At the same time, our active engagement is required to make it happen. Once we see what is at stake, each of us, individually, can choose whether or not to venture the leap across the divide.

What follows can be read as an extravagant thought experiment—a kind of "no holds barred" poker wager or roulette gamble played in the realm of ideas. To take this wager, no "belief" is required, and none desired—all I ask from readers is an open mind, a critical intelligence, and the willingness to consider things in a different light. While wars, economic recessions, and natural disasters are self-evident phenomena that can be acknowledged by anyone, a change in the nature of the psyche would, at least at first, be a subtle and invisible process. Eventually, however, such a mind-shift would have results in the physical world just as real as those headline-grabbing events that capture global attention for a fleeting moment or two. After all, in only the last two centuries, the surface of the Earth has been reshaped, paved over, penetrated by human thought—thought projected into material form by increasingly powerful technologies. If we were to experience a change in our way of thinking, the world could be transformed once again.

Approaching this subject requires "uninhibited fingers for the unfathomable," to borrow a phrase from the philosopher Friedrich Nietzsche. It was Nietzsche, that solitary visionary and destroyer of false values, who noted that much of what we take for knowledge—even in science—is actually a projection of our psychology onto the world. Our thinking tends to circle around established conventions whose basis is forgotten or obscure. Nietzsche proposed that the attainment of knowledge requires a "solid, granite foundation of ignorance" for its unfolding—"the will to knowledge on the foundation of a far more powerful will: the will to ignorance, to the uncertain, to the untrue! Not as its opposite, but—as its refinement!"

As I am seeking to define a change in the nature of consciousness, there is only one medium in which this change can be registered: within consciousness itself, the mercurial domain of our subjective and personal experience—the ground of all experience, in fact. For this reason, my book follows parallel tracks: On the one hand, I seek to build the philosophical scaffolding, drawing down an imprint for understanding this transition; on the other, I follow my own process of discovery—of hesitation and pondering and sleepless nights, of shamanic journeys in desert mesas and Amazonian jungles, of investigations made in English wheat fields and ruined cathedrals of Avalon, of visionary episodes that were often as wrenching as they were revelatory, of the effort made to overcome my own deep-seated "will to ignorance" and fear of the unknown.

THESE DAYS, an approaching or imminent "End of the World" has taken on the Styrofoam ambience of a cultural cliché. Who among our liberal intelligentsia would want to find himself another Montanus or William Miller, Meher Baba or Lubavitcher Rebbe, passed-over prophets and self-proclaimed messiahs of the past, or a deluded devotee of Heaven's Gate or one of the secret flagellants of Thuringia, impatiently anticipating world-dissolving rapture or the final curtain call of God's thunderous judgment? At the same time, the Apocalypse refuses to release its grip on the profane imagination of our culture. Movies and television shows spray the message of cataclysmic world destruction, by polar shift, alien attack, flood, flame, comet, or nuclear wipeout. On the level of the collective imagination, W. B. Yeats's famous poem "The Second Coming" is perpetually rephrased:

> *Things fall apart; the center cannot hold;*
> *Mere anarchy is loosed upon the world,*
> *The blood-dimmed tide is loosed, and everywhere*
> *The ceremony of innocence is drowned;*
> *The best lack all conviction, while the worst*
> *Are full of passionate intensity.*

For some, a literal Second Coming of Christ is eagerly anticipated. Millions of Christian Fundamentalists—in the American heartland and the White House—consider themselves Rapture-ready, living out the script of the book of Revelation. Best-selling novels predict the return of a Schwarzenegger-like Messiah, mowing down the armies of the faithless. It sometimes seems that our disinterest in environmental protection and climate treaties finds its ultimate basis in this shallow eschatology. Why reduce gas mileage or preserve resources when the End of Days is on its way? Why not, instead, accelerate the process? While the actions of Fundamentalists might seem to contradict principles of their sacred texts, they believe that no harm can come to those who are already saved—even the promise of the angels in the book of Revelation to "destroy them which destroy the Earth" causes them no sleepless nights.

The world-denying ambience and transcendent emphasis of our patriarchal religions has reached a kind of fever pitch. God is thought to cast his blessings on imperialist wars, territorial violence, and acts of terrorism; to condone free market capitalism and its exploitation of resources, to uphold the sanctity of the nuclear family, while opposing abortion and same-sex unions. God seems a convenient reference point for whatever regressive movement or lockdown of possibility any pundit, pontiff, or president chooses to announce in his name—for the increasing institution of a barbed-wired world.

Those on the other side of the fence, liberals and skeptics, secular believers in science and progress, still expect that the future will follow the pattern of the past century: sleeker machines, more immersive virtual worlds, longer life spans, further ecological deterioration. They accept economists' projections of limitless growth and ignore other figures indicating depletion and devastation. Faced with terrorist threats, most are willing to forfeit a degree of freedom for a measure of security. They ignore or avoid the underlying aspects of our situation that contradict their hopes and plans for the future.

Beyond the Apocalypse and blue-eyed Jesus awaited by the faithful, another "rough beast" slouches toward us: a biospheric crisis of such magnitude that it will touch the life of every person on Earth: "We are being confronted by something so completely outside our collective experience that we don't really see it, even when the evidence is overwhelming. For us, that 'something' is a blitz of enormous biological and physical alterations in the world that has been sustaining us," writes Ed Ayres, editorial director of World Watch, in his 1999 volume *God's Last Offer,* a compelling analysis of our global plight. *God's Last Offer,* like many such tomes, offers statistical support and rational argument for sounding the alarm. Its sobering predictions were ignored by a populace hungry for sex advice, celebrity gossip, and literary distraction. Ayres identifies four accelerating developments, or "spikes," that threaten our immediate future: population growth (though slower than once projected, global population increases by eighty million each year), consumption of resources, increasing carbon gas emissions, and the mass extinction of species (a systemic loss of amphibians, birds, mammals, insects, plants, and marine life worse, by far, than the sudden crash that eliminated the dinosaurs a while back).

"Global warming and bio-extinctions are all-encompassing, and the expansions of population and consumption trigger chain reactions that cannot be stopped by shores or borders," Ayres writes. The trends create feedback loops that further increase the rate of change—for instance, climate shift makes forests more vulnerable to fires, like the vast conflagrations that have swept the West in recent years, releasing more carbon gas while further reducing the number of carbon-absorbing trees. Ayres picks apart the mechanisms of dissimulation and deception—as well as the much more powerful denial, or will to ignorance—that have obscured the massive changes now taking place around us. "It's likely that a general pattern of behavior among threatened human societies is to become more blindered, rather than more focused on the crisis, as they fall," he suggests. The current world situation is like a car whose driver responds to the fact that his gas is running out by flooring the accelerator. The naughty thrill of breaking the speed limit distracts from any worry that the ride will soon be over, or any preparation for how it might end.

Within forty years, at the current rate of deforestation, there will be no tropical forests left on the Earth. Almost all of the world's delicate ecosystems, from polar ice caps to coral reefs, are facing a similar speedy decline. Our resources are dwindling at the same pace. As the 2005 Millennium Eco-Systems Assessment, prepared by the United Nations, states: "Nearly two-thirds of the services provided by nature to humankind are found to be in decline worldwide. In effect, the benefits reaped from our engineering of the planet have been achieved by running down natural capital assets. In many cases, it is literally a matter of living on borrowed time." The UN report tells us our oceans—until recently deemed vast beyond human influence—are ninety percent fished out, with entire food chains of aquatic life disappearing.

Geologists have proposed that depletion of the Earth's fossil fuel reserves could lead to shortages and steep jumps in price within the next few years. Imminent "Peak Oil" scenarios suggest a global recession—or far worse. In the meantime, we continue to pump more than six billion tons of heat-trapping carbon into the atmosphere annually. As climate change accelerates—along with the thinning of the ozone layer, desertification, the increasing poisoning of the biosphere by industrial chemicals, and so on—almost every year sets records for the number of earthquakes, wildfires, droughts, hurricanes, and other deluges.

Corporate globalization, a system aimed at short-term profit over long-term consequences, runs a going-out-of-business sale of the Earth's resources. What the West has started, other nations may finish: Studies have shown that, due to rapid industrialization and continued population growth, China alone would need an extra 200 million tons of grain per year by 2030, an amount equal to the export capacity of the entire world in 1997—and agricultural yields are falling, due to topsoil depletion, droughts, and unforeseen consequences of industrial farming. "In addition to twenty-six or more countries that are now water-deprived, hundreds of additional regions within countries, including large areas within China, India, Mexico, and the United States, are water-scarce," Ayres writes. Water and food shortages would lead to widespread migrations and social unrest, inciting short-term authoritarian regimes, followed by global chaos: "Leaders, unable to deliver relief, would likely be toppled, and governments would lose their ability to maintain order," Ayres notes. "Millions of people would depart as refugees, spilling over borders in diasporas too large to either control or support."

The long-term health of human societies cannot be separated from the health of the biosphere. We have reached the end of that relatively short period when the momentum of mass material progress could obscure the rapidity of global decimation. If one takes the time to reflect on our plight, as Ayres and others have done, it seems likely that the structures currently supporting our civilization, such as the sovereign nation-state, will splinter along with the biosphere. In fact, it may be the case that this process is already under way. The shrill tone of the mass media and the disregard for science and fact shown by our government—the "will to ignorance" imposing a kind of knowledge blockade—foreshadow the deeper crisis just ahead. Many of us recognize the dangers facing us but refuse to consider them carefully, frozen by anxiety over our own fate and the fate of our loved ones. It is possible, however, that this dread must be overcome, or we risk bringing about the result we most fear.

To make sense of this situation, each person has to go through their own individual process, confronting their habitual mechanisms of avoidance and denial, overcoming their fear and conditioned cynicism. This process happens in stages. My own awareness of the global crisis was sharpened by a visit to the Amazon rain forest in Ecuador, several years ago. I went to Ecuador for ceremonies with the shamans of the Secoya, an ancient Amazonian tribe with a complex language and culture that was almost annihilated during the last century. From thirty thousand Secoya at the end of the nineteenth century, about 750 survive today. A parallel process to mass species extinction has sharply reduced the diversity of human cultures. Of six thousand known cultures on Earth, more than half are currently threatened with elimination.

To reach the Secoya, we traveled through areas that had been abundant rain forest only a decade before—but were now reduced to parched grassland where mestizo farmers scratched out meager livelihoods. Each vast tract of millions of acres of jungle, opened up for exploitation by the oil companies, then logged and clear-cut, yields enough oil to sate the U.S. demand for fuel for, on average, three to five days. Once gone, the rain forest does not regenerate. I had read about the cutting down of the rain forest and understood it in the abstract, but confronting it viscerally was a shocking event in my own psychic life—a personal tipping point.

When I was young, I looked forward to a future as a writer and editor in an essentially stable culture that, despite change, would endure long past my life. Literature and art seemed, in themselves, of enduring value—a refuge for subtle perceptions and subversive ideas that could influence the broader culture over time. Like most people, I accepted the concrete solidity of modern civilization and believed its institutions would remain in place. I no longer have that perspective. Mulling over the facts, considering our situation over time, I concluded, sadly, that our current civilization is not a machine built to last.

The cultural critic and mystic William Irwin Thompson thinks we are approaching an imminent "catastrophic bifurcation" of the species:

> This hominid moment is not a state that can be perfected; it is a process, and to arrest that process is probably not possible. . . . Once we were prokaryotic bacteria, then we were dinosaurs, and now we are humans about to become, through a catastrophic bifurcation, subhuman and posthuman, or God only knows what else. . . . From the greenhouse effect to the ozone hole, or from sex, drugs, and rock and roll to fundamentalist purifications, or from genetic engineering to artificial intelligence, everything we

like to call human and home, even the planet as we have known it, is being taken away from us by our own actions, conscious and unconscious.

A former professor of the humanities at MIT and practicing yogi, Thompson is one of a small number of original thinkers who not only understands our present impasse but realizes it is not the whole story. Something else is taking place as well—a sidereal movement of consciousness returning us to levels of awareness denied and repressed by the materialist thrust of our current civilization. Essential to this process, according to Thompson, is a change in our understanding of myth. We can shift "from a postmodernist sensibility in which myth is regarded as an absolute and authoritarian system of discourse to a planetary culture in which myth is regarded as isomorphic, but not identical, to scientific narratives."

But this return to myth—what could it mean? Would it mean regressing to the superstitious mind-set of the past? Does it imply rejecting the scientific view of the world and the empirical knowledge gained over the last centuries? Or could it be a movement forward toward a more holistic and integrated understanding? And to what extent are the postulates underlying our current science-based or materialist worldview actually mythical, or metaphysical, ones?

Every culture is based upon a mythological substrate that provides a particular basis for understanding and interpreting reality. The ancient Greeks gave us Zeus, Athena, Hermes, and a pantheon of quarrelsome Olympians who interject themselves into the human world as a kind of aristocratic blood sport. The Hindus embody their cosmic principles in colorful multi-armed divinities such as Brahma, Vishnu, Shakti, Kali, and Shiva—creators, protectors, destroyers, and serene contemplators. The Mesoamerican cultures personified their metaphysic in a dizzying array of figures who enfold and interpenetrate one another: Omeototl, Quetzalcoatl, Itzpapalotli, Mixcoatl, Tezcatlipoca, to name but a few. Christians worship the Trinity of Father, Son, and Holy Ghost, embodying the feminine principle in Mary as well as the long-neglected Sophia, goddess of wisdom known to the Gnostics. Is it far-fetched to suggest that the deities of our secular age include the superstrings, selfish genes, Black Holes, and

Big Bangs described by our scientists, that define the limits of the materialist worldview?

Myth imparts a structure to space and time; myth weaves a world into being. According to the perspective of the Swiss psychoanalyst Carl Jung, the structure of myth is buried deep in hidden processes of the psyche, and this structure recurs universally in individuals and across all human societies. From his own dreams, cross-cultural studies, and the material provided by his patients, Jung developed the theory of a "collective unconscious," a repository of myth, symbol, and archetype that emanates from a source beyond the individual mind. Jung described the archetypes of the collective unconscious as "spontaneous phenomena which are not subject to our will, and we are therefore justified in ascribing to them a certain autonomy." A mythological or archetypal complex—such as the Judeo-Christian Apocalypse—is, from this Jungian perspective, ultimately a psychic event that can take material manifestation, like a collective dream coming to life.

The myth-based civilizations that preceded the rise of the modern world lived according to a model of time that differs from our contemporary awareness of it as a one-way linear extension. They believed that events and epochs inevitably followed a cyclical or spiral pattern, and the development of human societies and human thought was integrated, synchronized, with the immense gyrations of planets, stars, and constellations. For the Maya and Egyptians and the architects of Stonehenge and Chaco Canyon, astronomy was a sacred science. They built their pyramids and monuments as calendars and observatories, anchoring themselves in relation to the observable cosmos.

Periods of decline and dissolution represent necessary phases within this cyclical perspective. In Hinduism, for example, our present epoch is considered to be the last of four ages—corresponding, roughly, to the Golden Age, Silver Age, Bronze Age, and Iron Age of the Greeks. It is called the Kali Yuga, named for the destroyer goddess and dark mother, Kali, usually depicted dancing on a corpse, brandishing weapons, and wearing a garland of skulls, strumpet tongue stuck out, blood dripping from bared fangs. According to the French scholar René Guénon, writing in the early years of the twentieth century: "We have in fact entered upon the final phase of this Kali Yuga, the darkest period of this dark age, the state of dissolution from which there is to be no emerging except through a cataclysm, since it is no longer a mere revival which is required, but a complete renovation." Grim as this sounds, the good news is that the end of the Kali Yuga would mean a return to the Golden Age, the Satya Yuga, when Kali's rages give way to Shakti's ecstasies. Similarly, a principle of Chinese Taoism is that "reversal is the movement of the Tao." The Taoist symbol of the yin yang shows extreme opposites containing each other in embryonic form.

The Aztecs believed that the cosmos dissolved and regenerated itself from time to time. They called this world the age of the Fifth Sun, and feared the transition to the next epoch. During the last years of their empire, they sacrificed an estimated seventy thousand people a year in order to keep the Fifth Sun alive—a textbook case of the destructive mania that overtakes empires when they decline and turn decadent. Similarly, the Hopi, an indigenous tribe of Arizona considered to be the original inhabitants of the North American continent, descendants of the Anasazi who built the monuments of Chaco Canyon, say we are currently in the Fourth World, the fourth incarnation of the Earth. According to their oral prophecies, we are approaching the transition to the Fifth World. However, in the Hopi understanding of reality, there is no past or future as we think of them: "All time is present now," and events manifest according to their predestined pattern.

While followers of New Age trends often idealize indigenous tribal cultures for their connection to natural forces, modern skeptics dismiss them—our superiority seems proven by our technical acumen, our efficiency in manipulating resources, exploring the tiniest quanta of matter, or shooting rockets into deep space. An alternative perspective withholds judgment. We might consider the ways that the current world situation—billions of people crowded on an increasingly despoiled world, facing an increasingly uncertain future—is both an improvement and a decline from the previous epoch—tribal societies with populations stabilized at a low level, living in balance, more or less, with the natural world. Similarly, the modern materialist sees myth as antiquated and simpleminded—at best, metaphoric or symbolic. We consider it a way of thinking we have outgrown. This book will argue, instead, that the mythic form of thought known to indigenous and traditional cultures remains a valid orientation to reality, even a form of knowledge we need to reconcile with our own.

Prophecy might be based, not on haphazard fantasy, but on attunement to subtler levels of reality—on forms of awareness that the modern West forfeited, temporarily, in order to develop empirical and rationalist thought. As Armin Geertz, an anthropologist of the Hopi culture, writes: "Prophecy is not prediction, even though it purports to be so. Prophecy is a thread in the total fabric of meaning, in the total worldview. In this way it can be seen as a way of life and of being."

FOR OVER A THOUSAND YEARS, the geniuses of the Mayan and Toltec civilizations that preceded the Aztecs were fixated on codifying their understanding of time, memorializing it in their stone architecture and sculpture. According to their precise calculations, the Earth was approaching a cosmic conjunction that represented the conclusion of a vast evolutionary cycle, and the potential gateway to a higher level of manifestation. As we will discuss in more detail later, this approaching milestone also marks a rare astronomical event: the alignment of the Earth and the Sun with the "dark rift" at the center of the Milky Way on the winter solstice of December 21, 2012. The particular date is less important than the Mayan perspective, as the independent Mayan scholar John Major Jenkins puts it, "that around the year 2012, a large chapter in human history will be coming to an end. All the values and assumptions of the previous World Age will expire, and a new phase of human growth will commence." He suggests that "the Maya understood this to be a natural process, in which new life follows a death."

As Thompson noted, mythic thought might be considered "isomorphic," rather than contradictory, to modern science. In the last centuries, science has exposed the elegant logic in all natural processes, from the implosion of dying stars to the intricate coils of DNA ceaselessly shimmering within each of the hundred trillion cells that glom together to make up our bodies. Unlike modern industrial manufacturing, nature does not create waste. The timing of natural processes—from the chemical signals transmitted during fetal development to the seasonal blossoming of a flower—is usually flawless.

The physicist Werner Heisenberg wrote: "Natural science does not simply describe and explain nature; it is part of the interplay between us and nature." It is an interesting paradox that as science has studied the world, attaining deeper levels of understanding, the technology produced by our scientific knowledge has simultaneously pulled it apart. The biospheric crisis we have unleashed is entirely man-made. We are the ones who have interfered with the Earth's self-regulating systems, releasing pollution, garbage, chlorofluorocarbons, depleted uranium, and other streams of toxic spew. Could this planetary emergency also be "part of the interplay between us and nature"?

Myth-based civilizations and traditional cultures believe that human beings are inseparable from natural laws and cosmic cycles. The pattern of the growth of human societies might be viewed as an aspect or expression of nature. Even the development of our individual capacities and the evolution of our self-awareness unfold from capacities contained within the potential of the natural world. We are capable of sensible, intelligent thought because nature is intelligent and sensible. The aboriginal mentality recognizes no ultimate distinction between matter and spirit, no dualistic split between mind and world. As we shall see, at least one cutting edge of thought in modern physics offers support for such a worldview.

By this logic, an evolutionary process on the level of human consciousness should be comprehensible to us, if we can stretch our minds toward it. From the perspective of the no-waste, systemic logic of nature, it would be pointless—and, indeed, a cruel joke—to engender a complex world system such as the one we currently inhabit, only to smash it to bits with no gain and no purpose. If, as my research suggests, we have reached the time of prophecy, then this process should have a logic behind it—even an empirical basis. It should conform to natural laws. Of course, this logic and these laws could be radically different from the ones upheld by the narrowly defined materialist paradigm. What might help is a perspective that is open to radical possibilities, yet at the same time is logical, sensible, and calm.

It is my thesis that the rapid development of technology and the destruction of the biosphere are material by-products of a psycho-spiritual process taking place on a planetary scale. We have created this crisis to force our own accelerated transformation—on an unconscious level, we have willed it into being. Human consciousness, the sentient element of this Earth, is in the process of self-organizing to a more intensified state of being and knowing—what the Russian mystic G. I. Gurdjieff called a "higher octave." When the Hopi talk of a Fifth World, or the Aztecs anticipate a Sixth Sun, when St. John foresees the descent of the Heavenly City or New Jerusalem, they are describing the same thing: a shift in the nature of consciousness.

The philosopher Martin Heidegger made a distinction between "world," the conceptual framework that imparts meaning and structure to our lives, and "Earth," the physical ground of our being. In his phenomenological thought, "world" represents a set of subliminal assumptions, a way of relating to history, to time and space, protected by deep-buried processes in the collective psyche. "The *world worlds,* and is more fully in being than the tangible and perceptible realm in which we believe ourselves to be at home." A world allows for certain possibilities while denying others:

> By the opening up of a world, all things gain their lingering and hastening, their remoteness and nearness, their scope and limits. In a world's worlding is gathered that spaciousness out of which the protective grace of the gods is granted or withheld. Even this doom of the god remaining absent is a way in which world worlds.

The process of transition from one world to another could be one of simultaneous creation and destruction. The "opening of a world" might be an exquisitely timed process, like the stages of fetal development, but taking place within the psyche. It might even be necessary that this process remain shrouded in mystery until its final stage. As Heidegger also wrote: "All coming to presence . . . keeps itself concealed to the last."

I FIND IT VERY STRANGE—yet strangely familiar—to be making this investigation. I grew up a secular atheist in Manhattan. As a journalist in my twenties, I drank to excess at cocktail parties and wrote magazine profiles of celebrities, shoe designers, and artists. I never imagined I would recover a mythic dimension to existence. I never expected to be fascinated by visions and dreams and synchronicities, to seriously scrutinize biblical passages and indigenous prophecies as if they were the hottest of stock tips during the dot-com boom. Often I have felt less like a person than a convenient intersection for ideas to meet and mesh, a magnet or strange attractor, compelled or fated—perhaps tragically misguided—to draw together Jungian psychology, quantum paradox, Frankfurt School critique, anthroposophy, and Mayan mythology with explorations of such seemingly outré subjects as crop circles, alien abductions, Amazonian shamanism, and the end of time. Whatever the fate of this work, I feel grateful—as well as humbled—to have received the chance to explore such awesome mysteries.

I offer this book as a gift handed backward through space-time, from beyond the barrier of a new realm—a new psychic paradigm that is a different realization of temporality, a reordering of thought that embraces prophetic as well as pragmatic dimensions of reality. My aim is to help the reader share in my understanding. But a text can only act as a scaffolding of concepts, a ladder for others to climb. Real knowledge of what I am saying must be earned, and lived, by each individual, in his or her own way.

If we were to conclude, after careful consideration, that our modern world is based upon fundamentally flawed conceptions of time and mind, that on these fatal defects we had erected a flawed civilization—like building a tower on an unsound foundation that becomes increasingly wobbly as it rises—then logic might indicate the necessity, as well as the inevitability, of change. By closing the gap between science and myth, rationality and intuition, technology and technique, we might also understand the form that change would take. Such a shift would not be the "end of the world," but the end of a world, and the opening of the next.

PART ONE

A UNIVERSE IN RUINS

Moloch! Moloch! Nightmare of Moloch! Moloch the loveless! Mental Moloch! Moloch the heavy judger of men!

ALLEN GINSBERG, *Howl*

CHAPTER ONE

Throughout my early life and into my thirties, I lacked a metaphysical view of any sort. What I had, however, was an intense and even anguished yearning that deepened as I got older. Growing up in Manhattan, a churning whirlpool of cultural and erotic distractions, I could not shake the feeling that something was absent from my life and the world around me, something as essential as it was unknown, and virtually inconceivable. I belonged to an artistic but deeply atheistic New York City milieu; my father was an abstract painter and my mother was a writer and editor who had been part of the Beat Generation in the 1950s. Both of them had rejected the religions of their ancestors, and internalized the scientific view of a world lacking the sacred. The edifices of organized religion—the gray stone synagogues and massive churches around my neighborhood, the Upper West Side—seemed like somber remembrances and archaic vestiges of the past. I found it almost unbelievable that people would still participate in that kind of worship to a god that had clearly died, as Nietzsche put it, or absconded from the scene long ago.

Culture was my religion—at least its nearest approximation. I recognized myself in Allen Ginsberg's howls, in T. S. Eliot's wasteland of fossils and fragments, in James Joyce's "silence, exile, and cunning." I felt, with Rainer Maria Rilke, the inaccessibility of the angels hovering over our "interpreted world." I identified with my father's tenacious and uncompromising struggle, waged in his cavernous SoHo loft, to pull out of paint and canvas some pure symbol of his longing for being, his awkward authenticity a stance beyond what the market could bear. I identified, also, with my mother's quiet desperation, the attention to detail in her prose an effort to rescue from life some small but enduring essence—so different from the tidal throes of the Beats, their self-mythologizing mysticism, yet connected to them by a secret desiring strain, as well as shared history.

From my parents I inherited the discipline of solitude, the habit of contemplation. They had found their identity in fleeing the conformity of their parents' world for the comparative freedoms of bohemia and the life of the mind—this escape from the ossified past shaped their worldview. They belonged to the generational movement of the postwar era, embracing modernism, shedding old conventions.

New York in the 1970s was a city of old ghosts and frayed nerves. My mother and I lived on stately, gloomy West End Avenue, in a turn-of-the-century apartment building lacking a thirteenth floor. I attended the progressive Bank Street School down the street from Columbia University—its focus was on "learning how to learn," rather than learning anything in particular. Most of our teachers were leftists with a certain slant on history. They had us read Arthur Miller's play about the Salem witch trials, *The Crucible,* several years in a row, teasing out the paranoid strain in American life that resurfaced as the "Red Scare" of the 1950s. To this day, I know no other languages, can barely do long division. My early education may have incapacitated me in certain ways, but it helped shape me as a generalist, a perceiver of pattern rather than a delver into detail.

My grandmother lived a few blocks away from my school. In her apartment with the upholstered furniture covered in vinyl slipcovers and the piano that nobody had played in decades, I spent the most stifled afternoons of my life. An ambience of irreparable loss seemed to linger over her and her two elder spinster sisters who lived upstairs. I watched chunks of the Watergate hearings while lounging on her couch. Out of some sense of civic duty, she followed that spectacle of national disgrace from opening gavel to closing bell.

Everyone I knew, adults and children as well, seemed to me to be fleeing from unspoken trauma. I suppose the trauma was history itself, its wars and diasporas, pogroms and Final Solutions—that long-playing nightmare from which Stephen Dedalus seeks to awake at the beginning of *Ulysses.*

My grandmother and her sisters had been marked by the lean years of the Depression, and, before that, by the death of their father, Samuel Rosenberg, a gentle rabbinical student from Poland who could not make it in the New World, putting his head inside a gas oven decades before Auschwitz decreed a similar fate for our relatives who stayed behind. I know nothing about them, not even their names. In the aftermath of my great-grandfather's suicide, the family burned his papers and poems, never spoke of him again. Around this repression, like an original wound, a scar tissue of secrecy and sorrow had grown. More than half a century after the fact, they remained pierced by the shame of it.

That sense of unreeling trauma was also, I now realize, my own psychological response to my parents' separation. I was five when my mother and I moved to the Upper West Side, after years of screaming fights between her and my father, caused by his shoddy and unsupportive behavior. Much later, I learned he had never wanted a child, acquiescing uneasily to the situation.

When I was eleven, I became very sick, with osteomyelitis, a bacterial infection of the bone that in my case announced itself as a sharp and unyielding pain in my lower back. I spent eight months in hospitals, much of that time encased in a body cast at a children's residential ward in upstate New York. My disease did not seem like an accident to me. It seemed like a fateful culmination. All the wrongness of things pointed toward it, like a giant accusing finger. My sojourn in the world of the sick left me feeling semidetached from my body for a very long time. I acutely felt, and feared, the nearness of death, the onset of nonbeing. My mother's first husband—another abstract painter of Irish Catholic descent—had died, drunk, in a motorcycle crash, and her anguish over that traumatic loss had condensed into anxiety that weighed on me, keeping me away from physical risk. The scoliosis I developed in the wake of my disease seemed a fulfillment of her forebodings. In my own mind, I felt the damage done to me made me unsuitable for adolescent attractions, and for a long time I feared sexual contact, as if it might endanger me by breaching my interior exile, the severity of my self-consciousness.

Growing up in New York City was a teaching in impermanence. A bookstore, a movie theater, a cafe would arrive, as if to crystallize a certain idea of culture—down my block was the New Yorker bookstore, with its worn wooden spiral staircase and its encyclopedic assortment of science fiction; another block away was Griffin, a used book store whose diffident attitude to its dusty wares seemed designed to encourage shoplifting—and then vanish like a twig carried off by the rushing torrents of the river of oblivion, named Lethe by the Greeks. My friends and I spent much of our high school years in revival houses, making diligent dilettante studies of Godard, Fassbinder, Antonioni, Woody Allen, Kubrick, adding their phrases and poses to our lexicon.

Our parents had participated in the radical shifts of the 1960s, escaping the stilted past—that past embodied, for me, in the almost luridly repressive force of my grandmother, for whom culture had stopped with Schubert's quartets and the *lieder* of Schumann. No such freeing gesture seemed possible for my friends and me. We had the permanent presentiment that we had arrived too late—there would be no new underground, no French Resistance, no Summer of Love, not even another cleansing scrub of nihilism like the Punk Era that exploded in the late 1970s and immediately collapsed on itself. The revolution—any revolution, or movement, or meaning—was over. It had ended in failure, and we had lost. The butterfly lay crushed under the tank tread. History had snapped shut its traps, and we were exiles in a time after time—the permanent three A.M. of those who "hope without hope," in T. S. Eliot:

Waking alone
At the hour when we are
Trembling with tenderness
Lips that would kiss
Form prayers to broken stone.

In place of revolution, we witnessed a repetition of gestures and pseudo-rebellions that quickly revealed themselves to be marketing strategies. We were subjected to the AIDS plague, Reaganomics, Star Wars, and Gulf War syndrome, to the endless banality of cynical glitz paraded on television as if it mattered, or made sense. We learned to avoid the dangerous white-eyed stares of crack addicts, step over armies of the homeless, and dodge brazen hostility from poor minority kids from the nearby housing projects, who mugged us out of boredom.

No film summed up the menacing and claustrophobic mood of our New York better than *Taxi Driver*; it was one of the films we returned to again and again, my friends and I. It was a touchstone for the chaos we felt around us, the frustration that permeated the air like a nearly palpable force. In the ever-tightening gyres of Travis Bickle's psychotic mind, fixated on yet denied reality by the neon-lit spectacles of porno and politics, of trash-talking human beings who were like living pieces of trash, we could not help but recognize an echo of our own predicament. We almost lived more in the movies, books, video games that we absorbed than we did in the reality surrounding us, which seemed its own dark movie, its own fable of futility.

We were humanists with little interest in science, yet science and its technological expressions were the stabilizing force, the glue holding together our drab and doomed world. Materialism seemed iron-clad; evolution told us how species arise and die out—even the sun would flare out and collapse someday; entropy was the inevitable rule bringing an end to all things. "The whole temple of Man's achievement must inevitably be buried beneath the debris of a universe in ruins," wrote the philosopher Bertrand Russell, encapsulating this modern vision. "Only on the firm foundation of unyielding despair, can the soul's habitation henceforth be built."

Our self-conscious inwardness was a peculiar epiphenomenon that happened to pop up in some complex living systems. It was a by-product of the opposable thumb, an adaptation made by some clever primates trying desperately to avoid getting eaten by fiercer creatures. We could chalk up the success of our species to the law of the jungle, genetic mutation, and the survival of the fittest. We implicitly accepted that our identity and memory, feelings and ambitions, were, as DNA researcher Francis Crick confidently proclaimed, "no more than the behavior of a vast assembly of nerve cells and their associated molecules." Or, in the assertion of academic philosopher John Searle, "It is just a plain fact about nature that brains cause consciousness."

No scientist, as of yet, had figured out how consciousness emerged in the brain—but we were assured that it was only a matter of time before that last detail was ironed out and our three-pound jelly of gray matter yielded up its ultimate secret. Our best minds were working on it round the clock. We could rest assured, as well, that there was no life after death, no continuity of soul or flight of spirit. All that was superstition. What was not superstition was what was factual, quantifiable, tactile; whether nuclear bomb, body count, or skyscraper. Oblivion was not superstition. It awaited us after the music of our allotted time ended and we sucked our last breath.

I dropped out of college after two years—Wesleyan University in snowy Connecticut was a playful paradise for some, but a combination of intellectual boredom and erotic failure poisoned the atmosphere for me—and entered the working world, using my mother's connections to secure editorial jobs in magazines. For a few years, I flourished in the hectic hothouse and pinchbeck glitter of the "roaring eighties." By trial and error, I learned the overuse of alcohol to overcome my own shy resistances and the defenses of women. I was feted by publicists, met rock stars and movie actors, was flown to Munich to discuss Bambi, banality, and the baroque with the pop artist Jeff Koons and his wife, Cicciolina, Italy's famed porn star and politician. As some kind of internal reaction to this submersion in slickness, in the new and the now, I also became obsessed with chess and played in weekend tournaments at the Marshall Chess Club on Tenth Street, its paneled chambers smelling of stale pipe smoke and male sweat, draining my brain in losing contests with weird Russian adolescents, sporting taped-up glasses and misbuttoned plaid shirts, who melted down my defenses, drip by drip, over four or five hours. Life almost seemed a kind of chess problem; office politics and relationships required strategic positioning and tactical response. I interviewed the World Chess Champion Garry Kasparov in his Fifth Avenue hotel room, preparing for a match against IBM's Deep Blue. The computer was catching up to his game, and I could feel his strain as he sought to defend his turf from the accelerated onslaught of integrated circuits that would soon rip the crown off his head. It seemed symbolic of our time: The approaching age of Artificial Intelligence was being promoted in the glossy pages of *Wired* magazine, whose pundits thrilled that our machines would soon surpass us. I stopped playing chess soon after that meeting.

Out of inherent idealism, I also started a literary magazine and small publishing company with friends while in my early twenties. I drew inspiration from the Beats and earlier modernist movements such as the Surrealists, Dadaists, and Vorticists. I had grown up with stashes of slowly yellowing literary magazines from the 1950s and 1960s stored on my mother's bookshelves, such as *The Evergreen Review, Paris Review,* and *Floating Bear,* a stapled-together, mimeographed broadsheet edited by the poet LeRoi Jones, and always loved the form. Publishing such journals seemed essential for nurturing and inspiring a burgeoning underground scene, advancing the edge of the culture. It was also, I discovered, a great way to meet girls. Through our parties, we gained a low-level notoriety—written up in *Vogue* and *New York Times* trend pieces—and an increasing deluge of submissions, almost all of them terrible.

Seeking shadowy currents of vitality, we published stories of sexual malfeasance and personal dissolution, amidst more normal fare. But even the most imaginative acts of perversion, artful cries of despair, or exquisitely rendered relationship stories ceased to thrill me after a while. Instead of defining new forms of perception, we seemed to be stuck on repeat. The works of new fiction as well as art proclaimed masterpieces by the mainstream also left me cold. I began to suspect that our culture, whether experimental or conventional, obscure or acclaimed, was only echoing past achievements, not breaking new ground.

My loss of interest in contemporary culture was only one symptom of a deeper malady—a disavowal of almost everything that had once fascinated or at least entertained me. Some crisis was announcing itself, a "dark night of the soul" carefully prepared by my history, it seems to me now, though I did not see it that way at the time. Stalking the all-too-familiar pavement of New York City, I felt I was skating across the thinnest coating of ice, and beneath that slick crust the void was waiting to claim me, to crush me in the impersonal oblivion that had terrified me as a child. A simple question confronted me—*"Is this it?"*—and kept intensifying its mocking force, whispering that my life was a lie.

I had always been prone to occasional plunges into desolation—bat-like swoops into interior caverns of the psyche where life seemed without redemption—but as I approached the end of my twenties, the darkness deepened until it obscured everything else. Sometimes I felt I was already dead, wandering in some Hades or Tibetan bardo zone where the shades repressed the disquieting thought that they were no longer alive by engaging in a make-believe *danse macabre* of frantic activity. I did not know, then, about the astrological concept of the "Saturn Return," the orbital cycle that brings, every 28.6 years, the ringed planet of melancholy and density back to the place it occupied at one's birth, known to incite, for some, an intense period of existential reappraisal—what others might call a "major bummer."

Like some embittered Dostoyevskian renegade, I stopped caring about respectability, normalcy, or the literary and journalistic worlds in which I had tried to make my name. I lost interest in the stories that my culture was willing to tell me about the world—far more dangerously, I lost interest in my own story, which desperately needed a new plot. There was also, as secret undercurrent, my familial connection to the Beats. My mother had dated Jack Kerouac in the late 1950s; she had been with him when *On the Road* came out, catapulting him to disastrous fame. She sat nervously backstage during the taping of the TV talk show where a drunken Kerouac proclaimed, "I am waiting for God to show me his face." He broke up with her, then phoned her years later, from the depths of his slow alcoholic dissolution, to tell her tenderly, "All you ever wanted was a little pea soup."

In my high school yearbook, my mother took out an ad to me with a short quote from *On the Road*: ". . . mad to live, mad to love, mad to be saved . . ." This fragment had remained lodged in my mind like some dangerous splinter. Unlike the people I knew, the Beats had taken their chances, racing across continents in search of mystical visions, "starrynight ecstasies," hitchhiking and freight-train-hopping and bebopping to the tune of their own mythic destinies. Of course my mother wasn't suggesting to me that I should live like that—was she?—she was only reminding me, for some unknown reason, that somebody had.

My friends and I also pursued the "road of excess," to a certain degree, but it did not lead to any "palace of wisdom." It led to overindulgence in heroin and cocaine and alcohol, to an obsession with alternative rock bands such as Nirvana and Pavement, to blotted-out memories of inexcusable behavior—to drastically curtailed lives for some. Ultimately, it increased rather than diminished my claustrophobia, my dread of grim fate. "Drugs"—

at least the drugs we were taking, and the way we were using them—seemed another dead end, another consumerist substitute for true feeling or communion. Mirroring our culture, we had chosen to pursue altered states in a destructive way—one that accorded with our unconscious nihilism.

In my crisis, I scanned my past for a clue I had lost, a glaring wrong-turning. What was the missing piece of the puzzle? Was there even a missing piece?

It seemed to me that there might be lost modalities of consciousness, inconvenient possibilities of being, suppressed by the swarming distractions and anxieties of contemporary life. Out of desperation, I found myself recalling the handful of psychedelic voyages I had taken at college. Those long-ago trips—along with one surprising session during a course on Zen meditation when time seemed to disappear for a while—were my only memories of entering levels of consciousness that seemed more insightful or advanced, in some ways, than my normal state. They were more educational than my classes. On mushrooms and LSD, the world seemed temporarily renewed, restored to a level of sensorial acuity and openness I last knew as a child.

Those early hallucinogenic wanderings around my golden-leaf-speckled campus remained indelibly stamped on my memory. I recalled how, after eating a fistful of dried mushrooms for the first time, I had gone into a deli and pulled a crumpled wad of green bills out of my pocket at the counter, to find myself astounded that these dirty bits of paper were enshrined at the center of our civilization's entire system of values—there was something shameful, perversely pornographic, about this. The deli workers and the townspeople we passed seemed dazed or hypnotized, wired into a fraudulent construct of work and worry they had been hoodwinked into mistaking for reality. The built environment and all of the social codes surrounding us seemed cumbersome artifices—human creations—rather than permanent fixtures. During those forays, it was as though a thick coat of shellac had been lifted off of things, returning them to their original truth, their naked innocence.

When I returned to psychedelics as an adult, I found the impressions of those early trips immediately reconfirmed. I discovered the writing of Terence McKenna, who had taken the Timothy Leary role of psychedelic proselytizer for my generation. Following McKenna's advice, I explored, eyes closed, the strobing patterns produced by psilocybin-containing mushrooms. My private explorations led me to agree with McKenna that there is a vast psychic domain—a visionary reality—available to us, if we have the courage to explore it. The explanations offered by science—high-speed synaptic firing creating patterns—seemed like pallid excuses compared to the detailed scenes and jeweled fairylands reliably invoked by the ingesting of a few shreds of dry fungus or teeny paper squares. With mushrooms and LSD, it seemed as if the substances themselves had a mind—an ancient mycelial wisdom that suspended normal linear time in favor of riddles of eternity, of creative immanence concealed within the present.

Once I had embarked on my shamanic quest, I found, like Alice, I wanted to go all the way down the rabbit hole: I went to the jungle of West Africa for initiation into the Bwiti tribe; visited shamans in the Mexican mountains of Oaxaca and the Ecuadorian rain forest; plunged into the "archaic revival" of the Burning Man festival in Nevada. I smoked the super-potent, short-acting hallucinogen nn-dimethyltryptamine (DMT)—a seven-minute rocket-shot into an overwhelming other dimension that seemed more convincing than this one—at a hotel in Palenque, and tested new psychedelics invented in modern laboratories. Each journey, each psychedelic trip, seemed to reconfirm the knowledge I had gained and to impel me further into a new understanding of the world—one that was radically different from what I once believed, or thought I knew.

Allen Ginsberg noted that the goal of the Beat Generation was "to resurrect a lost art or a lost knowledge or a lost consciousness." In pursuing my exploration of shamanism, I found I was following in the footsteps of the Beats, and met many other seekers along the way, collecting their stories. I discovered that the pursuit of a postmodern Western gnosis remained a secret current, running beneath the massive momentum of our corporate mono-culture's drive to dominate the Earth, its shuddering inhabitants and quickly disappearing resources.

According to Buckminster Fuller, evolution takes place through a process of "precession." He used the orbit of the Earth as one example of this phenomenon: The enormous gravitational force of the Sun compels the Earth toward it, where we would be extinguished in its fiery furnace; meanwhile, our escape velocity would fling us out into cold interstellar space. Perpetually pulled apart by enormous forces, we continue our stable and cyclical orbit. Over the billions of years that the Earth spins in its groove, incredible things happen—matter self-organizes into life, life self-organizes into mind. On the scale of recent events, the growth of the Internet reveals a similar process: Billions of dollars funneled into the military-industrial complex, for the creation of advanced killing machines, accidentally funds a powerful tool for advancing human knowledge and creating new social networks. Evolution and destruction enjoy dancing the tango together on the same razor's edge.

What if the evolution of consciousness, rather than an adaptive quirk of the brain, was actually the central drama, purpose and point, of our whole show? There are, we shall see, philosophers and psychoanalysts, mystics and physicists who propose that this is the case. Inner development is an eccentric process, advancing in sudden jumps, in revelatory sparks and fizzles—each person is his or her own private universe. Strengthened by suffering and crisis, consciousness does not reach a new intensity according to any predictably linear progress that can be graphed by the tools of modern science. It follows its own wayward path.

What if this deepening of awareness takes place in the margins, in the nooks and crannies of contemporary life, like a weed flowering out of the thinnest sidewalk crack? Could it be, as the somnolent masses and the professional classes press forward in the old direction, seeking the same old rewards, that the new thing self-organizes out of chaos and noise, asserts itself in fragility and silence, then takes root and vitalizes until it suddenly manifests as established truth? If something like this was the case, we would be surprised at first—even shocked—but then it might occur to us: Perhaps it has always happened this way.

OVER TIME, I realized that the shamanic terrain into which I had stumbled out of inner yearning could be of the most intense importance. Of all improbable people, I felt compelled to venture across it—this twisting labyrinth of visionary kitsch and doomsday prophecy, "machine elves" and "cosmic serpents," of Borgesian palaces of paradox and Escher-like puzzles for the mind. This was an upsetting discovery. Where were the experts in these matters? Where were the proper wizards and esteemed dialecticians of academic mastery? Surely it was not left to someone like me—a clearly deficient, half-dissolute figure, a "freelance journalist" of dubious repute—to try to make sense of all these fragments and figments, to permeate them with thought, to put them in a proper order?

As Terence McKenna put it, when he found that the entire area of psychedelic investigation, and the philosophic implications of the intense visionary flights reliably produced through these compounds, had been abandoned by the intelligentsia after the 1960s, closed like an old dusty tome and returned to the library shelf, and it was up to him and a few of his friends to blow off the cobwebs: "Nobody else wanted this." Of course, this is not exactly the case. As I will seek to make clear, a series of brilliant thinkers have explored and refined these ideas, not the least of them McKenna himself. My task has been one of assembling, contemplating, reconciling these disparate efforts—not into some reductive system, I hope, but into an expression of an implicit order that resembles an infinitely unfolding masterpiece of origami.

Hovering over this process, it would seem, gently if somewhat uncannily guiding my steps along this journey, perhaps, has been an ancient spirit and deity of Mesoamerican provenance, representing an entire complex of mythological thought: Quetzalcoatl. When I began this investigation, the name meant little to me—primarily associated with D. H. Lawrence's novel *The Plumed Serpent,* which I had read long ago and forgotten. Quetzalcoatl—the name unites the quetzal, a bird of Mexico renowned for its colorful plumage, flute-throated flitterer atop rain forest trees, and the serpent, coatl, that slinks on its belly along the Earth. Integrating what slithers, cunningly, in the dust and what soars, brightly, in the air, Quetzalcoatl as a symbol unifies perceived opposites—Heaven and Earth, spirit and matter, light and dark, science and myth. He is the god of wind and the morning star, dispenser of culture, with a special affinity for astronomy and writing and the planet Venus.

"He was the Attis, Adonis, Thammuz, Bacchus, Dionysius, Osiris, and quite possibly the Pan of the Western World," writes Peter Tompkins, in his *Mysteries of the Mexican Pyramids.* The Plumed Serpent's antithesis in Toltec myth was the dark lord of limitation, death, and human sacrifice,

Tezcatlipoca, "Smoking Mirror," associated with black magic, obsidian, and the jaguar. In Hinduism, the snake represents the spiraling life force of kundalini, the universal energy vibrating in the frequencies that make up all physical stuff, according to Erwin Schrödinger's wave equation of matter. Tompkins writes:

> Moving and undulant, the serpent in Mesoamerica symbolized life, power, planets, suns, solar systems, galaxies, ultragalaxies, and infinite cosmic space. The plumes, says Szekely, were an added symbol of the levity with which birds can overcome gravity better than other creatures. When the plumes were depicted folded within the circle formed by the snake's body, they signified matter in its latent form, potential, as it was before the creation of stars and solar systems. If the plumes were fanned out from the serpent, they represented the universe in manifestation, with all its created worlds, each plume symbolizing a basic element in the strength of nature—fire, earth, water, air.

The mystery of Quetzalcoatl deepens when one learns that, according to legend, the conquistadors under Hernando Cortés arrived in the New World on the very year—1519, "One Reed" in the Aztec calendar—prophesied for his return. This synchronicity demoralized the Aztec Empire, which quickly capitulated to the rapacious Spaniards without putting up much of a fight.

For the visionary Mayan scholar José Argüelles, Quetzalcoatl seemed "not just a god, but a multiple god; not just a man, but many men, not just a religion but a . . . mental structure." He came to see the Plumed Serpent as "not just a local affair," but "an invisible and immanent force underlying and transcending the mythic fabric of mechanization." Unlike many of the other deities in the Mayan and Aztec pantheon, whose ferociously leering visages greet us from the walls of ruined jungle temples, this cosmic creature was believed to have a compassionate affinity for the human species, to be a protector and intercessor on our behalf in the court of Omeototl, the Brahma-like creator-deity of the Toltec pantheon.

Quetzalcoatl—the alien music of the name, in itself, can lead the modern mind along strange and disparate pathways, into vaudevillian fantasies and historical pageant-pomps, if we allow it to do so. Something in the sound of it ineffably suggests a kind of trickster, a magician-jester whose pantalooned goofs conceal dead-serious intent, conjuring shaman of tin-plate vintage, wounded fisher-king or Zarathustra-like pretender, serpentine guardian and gnarled avenger, winged stalker between realms of gravity and light, always playful yet playing for keeps. To continue the riff, in imaginative forays one can sense some Quetzalcoatlian essence among ink-stained instigators, plotting revolution in the back rooms of Parisian cafes in the last days of the ancien régime; one hears in the odd syllables an atonal strain of sexual anarchism, peacock strut of the zoot-suited pimp; steely-eyed homicide detective from Fritz Lang film noir investigating the murder of reality; hint of futurist philosopher DJ sampling old memes into new epistemes; or one catches faint glimpses of the great beast flaring fantastical plumage as Dionysian Pope presiding over pre-Raphaelite fairy kingdoms. On the other extreme, when studying Quetzalcoatl's mythological place, one senses an inner connection to the Gnostic Christ, whose secret countenance, so dignified and grave, is still obscured from us by centuries of religious propaganda and the shrill pronouncements of moralizing zealots. Not to forget those human incarnations of Quetzalcoatl, wizard-kings of jungle palaces sporting serpentine headdresses—the last one enshrined in legend: tenth-century ruler of the land of Tullan. Ce Acatl Topiltzin Quetzalcoatl forbade human sacrifice and instigated a brief "Golden Age" before he was defeated by Tezcatlipoca's cynical sorcerers. Afterward, as myth has it, he wandered across Mexico as a brooding exile before unifying Mayan and Toltec cosmology in the magnificent temple complex of Chichen Itza, and finally disappeared to the West on a raft of serpents, vowing his eventual return.

But Quetzalcoatl in New York? Aztec redeemer-spirit swirling above smog-tinged Manhattan like an ethereal Chinese dragon? Quetzalcoatl biding his time as I shivered in mid-winter, walking home from grade school; or lay in my hospital bed with dripping IV by my side; or staggered shitfaced away from a preppy soiree at *The Paris Review*; or spun my baby daughter on the creaky tire swings of Tompkins Square Park? The Plumed Serpent stretching iridescent wings over the churning crowds and urban grids of our all-too-human world?—but I get ahead of things.

CHAPTER TWO

Long before I had the slightest fever dream of any feathery serpent, there were the alligators. For a glossy men's magazine, I reported on the declining human sperm count, down more than 50 percent in the past half century worldwide, apparently caused by residues of pesticides and chlorine that concentrate in the food chain. I visited a swamp in the Florida Everglades, where a dioxin spill had turned the local alligators into hermaphrodites. In a speedboat, I skimmed across the moonlit marshes at night with a biologist who was studying this brave new syndrome. He pulled wriggling baby saurians out of the water, skillfully cinched shut their jaws, tagged them, and investigated their confused nether regions. These estrogen-mimicking compounds, it seemed, were deforming beasts around the world, thinning the eggshells of birds in the Arctic Circle, as well as causing cancer and reproductive dysfunction in humans.

Back in New York, I tested my own sperm count, which was subpar. When the networks started to pick up on my story, the chemical companies parried it perfectly. They released a study that compared the sperm counts of men from Los Angeles and New York, showing New Yorkers to have a higher potency. This made for more amusing copy—the "will to ignorance" reasserting itself—and the ominous issues raised by my article were quickly forgotten.

On the proceeds from my sperm-count article, I took a trip to Nepal and India. My psychedelic visions inspired an interest in the multi-armed deities and geometric mandalas of Hinduism and Tibetan Buddhism. In Katmandu, I visited Tibetan monasteries where the deep-throated chanting and horns played by the monks seemed like some primordial timbre or tone called forth by the cosmos itself. In India, I went to Haridwar for Kumbh Mehla, a sacred festival in which millions bathe in the Ganges together, on certain auspicious days. Fearing a nightmare of crowds and bad sanitation, I almost didn't go—but, luckily, I overcame my trepidation. The gathering was surprisingly orderly, the water fresh and clean from its source in the nearby Himalayas. Flower-bedecked gurus paraded with their orange-robed followers through the ancient town. Encampments were set up like festive amusement parks up and down the banks of the river. I ended up joining the ecstatic throngs bobbing beneath the current on the last holy day—we walked all night to find ourselves facing hundreds of naked naga babas, ascetic saddhus and cave dwellers with wild beards and staring eyes, accorded rock star status during the event.

According to Hindu lore, those who participate in the festival wash away bad karma accumulated during past lives. Of course, I didn't believe in karma or reincarnation. Those ideas seemed exotic anachronisms of an archaic belief system. But my immersion in this ritual—heart of a sacred culture continuing for thousands of years—filled me with a new and unfamiliar joy.

Returning home, I deepened my study of psychedelics, learning about plants used in indigenous ceremonies and chemicals invented in modern laboratories. I tried ayahuasca—sacred potion of the Amazon Basin, introduced to modern bohemia by William Burroughs, who sought out the brew as a possible cure for his addiction to junk in the early 1950s—in an East Side apartment, guided by pseudo-shamans from California. My visions were scarce—I saw green vines in front of a waterfall and visualized, for one startling moment, how thoughts take shape in the brain like clouds of synaptic connections—but interesting enough to make me want to go further.

For a hip-hop magazine, I went to the jungle of Gabon, a small country on the equator of West Africa, to take iboga, a psychedelic root bark that is the center of the Bwiti cult—my desire to connect with some spiritual source overwhelmed any fear of malaria, Ebola, or tribal violence. Even before the trip began, I seemed to enter a zone of hyperreality. My visa from Gabon was inexplicably postponed, arriving on my doorstep just hours before my flight to Paris. In France, I learned my connection to Gabon had been canceled, and I spent the night in an airport hotel, wondering if my guide would still be waiting when I arrived. Luckily, he was there for me, and we traveled from hot, oppressive Libreville into dense jungle. Out in the shaman's tribal village before the ceremony, more money was demanded from us by the screaming Bwiti—I had paid $600 and hardly had a penny left to my name. Finally, they agreed, gruffly, to put me through the initiation anyway. They forced me to strip naked and bathe before the men of the tribe in the local stream, gave me a Bwiti costume to wear, then fed me a huge amount of vile-tasting root bark powder.

At the beginning of the night-long ordeal, while the tribe drummed and sang around me, I saw, open-eyed, a golem-like figure made of rough tree branches sit down on a bench, cross his legs, and lean forward, observing me curiously. I was later told this was the spirit of Iboga, coming to meet me. Afterward, I watched Scrabble-like letters turn in the air to spell out a curious phrase: "Touchers Teach Too"—one of a series of hints that seemed vaguely prophetic. For much of the night, I was taken on a detailed tour of my early life. Many reports of iboga trips describe such a biographical survey, though nobody knows how a complex alkaloid molecule can unlock such deep doors in the psyche, or how neurochemical reactions can create the palpable sense I had—reported by others as well—of a presence guiding me through the process.

I reviewed my childhood, confronting old terrors. I saw how my parents' split had impacted my psyche, marking me with guilt feelings of responsibility. I was shown my misuse of alcohol—after the trip, I cut down on my drinking permanently. I had heard iboga described as "ten years of therapy in one night," and there seemed to be some truth to this. Iboga was like a stern but just father figure, pointing out all of my faults. At the same time, it imparted an exhilarating sense of possibility. Despite my conditioning and the forces that shaped me, Iboga whispered to me, I was free to reinvent myself, if I could find the will to do so.

A few nights later, we attended another iboga ceremony with a friendlier Bwiti sect in another jungle village. For hours we sat around a fire, and I observed how the members of the tribe tended the flame, adding wood or damping it down at just the right moment, without uttering a word. I realized they cared for each other in the same way; this was an insight into tribal life, a shared sense of purpose, a trust and fierce pride that we in the modern world have forfeited. I felt the deep loss of it. During that ceremony, one of the shamans—a powerful jet-black-skinned man with eyes bright from eating iboga powder—said he saw my grandmother hovering over me.

"She loved you very much," he said through a translator, "but now she is dead, and she doesn't want to let you go. Her spirit is hanging over you. She is stopping you from seeing visions, from visiting the other world."

My grandmother had died recently. It did not seem accidental that the Bwiti was so specific about her, but how could he have seen this? I did not believe in "spirits." However, if there were such things, my grandmother would be the type who hung around. She had clung to life tenaciously, as if awaiting some hope that life denied her.

Hyperreality continued on my return trip to the United States: I had a one-night layover in Paris, where a friend was lending me his apartment. I walked into a crowded cafe to watch France win the World Cup on television, then wandered all night as the city erupted, in delirious fountain splashings and climbings of monuments, that seemed to me, coming out of the primordial jungle, peculiarly histrionic and unreal.

A few months after returning home, I dreamed of my grandmother rattling around my apartment, going through my things, looking for "papers." I screamed at her, throwing her out of the house. When I awoke, I felt strengthened, as though I had somehow cleared my psychic premises of a lingering ghost. This was not the only odd correspondence: Daniel Lieberman, the young Jewish botanist who brought me to the Bwiti, told me, on the way into the jungle, that he had learned, during his initiation, he wasn't going to live very long. Two years after our journey, I received an e-mail that he had died in a freakish car accident, on his thirty-third birthday, while traveling across South Africa.

The modern perspective rejects the legitimacy of psychic phenomena. We base our certainty on the materialist paradigm that insists consciousness is a manifestation, or epiphenomenon, of the physical brain. Materialism institutes a strict separation between minds, and between mind and matter. But our curt rejection of such phenomena as telepathy, telekinesis, and clairvoyance is called into question by the long history of psychical research. Although the fact is little known, psychic effects of various kinds have been demonstrated in controlled scientific experiments. The influence of directed thought causes significant statistical deviations from random variation in many areas, including casino games and experiments where images or feelings are transferred between subjects who are not in contact with each other. Dean Radin, director of the Consciousness Research Laboratory at the University of Nevada, has compiled and analyzed the statistical evidence for "psi" phenomena, presenting the data in his 1997 book, *The Conscious Universe: The Scientific Truth of Psychic Phenomena*. According to his meticulous study, thousands of experiments in telepathy, precognition, and clairvoyance have fulfilled the scientific requirements of verifiability and repeatability, indicating that these phenomena do, in fact, exist, and can be measured.

It is not just a small coterie of cranks who approve these results. The data was combed through by U.S. government panels of scientific experts, including professional skeptics, in the 1980s and 1990s. The Congressional Research Service, in 1981, concluded: "Recent experiments in remote viewing and other studies in parapsychology suggest that there exists an 'interconnectiveness' of the human mind with other minds and with matter. This interconnectiveness would appear to be functional in nature and amplified by intent and emotion." According to the Army Research Institute (1985): "The bottom line is that the data reviewed in [this] report constitute genuine scientific anomalies for which no one has an adequate explanation or set of explanations . . . their theoretical (and eventually, their practical) implications are enormous." The weight of evidence impelled the American Institutes for Research, reviewing declassified studies in psi research performed by the CIA, to recommend to the U.S. Congress, in 1995, that "future experiments focus on understanding how this phenomenon works, and on how to make it as useful as possible. There is little benefit to continuing experiments designed to offer proof."

In *The Conscious Universe,* Radin examines the underlying biases, ingrained prejudices, and multiple mechanisms that cause the evidence for psychic phenomena to be ignored and suppressed. Some of these mechanisms are psychological and some are professional—scientists, like those in other fields, contend for research grants and academic positions, and those who take on unusual or heretic causes are often rejected by the system. On the most fundamental level, psychic phenomena may not be dismissed for rational reasons, but for irrational ones. The acceptance of an interconnectivity between minds, or between mind and matter, would shatter fundamental postulates of the materialist worldview, forcing a paradigm shift. If, as successful scientists and academic philosophers such as John Searle insist, "the brain causes consciousness," and consciousness is limited to the brain, what mechanism would allow for telepathy, precognition, remote viewing, and the host of other effects that have been documented? By shielding their subjects or separating them by large distances, experiments in psychic phenomena have demonstrated that these phenomena cannot be caused by any type of physical "waves," whether consisting of electromagnetism or any other form of energy.

Faced with such paradoxes, most of us choose to accept the default setting that makes life easiest, while reducing our exposure to ridicule. The contemporary world besieges us with worries and work that seem immediately important to our success and survival. Few of us have time to make our own investigation of such abstruse realms as psychic research—and why would we bother, when our "experts" assure us they have the situation wrapped up? Without giving the subject our careful attention, we accept the materialist view that does not allow for psychic capabilities—even if we have experienced psychic events, in some form or other, in our own lives (how many men have explored the ability to stare at a woman's back until, sixth-sensing this pressure, she abruptly turns around? How many of us are tickled by the occasional synchronicity that seems to defy even the most extravagant dice rolls of probability?). This is not usually a considered decision, but a subliminal response designed to reduce our exposure to "cognitive dissonance."

The materialist view is supported not just by the mainstream media, which tends toward knee-jerk dismissal of psychic phenomena, but by a hard core of skeptical scientists who continue to assert that any evidence for psi is the result of "bad science"—even when those results are published in peer-reviewed science journals that support the same science they consider to be "good." Another response from skeptics, when faced with the statistical evidence, is to argue that, while something unusual seems to be happening in these experiments, it is not important enough to merit our attention. Radin disagrees, noting, "effects that are originally observed as weak may be turned into extremely strong effects after they are better understood. Consider, for example, what was known about harnessing the weak, erratic trickles of electricity 150 years ago, and compare that to the trillion-watt networks that run today's power-hungry world." Once we accept the reality of psychic effects, Radin suggests, we may experience a rapid evolution. Our ability to utilize these capacities—for healing, telepathy, telekinesis, and other purposes—could develop rapidly.

Before I began my study of shamanism, I was agnostic about psychic phenomena, and did not really care much about the subject. As I deepened my psychedelic explorations, psychic events seemed to multiply in my life—at the same time, my dreams, previously gray and unremarkable, became colorful tapestries that demanded my nightly attention. My initiation into the iboga cult of the Bwiti seemed to have long-lasting effects on my inner life.

The Bwiti believe that iboga opens you to the "spirit world." My own trip seemed inconclusive on that front. But after my journey to Gabon, I would often dream of people I knew who had recently died. In these dreams, they seemed lost, confused, looking to me for advice. My dream-self told them that they were no longer alive, and that was the reason for their confusion. While the modern perspective dismisses such dreams as projections, I couldn't shake the eerie feeling that I was, indeed, meeting with spirits. In indigenous cultures, shamans are believed to take the role of the "psychopomp," the hermetic messenger who guides the souls of the dead to their proper place in the underworld. Since our culture lacked shamans, my Bwiti initiation and shamanic study, perhaps, made me the closest available substitute. This is how it felt to me intuitively—however, I had no rational framework for conceiving any type of afterlife.

Like the accumulating evidence for psychic effects, study of psychedelics has been suppressed and dismissed in the modern world. Most hallucinogens were outlawed in the late 1960s, when the U.S. government feared their popularity was contributing to social unrest. Once considered extraordinary tools for studying the mind—not just by Beat poets and rock stars, but by some of the best minds in psychiatry—psychedelics were decisively shoved to the cultural margins and stigmatized. Among my literary and artistic crowd in 1990s New York, the attitude toward psychedelics was one of contempt. Tripping was once considered "mind expanding," but expanding the mind in this way was, clearly, no longer a hip thing to do. Although they were illegal, psychedelics lacked the outlaw chic and dangerous allure of heroin and cocaine—they were "hippie drugs," a "sixties thing," outmoded and obscure; at best, another way to party.

Dr. Rick Strassman, author of *DMT: The Spirit Molecule* and the conductor of the first government-approved study of a psychedelic compound with human subjects in more than two decades—testing the effects of nn-dimethyltryptamine (DMT), a short-acting hallucinogen produced by our own brains as well as many plants—believes that the medical and psychedelic professions were traumatized by the curtailed ending of psychedelic research: "The most powerful members of the profession discovered that science, data, and reason were incapable of defending their research against the enactment of repressive laws fueled by opinion, emotion, and the media. . . . Psychedelics began as 'wonder drugs,' turned into 'horror drugs,' then became 'nothing.'" By the time he attended medical school for psychiatry, in the mid-1970s, they had been almost completely excised from the curriculum. The success of antidepressants offered a different paradigm for treatment, one that fit more comfortably with the culture's underlying biases.

Much of our culture continues to reject psychedelics, which can act as psychic amplifiers, while dismissing psychic phenomena as either nonexistent or meaningless. This rejection may be based on deep-rooted psychic processes, protected by subconscious motivations. Our cultural conditioning tends to support a willful ignorance, based on the impulse to preserve the materialist worldview, and the system of values it supports, from any danger. The possibility of establishing a radically new understanding of the nature of the psyche, supporting age-old beliefs, threatens the underpinnings of a culture obsessed with acquiring wealth, goods, and status. If we were to discover that other aspects of reality deserved our serious consideration, we would have to reexamine the thrust of our current civilization; entire lives and enormous expulsions of energy could seem misdirected or even wasted. As I deepened my own explorations, as layers of conventional beliefs fell away, I found I was finally approaching the questions that had eluded me during the course of my life—questions I had not even believed I had the right to ask. Even if it required isolation from the mainstream, I preferred to sacrifice my beliefs and preconceptions, along with the comforts and status they afforded, rather than cling to a set of inherited values that I increasingly suspected to be false.

CHAPTER THREE

Friedrich Nietzsche found the will to superficiality—an embrace of the trivial and an instinctive avoidance of anything troubling, profound, or anomalous—to be a healthy impulse and innate tendency in the ordinary human psyche. He believed that this instinct was also hidden beneath most of the confident postulates of science. "Here and there we understand and laugh at the way in which precisely science at its best seeks most to keep us in this *simplified,* thoroughly artificial, suitably constructed and suitably falsified world—at the way in which, willy-nilly, it loves error, because, being alive, it loves life," he wrote in *Beyond Good and Evil.* This instinct toward the false and the flighty protected against the chance "that one might get a hold of the truth *too soon,* before man has become strong enough, hard enough, artist enough" to handle it. Nietzsche proposed that the "seeker after knowledge," the "opposite-man," was "secretly lured and pushed forward by his cruelty, by those dangerous thrills of cruelty turned *against oneself.*" The insistence on truth was "a violation, a desire to hurt the basic will of the spirit which unceasingly strives for the apparent and superficial."

A civilization relies upon a set of unconscious agreements as to what constitutes meaning and can be allowed into discourse. When faced with information that falls outside these parameters, cultures and individuals alike forget or neglect, or actively suppress, the ill-fitting data. Yet the repressed elements return to haunt us eventually, psychologists tell us, as dissociated projections of our psyche. Ultimately, the only model of reality that can sustain us is one that accounts for even the most intractable and seemingly anomalous aspects of our experience. Carl Jung wrote:

> We are living in what the Greeks called the *kairos*—the right moment—for a "metamorphosis of the gods," of the fundamental principles and symbols. This peculiarity of our time, which is certainly not of our conscious choosing, is the expression of the unconscious man within us who is changing.

Despite what we might like to be the case, we may have no choice but to become "hard enough, strong enough, artist enough" to assist in this change.

IN THE SUMMER OF 2000, I attended the Burning Man festival, in the Black Rock desert of Nevada, for the first time. In a completely flat plain of dry dust, without plants or animals of any type, ferociously hot in the day and cold at night, thirty-five thousand people construct an anarchist utopian city that stands for one week, bringing food and water with them and removing all trash at the end, returning the desert to its pristine emptiness. From pictures I saw on the Internet, I did not have great expectations for this event, which looked contrived and self-consciously freaky. But the reality of it blew me away.

A congregation of seekers, streakers, and neo-shamans, Burning Man revealed secret currents of underground progress since the 1960s. The festival celebrated an evolution of countercultural consciousness, under the radar of the mainstream. Its drum circles, parades, and spontaneous rituals reminded me of the ecstatic spectacles I had found at Kumbh Mehla. I was astonished by the sculptures and camp dioramas lovingly crafted for the festival, most of them torched at the end of the week. In New York galleries, artworks were highly prized and fetishized commodities meant to be enshrined in museums. At Burning Man, art was created to be enjoyed by the community, then released, without attachment.

During my week at the festival, I often thought of my father in his loft, surrounded by stacks of paintings that lacked an audience outside his friends. It was an old habit of mine to keep up an internal dialogue with him—about the meaning of art, the value of persistence, against all odds, in the face of the world's indifference. Burning Man seemed to define a new context for art-making, reconnecting the inspiration of the individual with communal joy. I spent the last night of the festival sitting in front of a large sculpture of a heart that was also a furnace, welded together from sheets of metal. Filled with wood, it beat bright red, shooting out sparks, spreading warmth across the cold desert ground.

When I returned to New York, I played my answering machine to discover a series of increasingly worried calls from friends of my dad's who could not get in touch with him. With gloomy foreboding, I walked from my partner's house across the boutique-strewn streets of SoHo—the vast bulk of time my father and I had spent together was in that area, which had transformed, over decades, from a quiet neighborhood of solitary artists in paint-splattered jeans and old tool and die shops to its present manic glitz—to his Greene Street loft. I climbed the stairs to his door, where an acrid smell suggested the worst. From his neighbor's apartment, I called the paramedics, who went inside to remove his body. At some point during the past days, he had died of heart failure in his bathroom. I knew that he took medicine for a defective heart valve, but he had concealed the gravity of his condition from those who loved him. The exact date of his death could not be determined—but it was likely he died during the night I spent on the cold ground, warming myself before the heart sculpture.

TWO MONTHS AFTER MY FATHER'S DEATH, I traveled to the Amazon rain forest in Ecuador to take ayahuasca with the shamans of the Secoya, an indigenous tribe. This was another plunge into hyperreality. Ecuador was a study in extreme contrasts. Its economy was in ruins and its population was desperately poor. Machine gunners stood outside the tourist spots in Quito, the capital, protecting travelers from the threat of kidnapping. We spent one night in Lago Agria, a Dodge City–like boomtown that owed its rapid development to the oil industry, where whorehouses blared tinny pop music from loudspeakers around the clock. From there, we took a boat into the lush, quiet rain forest, to meet the Secoya, the last of an ancient tribal lineage for whom the surrounding forest was a vast encyclopedia of living knowledge.

During all-night ceremonies, we drank bitter cups of ayahuasca and lay back in our hammocks. We listened to the songs, the *icaros,* that the old shamans called out to the "heavenly people," the spirits, whom we could see as flickering phantasmal forms while under the influence of their medicine. The rituals seemed to weave subtle worlds into being, carrying us into a magical reality. According to Don Caesario, the lead shaman, in the past, when the tribe had been larger and more unified, at the end of an all-night ceremony he would occasionally find an unknown seed or sapling in the palm of his hand. This would blossom into a new medicinal plant the tribe needed—a gift from the spirits. Of course, I assumed this was fable or fantasy.

The Secoya's old way of life was quickly vanishing. The adolescents of the tribe wore sneakers and Nike knockoffs. Schooled by missionaries, many of them had never participated in the ceremonies that were once an integral part of tribal life. Our guide for this journey—another Jewish ethnobotanist obsessed with tribal life—had gotten grants to preserve aspects of Secoya culture that were quickly disappearing from memory. He took us downriver to visit a re-creation of one of the tribe's original dwellings, an oval-shaped building, one enormous room made of rain forest logs lashed together with vines, where a large family would have lived together, in their old ways. Only a few of the oldest members of the tribe recalled how to make such a structure. Lying inside it, I studied its intricate lattice-like design, watched patterns of sunlight play across the ground, and heard cries of birds and distant animals. I found it one of the most harmonious and peaceful shelters I had ever entered—not a primitive hut, but an Amazonian cathedral, perfectly melding form and function. The quickly disappearing rain forest culture of the Secoya did not seem rudimentary; it possessed extraordinary subtleties and refinements.

On our shuddering bus ride back to Quito, a long pipeline for oil snaked beside us along the highway, sucking out the lifeblood of the Amazon. Ayahuasca was frequently associated with snakes—I had seen swirling serpents during my visions. In his book *The Cosmic Serpent,* anthropologist Jeremy Narby argues that the ayahuasca vine communicates the living intelligence of the natural world, sending images to us through the crystalline receptors and photons emitted by our DNA: "The global network of DNA-based life emits ultra-weak radio waves, which are currently at the limit of measurement, but which we can nonetheless perceive . . . in hallucinations and in dreams." I was struck by the conjunction of the oil pipeline, coiling out of the rain forest, and ayahuasca, the cosmic serpent, also seeking to escape from the jungle, to transmit its urgent, healing messages to the modern world.

Other correspondences compelled my attention. We had entered the rain forest at the end of October 2000, assuming that, when we exited, we would find out who won the U.S. presidential election. To our surprise, when we returned to "civilization," we found the result was still in doubt, the contest hanging by a chad.

I'd never cared for American politics, which seemed a cynical puppet show, stage-managed to preserve the status quo. In the jungle I learned that the U.S. oil company with the worst track record in Ecuador was Occidental Petroleum—known as "Toxidental" through the Amazon. Occidental had free rein through the 1990s, despite President Clinton and Vice President Gore's avowed environmentalism. From that foreign vantage point, the differences between Republican and Democratic regimes seemed insignificant compared to their overriding commitment to U.S. military might and economic hegemony. My experience of ayahuasca, shamanism and rain forest destruction, and the sudden change in the dynamic of American political life represented by the 2000 election, seemed accidentally juxtaposed in time—yet the synchronicity of these events haunted me.

In shamanic cultures, synchronicities are considered to be teachings as well as signs indicating where one should focus one's attention. Such correspondences demonstrate the usually hidden links between the individual psyche and the larger world. Synchronicities express themselves through chance meetings and natural events as well as in dreams and supernatural episodes—for instance, among the Secoya shamans, a dream of seeing yourself in the mirror suggests that you will soon encounter a jaguar in the rain forest. Shamans see such temporal conjunctions as essential aspects of reality, revealing its ultimately dreamlike and magical qualities.

According to the materialist belief system I grew up with, such correspondences can only be accidents of probability. The brain's swarms of synaptical connections automatically strive to find patterns—our survival is based on our ability to recognize the difference between woolly mammoth or saber-toothed tiger, edible plant or poisonous shrub. As a by-product of our habitual pattern-seeking, we are neurologically programmed to seek deeper meanings in a world that is, at the most fundamental level, devoid of such things. Our belief that there are "signs" hidden within the chaos of events is an age-old survival mechanism, an attempt to endow our lives with importance and avoid the existential fact of our insignificance. Although I didn't realize it at the time, deep currents of twentieth-century thought, in the disciplines of physics and psychoanalysis, suggest this materialist perspective is a flawed one.

CHAPTER FOUR

While working at a Swiss patent office a century ago, the physicist Albert Einstein proposed, in his theory of relativity, that space and time were not separate domains, but deeply interrelated. He discovered that the gravitational force of physical objects actually curved space and bent time as well, and laid out this hypothesis in elegant formulae that contradicted the core of the Newtonian worldview, in which space and time were conceived of as absolute dimensions, with no connection to each other. In place of Newton's "absolute space" and "absolute time," Einstein defined a four-dimensional space-time continuum, in which no perspective is privileged. According to the physicist Mendel Sachs, "Relativity theory implies that the space and time coordinates are only the elements of a language that is used by an observer to describe his environment." The shock of this discovery was quickly compounded by other shocks.

The roots of physics can be found in the thought of the ancient Greeks, who made inquiries into *"physis,"* the essential nature of things. As modern physicists developed the analytical and experimental tools to probe deeper into the fundamental building blocks of matter, they were surprised—at times, appalled—by what they found. They found that matter was largely composed of empty space. If you were to blow up an atom to the size of St. Peter's Cathedral in Rome, the neutron at its center would be the size of a grain of salt. The electrons whizzing around that neutron cannot be considered objects in the traditional sense of the term. They do not exist the way matter exists, but only evince "tendencies to exist."

At this quantum level, physicists discovered that their attempts to measure the phenomena they were studying affected that phenomena, which led them to realize that consciousness had to be integrated into their understanding of matter. The perceiving subject could no longer be separated from the objects under investigation. The physicist Werner Heisenberg codified this understanding in his Uncertainty Principle: With quantum objects such as photons and electrons, it was impossible to determine both their position and their momentum. If the scientist chose to observe momentum, the quantum object appeared as a wave. If the scientist chose to determine position, the quantum object appeared as a particle. But in actual fact, it was neither, or both at the same time—or it could be considered a transcendent "wavicle," only existing in the Platonic realm of ideas. "The path of the electron comes into existence only when we observe it," wrote Heisenberg. According to the physicist Niels Bohr, "Isolated material particles are abstractions, their properties being definable and observable only through their interactions with other systems."

Physicists discovered that the quantum world seemed to disregard the rules of classical physics in a number of startling ways: As probability waves spreading through space, photons and electron "wave packets" are found at more than one place at the same time, only manifesting or collapsing to a particle when an observation is made. The physicists were also confronted with "quantum jumps": electrons vanishing from one point and appearing at another, without passing through the space in between. Experiments also established quantum nonlocality or Action at a Distance: Once-correlated quantum objects remain linked even when separated by vast distances. If the probabilistic wave of one object is collapsed to make a particular observation, the other object is affected as well. The change happens immediately, with no time lag for a message to be transmitted through space, indicating that the objects are connected through a transcendent domain.

Bohr once declared: "Those who are not shocked when they first come across quantum theory cannot possibly have understood it." In the world of "quantum strangeness" revealed by quantum mechanics, time, space, and consciousness are intimately interrelated and inseparable, and there exists a higher dimension, outside our perceptions of space-time, in which everything is interconnected.

According to the Princeton physicist John Archibald Wheeler, in the universe postulated by quantum mechanics, there can be no such thing as an observer: "'Participant' is the incontrovertible new concept given by quantum mechanics. It strikes down the 'observer' of classical theory, the man who stands safely behind the thick glass wall and watches what goes on without taking part. It can't be done, quantum mechanics says it." That consciousness is embedded in the processes it perceives, continually changing them while it is changed by them, was an insight conveyed to me, and many others, during psychedelic trips.

Wheeler was responsible for one of the most mind-bending explorations of this new realm: the "delayed choice" experiment. Wheeler utilized mirrors to split a beam of light into two paths that crossed each other, and created an apparatus that could register the photons as particles that take a single pathway, or as probabilistic waves that travel both routes at the same time. Once the light beam had passed the point where it split into two, experimenters made the decision whether or not to measure the wave aspect. This choice seemed to have a retroactive effect on the nature of the beam, which still revealed itself either as wave or particle, depending on the scientist's choice. The experiment demonstrated that quantum phenomena exist only in potentia, until a decision is made, by conscious choice, as to how they are to be perceived—even if this choice is made retroactively. Although this immediate and nonlocal effect happens beyond the speed of light, Heisenberg realized it does not violate the laws of causality because no signal can be transmitted in such a way. "May the universe in some sense be 'brought into being' by the participation of those who participate?" Wheeler wondered.

The existence of a four-dimensional space-time continuum means that what we perceive as the linear direction of time is only an illusion created by our particular perspective. As the physicist Arthur Eddington put it, back in the 1920s, "Events do not happen; they are just there, and we come across them." Elaborating on this concept, Louis de Broglie wrote:

> In space-time everything which for each of us constitutes the past, the present, and the future is given en bloc. . . . Each observer, as his time passes, discovers, so to speak, new slices of space-time which appear to him as successive aspects of the material world, though in reality the ensemble of events constituting space-time exist prior to his knowledge of them.

Such a perspective is identical to the mystical or shamanic understanding of reality. It matches the Hopi perspective on time, in which "All time is present now," and events unfold according to a preset pattern. The model of space-time presented by relativistic physics is a "timeless space of a higher dimension," wrote Fritjof Capra, who found this idea echoed by many mystical traditions. "All events in it are interconnected, but the connections in it are not causal." Instead of causal connections, the model of events in quantum mechanics is probabilistic and discontinuous, defined by the perceptions of a conscious observer.

Only a few Western philosophers had made the daring leap to such a perspective, rejecting materialist dualism and linear causality for a world created by our participation in it, among them Nietzsche:

> In the "in-itself" there is nothing of "causal connections," of "necessity," or of "psychological non-freedom"; there the effect does *not* follow the cause, there is no rule or "law." It is *we* alone who have devised cause, sequence, for-each-other, relativity, constraint, number, law, freedom, motive, and purpose; and when we project and mix this symbol world into things as if it existed "in itself," we act once more as we have always acted—*mythologically.*

As Wheeler put it bluntly: "There is no space-time, there is no time, there is no before, there is no after. The question what happens 'next' is without meaning." As if anticipating Wheeler, Nietzsche also wrote: "Every power draws its ultimate consequences at every moment."

After my father's death, I inherited his life's work as well as his library. Among the books on his dusty, homemade shelves were many on art history, philosophy, and physics—these last including a number of once-popular titles comparing the findings of quantum mechanics with the basic concepts of ancient spiritual traditions. These included the 1974 best seller *The Tao of Physics* by the physicist Fritjof Capra, and *Mysticism and the New Physics* (1983) by Michael Talbot. Despite its concern with "the essential nature of things," the subject of physics had always seemed abstruse to me, and I had never pursued this area of inquiry.

The modern fragmentation of knowledge into many disciplines, each with its own specialist discourse, gives us the belief, or illusion, that we cannot attain an integrated understanding of our reality. As the poet Wallace Stevens put it, "The squirming facts exceed the squamous mind." I assumed that the concepts of physics, based on complex equations and experiments with super-accelerators, could not be reduced into ordinary language, made sensible to a generalist such as myself, without deforming them. I knew, however, that these concepts had exerted a powerful effect on my father's vision, taking permanent hold of his imagination. His early works depicted simple rectangles on a flat picture plane, calling to mind the early nonobjective explorations made by Russian Suprematist painters such as Kasimir Malevich. In his later work he was obsessed with volume and mass, painting gigantic menhir-like shapes of shimmering color that curved and fused with the fields around them. He was seeking to capture the space-bending effects of gravity, the interpenetration of diaphanous forms suggesting energy in constant transformation.

In *The Tao of Physics,* which developed out of the author's initial insights during explorations of psychedelic "power plants," Capra explores various concepts in Hinduism, Taoism, and Buddhism and correlates them with the worldview implied by the discoveries of quantum physics. He finds that these ancient traditions reflect the quantum understanding of interrelated and inseparable phenomena, of the centrality of consciousness to the world, and the unity of all appearances in a domain that transcends space-time. Although Hinduism describes a pantheon of gods, all of these gods are ultimately expressions of a single principle, Brahman, the unitary consciousness that pervades everything, "the one without a second."

In Hinduism, the world of appearance in which we live is "lila," the divine play of the gods. "Brahman is the great magician who transforms himself into the world and he performs this feat with his 'magic creative power,'" Capra writes. This "magic creative power" was given the name "maya" in *The Rig Veda,* one of the most ancient Hindu scriptures. "The word 'maya'—one of the most important terms in Indian philosophy—has changed its meaning over the centuries. From the 'might' or 'power' of the divine actor and magician, it came to signify the psychological state of anybody under the spell of the magic play."

The space-time reality we perceive from our limited perspective is lila, manifested through maya, the magical power of the gods. The transcendent consciousness, beyond all conceptualization, outside of space-time, the source of space and time, is Brahman. In the 1920s, Arthur Eddington was one of the first physicists to propose that Relativity Theory suggested "the stuff of the world is mind stuff," and that this "mind stuff is not spread out in space and time; these are part of the cyclic scheme ultimately derived from it."

Enlightenment or awakening, in the Eastern traditions, is the individual's experience of reconnecting with this ultimate ground of being, the "suchness" (*Thathagatta*) that supersedes all dualisms. As the Buddhist sage Ashvaghosha put it two thousand years ago: "Suchness is neither that which is existence, nor that which is nonexistence, nor that which is at once existence and nonexistence, nor that which is not at once existence and nonexistence." Capra points out that such paradoxical descriptions sound identical to the physicist's attempts to grasp the slippery essence of quantum objects, which are neither wave nor particle, do not exist yet do not not exist.

Capra found it interesting to follow the "spiral path" of Western science's 2,500-year evolution, from the mystical philosophies of ancient Greece, to a materialistic dualism and mechanistic worldview that was in sharp contrast to Eastern thought, and now returning to the integrated perspective of the ancients: "This time, however, it is not only based on intuition, but also on experiments of great precision and sophistication, and on a rigorous and consistent mathematical formalism." The left-brain rationality of modern science, based on objective experiment, had confirmed the right-brain intuitions of mysticism, based on subjective experience.

Materialist critics of Capra's perspective argue that quantum effects are limited to the submicroscopic quantum level, and that "macro objects," the stuff we see in our world, continue to obey the rules of Newtonian physics. This, however, is not really the case. The laws of quantum physics, its quirks and quarks and discontinuous breaks, also govern the macro level. Macro objects obey the wave equation for matter, discovered by the physicist Erwin Schrödinger, their size reducing the probabilistic spread of quantum effects to an unnoticeable amount. The Newtonian equations still maintain their effect on the macro level, as a "special case" of the new relativistic physics, but they are fuzzy approximations. The world we live in is fundamentally a quantum one.

In our daily lives, do we experience quantum effects such as nonlocality, action at a distance, and so on? Is that what the statistical evidence for psychic phenomena—such as telepathy, remote viewing, and precognition—indicates? Is it possible that synchronicity is also one of these effects, revealing a deep-buried interrelationship between time, space, and consciousness? Is consciousness itself a quantum phenomenon?

In *Synchronicity: The Bridge Between Matter and Mind* (1987), the physicist F. David Peat makes this argument. "Synchronicities are the jokers in nature's pack of cards for they refuse to play by the rules and offer a hint that, in our quest for certainty about the universe, we have ignored some vital clues," he writes. A one-time collaborator of the physicist David Bohm, who developed the "holographic universe" theory, Peat finds, in his study of synchronicities, a link between physics and psychology.

The Swiss psychoanalyst Carl Jung, in his 1951 essay on the subject, defined synchronicity as an "acausal ordering principle." He noted that these episodes multiplied in his own life, and the lives of his patients, during periods of intense psychic transformation. In Jung's autobiography, *Memories, Dreams, Reflections,* he recalls numerous such episodes—such as awakening from a startling dream in which a bullet passed through his head, to learn the next day that one of his patients had committed suicide by shooting himself in the head during the night. Elaborating on the concept, he wrote:

> Synchronistic phenomena prove the simultaneous occurrence of meaningful equivalences in heterogeneous, causally unrelated processes; in other words, they prove that a content perceived by an

> observer can, at the same time, be represented by an outside event, without any causal connection. From this it follows either that the psyche cannot be localized in time, or that space is relative to the psyche.

Studying these events, Jung found it probable that mind and matter were, ultimately, "two different aspects of one and the same thing."

For Peat, synchronicities give us glimpses into the deeper patterns of nature, indicating an organizing intelligence underlying the seeming chaos of daily events. They are a part of an order of psychic reality that we recognize, intuitively, when they occur to us, as profound glimpses or sudden revelations of the nature of mind: "Synchronicities, epiphanies, peak, and mystical experiences are all cases in which creativity breaks through the barriers of the self and allows awareness to flood through the whole domain of consciousness. It is the human mind operating, for a moment, in its true order and moving through orders of increasing subtlety, reaching past the source of mind and matter into creativity itself."

Peat wondered if these episodes, instead of occasional reaches into a different order of cognition, could become a continuous flow of synchronic phenomena, experienced collectively. If that was the case, synchronicities would suggest "an intimation of the total transformation that is possible for both the individual and society." In such a shift—which Peat equated with the "ending of time" described by various spiritual traditions—different "time orders" would be experienced simultaneously; spontaneity, synchronicity, and creativity would become the rule, rather than the rare exception. The old model of linear and mechanistic causality should give way to one more accurately based on "transformations and unfoldings." In *Mysticism and the New Physics,* Talbot similarly concluded: "We may suspect a slow and continual change of axis from causality to synchronicity." He quoted the Rig Veda: "Without effort, one world moves into another."

My personal exploration of psychedelics and shamanism had attuned my sensitivity to such episodes—and this seemed to lead to an exponential increase in their frequency. I found this, at first, to be bizarre and destabilizing. From my previous mechanistic or materialist worldview, in which all events were accidents of physical processes colliding over time, I found myself catapulted into a realm of occult correspondences where signs seemed to multiply endlessly, suggesting the possibility of a schizophrenic break, in which the mind collapses from an overload of signifiers and uncanny signals—from an oversaturation of meaning. According to Peat: "It is as if the formation of patterns within the unconscious mind is accompanied by physical patterns in the outer world. In particular, as psychic patterns are on the point of reaching consciousness, then synchronicities reach their peak; moreover, they generally disappear as the individual becomes consciously aware of a new alignment of forces within his or her own personality. . . . It is as if the internal restructuring produces external resonance, or as if a burst of 'mental energy' is propagated outward onto the world."

If consciousness somehow cocreates such events, it is only through the subjective perspective of an individual that the episodes can be observed and correlated. Like a quantum object that can be registered as either wave or particle, such temporally resonant phenomena exist only through our activity of conscious discrimination. It is our psyche that determines if an event reveals a deeper order of significance, or if it is ignored as part of the quotidian flux. I found that attuning myself to synchronicities, learning to separate signal from noise, required the development of a kind of intuitive skill, and revealed an aesthetic dimension. Meaningful synchronicities, ones that suggested some new pattern forming, seemed to occur just beyond the thought or idea I was currently holding, and were accompanied by a psychic sensation akin to a key turning in a lock. Some of these synchronicities involved conjunctions between personal episodes and world events that seemed to me both numinous and inexplicable.

CHAPTER FIVE

What appears to be the established order of present-day civilization is actually only the inert but spectacular momentum of a high velocity vehicle whose engine has already stopped functioning.

JOSÉ ARGÜELLES, *Earth Ascending*

In the spring of 2001, I agreed to publish a book-length poem by a friend of mine through our small press. The author of the book, Michael Brownstein, was a poet and novelist who had been involved in the post-Beat counterculture of the 1970s, living in Boulder and teaching literature at the Naropa Institute, founded by the Tibetan lama Chogyam Trungpa Rinpoche, with Allen Ginsberg. Michael also explored psychedelic shamanism—he had introduced me to ayahuasca, and his poem included a long section describing an ayahuasca ceremony he attended in the Amazon, where the oil wells could be heard pounding in the distance. His manuscript was too radically downbeat to be considered by any mainstream publishing house—even I had a lot of trouble, at first, digesting its contents. More than a poem, it was an impassioned outcry against the current order, looking toward an imminent future of abandoned cities and dead lands. He attacked Pasteur's germ theory of disease, decried the media's systemic suppression of information—such as reports on the dangers from electromagnetic radiation emitted by cell phones—and incorporated ideas and quotes from left-wing critics such as Vandana Shiva, Jerry Mander, and Noam Chomsky. With the fist-thumping tempo of an old-fashioned fire-and-brimstone sermon, he decried corporate globalization and the oil companies and the poisoning of the biosphere, assailing

This age of manufactured mind.
This push to transform life into products.
This culture drowning the present in the name of the future.
This heart of darkness beating its fluid into every cell.

Michael envisioned the year 2012—focus of Mayan cosmology, and an increasingly popular meme in the counterculture—as the time when global cataclysm would consume our depleted, defeated world:

Centuries ago ancient Mayans predicted world upheaval for the year 2012.
The end of their sacred calendar's five-thousand-year Great Cycle.
Has 2012 come and gone?
The future everyone secretly fears, is it already here?
And the past—did it ever really happen?

One fall morning, I finally stopped procrastinating and started to edit Michael's poetic manifesto, opening his manuscript and spreading the pages across the dining room table. My partner was in our bedroom, breastfeeding our daughter, who was not yet a month old. Outside, we heard the roar of a low-flying airplane and then a loud metallic crunch. We opened the blinds of the loft and saw a flaming crater in one of the World Trade towers, as "9-11" dialed up our current state of planetary emergency. A few moments later, the second tower was hit.

Michael's book was already titled *World on Fire.*

GROWING UP IN SOHO during the early 1970s, I recall the construction of the World Trade towers. Even from my child's vantage point, the arrival of those sleek gray megaliths seemed to suggest a new order of things. Imposing themselves on the skyline of lower Manhattan, they offered an indelible image of the postmodernist technological future our society was supposed to be racing toward, where functionality replaced funk, where the handmade ambience of old-fashioned craft was abandoned for the sterilized swank of the airport lobby. The Twin Towers also represented a shift in economic paradigm. Like a tuning fork, they beamed out the shrill frequency of the rapacious corporate globalization that went into overdrive during the next decades—the accelerating movement away from a production-based economy to the ruthless transactional logic ruled by the speculations of the financial sector, where "futures" are traded by high-speed computer, and the economies of entire Third World nations can be gutted in a few hours. Their fall seemed, also, a movement into a new order of things—or perhaps a new disorder.

It was Jean Baudrillard, the Henny Youngman of contemporary French thought, who quipped that the World Trade towers were not actually destroyed, but committed suicide: "It is almost they who did it, but we who wanted it." So much has been declaimed about this event that it seems superfluous to offer more words. However, my own impression of 9-11 was of an almost overwhelming déjà vu, an inexplicable feeling of returning or recalling something I had only temporarily forgotten. "The tactics of terrorism are to provoke an excess of reality and to make the system collapse under the weight of this excess," Baudrillard wrote. And indeed, the attack did seem like an assertion of stark reality, a stern decree from the pointing finger of fate. Refracted through the prism of so many disaster movies and mediated spectacles, 9-11 was deeply shocking, but not surprising. Such a radical negation of the prevailing system of dominance seemed somehow built into the structure itself—a necessary corrective to its global hegemony, for which the Islamic ideology of the terrorists provided only a convenient excuse.

As the towers flamed, I left my house and walked downtown toward them, past dazed survivors covered in gray ash and police squadrons, past blow-dried TV anchors set up on street corners, through crowds rushing away or collected around radios and television screens, as if the media could tell them something more essential than what they could see with their own eyes. I wanted to feel the magnitude of the disaster, to absorb into my skin its biblical proportions, as well as its stage-prop-like unreality. The fall of the towers seemed to confirm William Irwin Thompson's perception:

> Some god or Weltgeist has been making a movie out of us for the past six thousand years, and now we have turned a corner on the movie set of reality and have discovered the boards propping up the two-dimensional monuments of human history. The movement of humanism has reached its limit, and now at that limit it is breaking apart into the opposites of mechanism and mysticism and moving along the circumference of a vast new sphere of posthuman thought.

Watching the chaos, I couldn't shake the uncanny feeling that I knew about this already—that I had been, in some obscure corner of my soul, even waiting for it to occur.

September 11 was the first event to be witnessed, in real time, by billions of people across the planet. On the level of the collective psyche, this episode had, astonishingly, measurable effects. The Global Consciousness Project at Princeton University conducts one of the most well-organized and ongoing experiments to measure psi as a worldwide phenomenon. In an attempt to take a kind of psychic EEG reading of the planet, Princeton researchers put fifty random number generators in cities around the world, monitoring their constant fluctuations. It has been substantiated repeatedly in psychic experiments that the activity of human consciousness influences random number generators, such as the roulette wheels and slot machines used at casinos. The Princeton experiment reveals a similar effect taking place on a global scale.

The Princeton project documents strong deviations from normal patterns of randomness during major world events and disasters. The most extreme deviation was registered on the morning of September 11, 2001. Even more interesting is the following: Although perturbations in the pattern peaked several hours after the planes hit the World Trade Center, deviations from the norm began a few hours before the catastrophic events. Roger Nelson, the project's director, notes:

> We cannot explain the presence of stark patterns in data that should be random, nor do we have any way of divining their ultimate

> meaning, yet there appears to be an important message here. When we ask why the disaster in New York and Washington and Pennsylvania should appear to be responsible for a strong signal in our world-wide network of instruments designed to generate random noise, there is no obvious answer. When we look carefully and discover that the [data] might reflect our shock and dismay even before our minds and hearts express it, we confront a still deeper mystery.

Nelson speculates that the Consciousness Project is witnessing the early phases of the self-organization of a global brain. He writes: "It would seem that the new, integrated mind is just beginning to be active, paying attention only to events that inspire strong coherence of attention and feeling. Perhaps the best image is an infant slowly developing awareness, but already capable of strong emotions in response to the comfort of cuddling or to the discomfort of pain." As of yet, Nelson and his team do not have an analytical framework for interpreting the data they continue to compile.

The Catholic mystic Pierre Teilhard de Chardin foresaw the development of a "new, integrated mind" of global humanity, calling it the "noosphere," from the Greek word *nous,* meaning mind. Noting that our planet consists of various layers—a mineral lithosphere, hydrosphere, biosphere, and atmosphere consisting of troposphere, stratosphere, and ionosphere—Chardin theorized the possible existence of a mental envelope, a layer of thought, encompassing the Earth. The "hominization" of the Earth had concluded the phase of physical evolution, during which species multiplied and developed new powers, leading to an entropic breakdown of the biosphere. This process, Chardin realized, requiring the tapping of the stored energy and amassed mineral resources of the planet, could happen only once. When physical evolution ended, the evolving stem of the Earth switched from the outer layers to the level of cognition, developed through human consciousness, containing the entirety of our thought, as well as the planet's future evolutionary program. Chardin proposed that the noosphere would eventually develop into "a harmonized collectivity of consciousnesses equivalent to a sort of super-consciousness."

The activation of the noosphere would be predicated on humanity's realization of itself and the Earth as constituting a single organism, followed by "the unanimous construction of a *spirit of the Earth*." Chardin considered this the logical and even necessary next phase of human evolution into a fully aware and self-reflective species. In his 1938 book, *The Phenomenon of Man,* he wrote:

> The idea is that of the earth not only becoming covered by myriads of grains of thought, but becoming enclosed in a single thinking envelope so as to form, functionally, no more than a single vast grain of thought on the sidereal scale, the plurality of individual reflections grouping themselves together and reinforcing one another in the act of a single unanimous reflection.

What would it take to activate the noosphere? According to Chardin, "A new domain of psychical expansion—that is what we lack. And it is staring us in the face if we would only raise our heads to look at it."

FOR SEVERAL WEEKS AFTER THE ATTACKS, the ruins of the Twin Towers continued to smolder—we could watch the effluvial trails from our windows. When the wind changed direction, it blew a metallic-tinged, acrid, gray doom-cloud of powdered fiberglass, concrete, corpses, office products, and other detritus in our direction. Despite the government's blithe assurances about the quality of the air, we did not want to subject our daughter's tiny lungs to this clearly poisonous fog. I was also alarmed by the bioterrorism threat. Anthrax had been found in the offices of several Democratic members of Congress and in the mailroom of *The New York Times* and ABC News, the most liberal outlets of the corporate media. It seemed possible that a widespread attack was on its way. Anything seemed possible, in fact.

To escape this ominous cloud of toxic possibilities, we fled the city several times, renting a car to drive to a friend's house in Connecticut. It was during one of our emergency getaways that my partner received a call from her mother and learned that her father, a magnate who ran a successful clothing company in Germany, had died the night before. He had been stricken with cancer during the previous winter. His death followed my father's by almost exactly a year.

The magnate was one of the most impressive men I had known, and certainly the most elegant. Along with my partner's mother, he had built up his business and then sold it at the peak of its success to focus on his passion, which was collecting contemporary art. The magnate was a man of forceful enthusiasms, a great believer in science, modern progress, and free enterprise. He loved to debate the issues of the day and to manifest his will through his projects, which were usually successful. He was six feet, three inches tall, with a long and handsome face, and a tremendous sartorial flair. His wife was his perfect match; she was beautiful, imposingly glamorous, and a trained art historian. In the mid-1990s, after the wall came down, they relocated to Berlin to participate in the revival of Germany's capital. They bought and renovated a building complex in the center of Mitte, the old Jewish quarter, installing their art collection on the top floors. Above this complex, they built an extra level, with glass walls, for their own bedrooms, guest apartment, and swimming pool—the transparency seemed to the point, as the magnate and his wife were proud to present themselves to the world as model citizens. On weekends, visitors were invited to tour the collection. In order to do so, they had to take off their shoes and wear the gray felt slippers that protected the wood floors from damage.

At their home in Berlin, they threw fabulous parties, dinners, and brunches in their main room, on long tables festooned with acorns and flowers, underneath bright-painted Frank Stella sculptures that projected jaggedly from the walls. My partner's parents accepted, and almost expected, eccentric or drunkenly wild behavior from artists, knowing that artists often suffered, embracing personal chaos or indulging in intoxication, to find inspiration. They were gracious hosts, and rarely offended.

I MET MY PARTNER at a party thrown by mutual friends in the art world. In my first memory of her, she stands on the corner of the dance floor, wearing a shiny blue down coat, her head bobbing back and forth, her eyes sparkling with curiosity. She was tall and thin and highly attractive, her brown hair cut in an angular swoop around her long neck. We spoke for a while. I found her accent hard to understand or to place. She was German but had lived in France for ten years before moving to the United States. After we met, we started to run into each other constantly—she lived a half block from the office I was using to write a novel. It seemed like whenever I left my work—whether 4 p.m. or 3:30 a.m.—I would encounter her on the corner. We started to spend time together. I learned that she had only been in New York for a year, moving from Paris to take a job as an editor at an art magazine. She had a wonderful, engaging smile.

We discovered a shared enthusiasm for the Austrian writer Thomas Bernhard, the master of the ceaseless complaint. Many of Bernhard's works take the form of extended rants from isolated narrators, mathematicians or philosophers or failed pianists, unfurling their alienation in long and repetitive sentences. Our attraction to this writer was no accident. My partner and I shared a Bernhardian ambivalence about existence; while this brought us together, it also made our relationship more difficult. Our ambivalences manifested in different ways and in frequent struggles. We flickered between phases of profound complicity and mutual exasperation.

A distinct pleasure of being a bohemian outsider is that you can sometimes subvert social hierarchies. Through my partner, I was ushered into her family's world of high-tension prestige and high-end culture. Spending long stretches at their house in Berlin, I felt like an accidental barbarian who had not only gotten through the gates but somehow ended up luxuriating in the inner sanctum.

I once saw the magnate by accident, across Prince Street in SoHo. I waved at him but he did not notice me. Wearing an elegant overcoat and a black Comme des Garçons suit, he was striding down the street purposefully, his carriage erect, with his eyes fixed on the distance, as if locked in on his goal. He was every inch the empire builder. I could not help but compare his confident gait with my own slouchy and digressive style of walking—and being—my eyes ever turning this way and that, observing people, especially women, and pursuing my own reflections wherever they led. There were aspects of him that, in my own way, I tried to emulate—and others I only wished that I could.

Only a few weeks before he died, we had gone to Berlin for the wedding of my partner's brother, a genetics researcher, who had compressed his engagement so that the magnate could preside over the nuptials. The ceremony gave us the chance to introduce him to his first grandchild. Sitting at his desk overlooking the church towers and rooftops of Mitte, he held her in his arms, and wept.

The wedding was held at the Alte Nationalgalerie, on Berlin's museum island, in the dark-wood-paneled main hall, just restored and not yet open to the public. During the service, the magnate rose carefully from his wheelchair to read a passage from the Bible, in German and then in English:

"Though I speak with the tongues of men and of angels, and have not Love, I am become as sounding brass, or a tinkling cymbal," he read, his voice trembling. "And though I have the gift of prophecy, and understand all mysteries, and all knowledge; and though I have all faith, so that I could remove mountains, and have not Love, I am nothing. And though I bestow all my goods to feed the poor, and though I give my body to be burned, and have not Love, it profiteth me nothing."

Although he read so movingly, he considered the text to be nothing more than great literature, expressing humanity's irrational and absurd yearning for transcendence. Like many men of his generation—like my father as well—he believed that science had disproved God, as well as mysticism, once and for all. When the magnate was healthy, I had been unable to impress him with my radical views on corporate globalization and shamanism. In the months before he died, I also failed in my fumbling attempts to open his awareness to the possibility of other realms of being, of bardo states and spirit guides.

One morning, a few years after the event, I suddenly recalled that wedding, and wanted to review the passage he had read that day. Being a biblical illiterate, I didn't know where to find it. When I checked my computer a few hours later, I found that a friend of mine had just sent me the entire text via e-mail—as if by noospheric delivery service. The passage was from St. Paul's Letter to the Corinthians. My friend, the daughter of a Protestant theologian, later told me she awoke in the middle of the night and felt compelled to do this—as if an invisible hand were pressing the back of her head. She wrote: "I don't think I have ever quoted scripture to anyone"—and I can't recall receiving such a scriptural message, before or since. The famous passage of 1 Corinthians 13:8 continues:

> Love never faileth: but whether there be prophecies, they shall fail; whether there be tongues, they shall cease; whether there be knowledge, it shall vanish away. For we know in part, and we prophesy in part. But when that which is perfect is come, then that which is in part shall be done away. When I was a child, I spake as a child, I understood as a child, I thought as a child: but when I became a man, I put away childish things. For now we see through a glass, darkly; but then face to face: now I know in part; but then shall I know even as also I am known.

I recalled the passage was about love, but I didn't realize it was also about prophecy—and its limits. My friend wrote in her e-mail: "I greatly admire your willingness to bear witness to your experiences and beliefs in such a radical and generous way. I will also say that I think the role of truth-bearer requires the purest of intentions. 'Do it with love' is good advice."

Although my father met the magnate only once—at an Easter brunch organized by my partner, they sat side by side on the couch, discussing art and current events—in my mind they are linked, not only by their deaths a year apart, but by their lives. Although one was a wealthy entrepreneur and the other an impoverished artist, on a deeper level, they were like long-lost brothers. They were archetypal embodiments of the twentieth-century male—driven and proud, unyielding in their will and prodigious in their efforts. They were progressives, existentialists, with attitudes to life shaped by hard childhoods spent in the ruined landscapes of postwar Europe. Accepting modernism and the scientific worldview, they had rejected not only religion but any possibility of access to what the Austrian visionary Rudolf Steiner called "supersensible knowledge." When I recovered the shamanic dimensions, I passed beyond the limits of thought that their lives and conditioning had imposed on them. To me, the conjunction of their deaths represented the closing of an era.

PART TWO

THE SERPENT TEMPLE

Does it really exist, this destroyer, Time?

RAINER MARIA RILKE

CHAPTER ONE

The English writer Patrick Harpur defines a "daimonic reality" of mercurial subtlety and subversion. The daimon, according to Harpur, takes pleasure in upsetting our logical categories, sabotaging our sciences, and collapsing our belief systems. By doing this, he represents the *anima mundi,* the soul of the world, which evades definitions and categories. "The soul prefers to body itself forth—to imagine—in personifications," he writes. "The Neoplatonists call these daimons. They are neither gods nor physical humans but inhabitants of the middle realm—nymphs, satyrs, djinns, trolls, fairies, angels, etc." These tricksters are not transcendent, hovering above us in some higher realm, but immanent, interwined with the Earth and our psychic life.

The daimonic realm, in between spirit and matter, fantasy and fact, is the realm of the soul—the third term, "the imp of the perverse," that undermines any dualism or dialectic:

> The world can be cut any way you like, but the inclination of Western culture has been to cut it in two: spirit and matter, mind and body, subject and object, God and Nature, sacred and secular and so on, it seems, forever. But the Platonic tradition describes a third realm, which neither Christianity nor science allow. This third realm, both mediating between the two halves of the world yet maintaining distinctions between them, is the realm of soul. It is neither spiritual nor material, neither inner nor outer, etc., but

> always ambiguous, always both-and. It is a "subtle" or "breath-body" which, Proteus like, can take on any shape.

According to Harpur, our attempts to banish the daimon only guarantee his return in some new guise, perpetrating some new form of mischief: "Daimons tend to disregard causality just as they ignore other laws, such as space and time, that we are pleased to impose on a world whose reality is quite otherwise," he notes. The Greek god Hermes—Mercurius, in Latin—conveying messages between gods and men, is the most daimonic of deities. "Mercurius is both earth spirit and soul of the world, volatile and fixed, Above and Below, psychic and hylic." This trickster god of thieves and liars, writers and alchemists, effortlessly slipping between words and worlds, between planes of reality and illusion, can be "harmlessly impish or seriously diabolical."

Through my shamanic explorations, I learned the hard way that the imagination can take on a palpable life of its own, producing daimonic effects I would never have considered possible, if I hadn't encountered them. In the spring of 2001, I field-tested a little-known psychedelic, dipropyltryptamine (DPT), a laboratory-created substance chemically similar to DMT. During our trip, my friend and I felt we were under psychic assault from imperious imps. We projected through purple and violet realms that appeared fully rendered in our visions, modulated by the moody techno music playing on the stereo. We explored decadent Edwardian mansions, jittering in hyperspace, where orgiastic banquets were in full swing. We encountered arrogant incubi that seemed to be the local inhabitants, rising around us like twirled smoke, unfurling bat wings and talons, mocking us for crashing their glamorous, incorporeal scene without an invite.

Although the drug wore off after several hours, my trip did not end all at once. For several weeks afterward, I picked up flickering hypnagogic imagery when I closed my eyes at night—honeycombed surfaces of distant planets, small dinosaurs darting through primeval jungles. In one scene, I entered a column of fire rising from the center of Stonehenge again and again, feeling myself pleasantly annihilated by the flames each time. It was as if I had inadvertently popped open a portal into another order of being, and the door didn't want to close.

I could not seem to get free of the grip of one of these phantasms, demons, or djinns. This entity—or thought-form or energy—haunted my dreams, always appearing as a mocking man in black beard and dark velvet suit, sometimes pursuing me through strange cities, or pounding me over the head with a pillow while giggling manically. In one dream, my stalker brought a crowd of rowdy friends into my house, stealing books and trashing the place. He told me, "I used to live here."

"Do you want to come back?" I asked him.

"Yes, I do," he said.

In other dreams, I sensed the presence of Iboga as a protective force. Once, my partner and I were being followed by my black-bearded antagonist in a European city. We entered a restaurant run by a large African woman that seemed a safe haven. In another dream, I visited a press conference for the World Heavyweight Champion Evander Holyfield, smiling confidently, before a boxing match. My daimon appeared as his opponent; a kind of cartoony Tasmanian Devil, black-colored, full of insolent rage, its arms and legs pinwheeling madly. I seemed to be identified with Holyfield's self-contained strength; I awoke feeling confident I could defeat or dispel my astral nemesis.

While my partner was in Berlin, I experienced poltergeist phenomena—mirrors falling off walls, strange bugs appearing in drawers—around our apartment. I had no experience with such things—and did not "believe" in the occult at all. Yet these apparitions seemed unambiguous signs. Luckily, my fellow tripper was something of a witch. She lent me a large black obsidian ball she had carried with her from Mexico. Obsidian, in shamanic practice, is thought to absorb negative spiritual energies.

Alone at night, I sat in front of this opaque orb, doing meditations. While concentrating, I sought to pull in whatever psychic energy had been unleashed through our journey. At one point, the obsidian ball, and then my entire field of vision, turned completely gray, as if covered in fog. I looked away, and my vision returned. I looked back at the ball, and the world turned gray once again. Eventually, I felt I reintegrated whatever shroudy adversary or aspect of my own ragged subconscious was struggling against me. For months afterward, I noted an upsurge of synchronicities, as well as other events—some of them horrific—that I felt, intuitively, were catalyzed by my unprepared, unprotected breach into arcane realms. At the same time, I felt significantly more intelligent, sharper, attuned to new patterns of information and ideas that had eluded me before.

I do not hope to convince anyone who has not had such experiences of the actuality of this episode. For me personally, the fallout from this journey caused a shocking expansion in my ideas of what is possible. I had been wrenched open and initiated into imaginal realms, attaining occult knowledge. I gained a deeper understanding of what Carl Jung meant when he insisted upon "the reality of the psyche," as well as William Blake's claim that "The imagination is not a State: it is the Human existence in itself." Stories I had previously dismissed as specious fantasy—such as the Secoya's description of creating magical plants through their seances, or the descriptions of Christ's miracles in the Bible—seemed far more plausible. I began to study Western occult systems and Qabalah. I also realized that the heedless pursuit of psychedelic gnosis had dangers to it, on a psychic and spiritual level, that I had not conceived of before. I altered the contents of the book I was writing to reflect this new understanding, approaching the subject with a new level of caution.

A deep level of connection between mind and matter seemed indicated by the episode, revealing the daimonic reality, of trickster subversion and in-betweenness, with "attributes of rapidity, luminosity, elusiveness" described by Harpur. Nothing in my personal history or subconscious seemed to prepare the way for such an encounter. The initial visions had been seductively Luciferic—as if referencing a whole realm of decadent art and literature, occult visions of Aleister Crowley and Joris-Karl Huysmans and Charles Baudelaire, that I had never taken seriously.

I continued to feel the presence of daimonic forces around me in a way I had never imagined before. Our world seemed a crude and grinding place compared to their subtle emanations. The pirouetting trail of smoke from an incense stick, the graceful unfolding of an iris, a peacock feather—such phenomena seemed to draw that other domain closer to this one. I saw how corporate symbols and signs—whether Nike swoosh, McDonald's arches, or policeman's badge—drew their energy from a negative occult realm, binding us in a lackluster paradigm that put prison bars around our possibilities. We were not innocent victims, however, but willing collaborators in this lockdown.

"THE STUPIDITY OF MORAL INDIGNATION," wrote Nietzsche, that fount of *ubermensch* wisdom, "is the unfailing sign in a philosopher that his philosophical sense of humor has left him."

In the winter of 2002, I fell prey to that stupidity. Obsessed with the global decimation that my visit to the Amazon had imprinted on my retinas, reinforced by the dull drizzle of despair falling from the daily newspapers, and the planetary emergency announced by 9-11, I turned glum. As I looked deeper into the global ecological and economic abyss, I feared I had made a big mistake bringing a child into this melting-down mess of a world. My reflections took lurid form when I went to sleep. In dream after dream, dread followed dread. The Earth was trashed, zapped by aliens, smacked by comets, sucked dry, blown to bits, dissolved in cartoon frenzy. The fallout from my forced entrance into Western occultism also intensified my anxiety—but I kept that experience under wraps.

When I tried to discuss the global situation with friends or acquaintances—"Oh by the way, have you ever checked out the *World Scientists' Warning to Humanity*? They seem to suspect the planet's life-support systems are approaching the point of collapse"—they tended to act like I was committing some inexcusable faux pas. As fast as politely possible, they returned to chattering about the latest art exhibit, romantic gossip, movie plot, or thrilling vacation plans. Some never spoke to me again. Others accused me of lecturing them—and, alas, I was. My partner was sympathetic at first, but soon found my continued harping on apocalyptic themes to be unbearable. I didn't disagree with her, but I couldn't stop. The more I tried and failed to open such discussions, the more I sulked.

Herbert Marcuse's *One-Dimensional Man* (1964) remains a surgical analysis of the fundamental "irrational rationality" of our system. He argued that industrialization and mechanization could—and logically should—have led to a reduction in labor time and the institution of a post-work and post-scarcity global society after World War II. The alternative to a conscripted social reality would be one that gave us new freedoms—freedom from work, freedom from propagandizing media, freedom to create and explore our own realities. The response to this deep threat to the controlling apparatus was the creation of "false needs" in the consumer, the perpetuating fear of nuclear war and terrorism, and the use of the mass media to enforce consensus consciousness:

> The union of growing productivity and growing destruction; the brinkmanship of annihilation; the surrender of thought, hope, and fear to the decisions of the powers that be; the preservation of misery in the face of unprecedented wealth constitute the most impartial indictment—even if they are not the raison d'être of this society, but only its by-product: Its sweeping rationality, which propels efficiency and growth, is itself irrational.

The great mass of humanity forfeits their inner freedom of thought, conscience, and will to participate in this system, which presents itself as inevitable, inescapable, and airtight.

Marcuse wrote: "Perhaps an accident may alter the situation, but unless the recognition of what is being done and what is being prevented subverts the consciousness and the behavior of man, not even a catastrophe will bring about the change." I kept thinking of this sentence in the wake of 9-11. For a few weeks after the attacks, the mood in the city was different—compassionate, contemplative. People gathered in public parks such as Union Square for all-night vigils and heartfelt exchanges. There seemed a potential for a genuine opening of awareness, but this new space of possibility quickly vanished. The police closed down the parks and ended the vigils, stripping away the homemade banners and artworks and candle-strewn memorials that had festooned the lawns and park fences.

As a German philosopher writing in the aftermath of the Nazi regime, Marcuse understood the sleep-inducing force of indoctrination, its power to make people forget and forfeit their own real interests. "The fact that the vast majority of the population accepts, and is made to accept, this society does not render it less irrational and less reprehensible," he wrote. "The distinction between true and false consciousness, real and immediate interest still is meaningful."

In the months after the attacks, the media repetitively droned the stereotyped jargon of the government, enforcing a consensus trance of nationalism and revenge fervor. Shopping was promoted as a patriotic act. Stopping to think about our current world, and America's role in it, was not. For Marcuse, this incessant bombast acts as a kind of hypnotic spell, compelling conformity without any need for introspection: "The new touch of the magic-ritual language . . . is that people don't believe, or don't care, and yet act accordingly." Soon, every deli seemed to have a framed photograph of the Twin Towers by its cash register, as that oversized tuning fork was elevated to the status of a religious icon.

The necessary movement into a truly rational and compassionate civilization would require entirely new systems—not a renovation, but a complete transformation. "Organization for peace is different from organization for war; the institutions which served the struggle for existence cannot serve the pacification of existence. Life as an end is qualitatively different to life as a means," Marcuse wrote. He realized that, with the increasing power of technology, the human imagination—rather than any abstract "necessity"—had become the determining force in creating social reality. "In the light of the capabilities of advanced industrial civilization, is not all play of the imagination playing with technical possibilities, which can be tested as to their chances of realization? The romantic idea of a 'science of the imagination' seems to assume an ever-more-empirical aspect." But in our present system, which Marcuse dubbed "totalitarian democracy," the imagination is denied its rightful place. Or, to put it another way, the imagination determining our social reality is fixated on dominance, security, the death-dealing powers of the military apparatus—on various projections of our shadows.

Technical efficiency is value-neutral. It could be applied systemically, to liberate humanity and heal the world. Writing before the environmental crisis entered its acute phase, Marcuse could see no pragmatic means to shift postindustrial civilization from its doom-orientation to a happier one. "The critical theory of society possesses no concepts which could bridge the gap between the present and its future; holding no promise and showing no success, it remains negative." He concluded with a quote from Walter Benjamin: "It is only for the sake of those without hope that hope is given to us."

From a shamanic perspective, the psychic blockade that prevents otherwise intelligent adults from considering the future of our world—our obvious lack of future, if we continue on our present path—reveals an occult dimension. It is like a programming error written into the software designed for the modern mind, which has endless energy to expend on the trivial and treacly, sports statistic or shoe sale, but no time to spare for the torments of the Third World, for the mass extinction of species to perpetuate a way of life without a future, for the imminent exhaustion of fossil fuel reserves, or for the fine print of the Patriot Act. This psychic blockade is reinforced by a vast propaganda machine spewing out crude as well as sophisticated distractions, encouraging individuals to see themselves as alienated spectators of their culture, rather than active participants in a planetary ecology.

"What is happening to our world is almost too colossal for human comprehension to contain. But it is a terrible, terrible thing," lamented Arundhati Roy, the Indian novelist-turned-activist, who documented and protested the enormous dam projects in India, orchestrated by the World Bank, displacing 30 million people from their homes, with little tangible result beyond the enrichment of multinational corporations and an increase in India's debts. "To contemplate its girth and circumference, to attempt to define it, to try and fight it all at once, is impossible. The only way to combat it is by fighting specific wars in specific ways." But among the people I knew in New York, there was little contemplation of the situation, and no courage, anger, desire, or will to fight against it.

"Kick a daimon out of the front door and it returns by the back," Patrick Harpur wrote. Perhaps the daimon was still around, but asserting itself through our collective capitulation to the powers that be? Harpur also noted: "The only way to get rid of anima is to bore her. In this case she manifests herself as absence, as loss of soul in the dead mechanical language of theorists who are bent on defining and categorizing what cannot be treated in this way." Is "loss of soul," absence of affect, the form taken by the daimon in our materialist world? Does anima express herself through our listless disinterest and shrugging disregard for the future of this planet?

The eighteenth-century mystic William Blake believed humanity was bound in "mind-forg'd manacles" of its own devising. Writing at the dawn of the Industrial Revolution, he decried this restricted vision that denied the primacy of the imagination, and foresaw its inevitable results with mystic rage denied later commentators:

> Thought *chang'd the infinite to a serpent, that which pitieth:*
> *To a devouring flame, and man fled from its face and hid*
> *In forests of night; then all the eternal forests were divided*
> *Into earths rolling in circles of space, that like an ocean rush'd*
> *And overwhelmed all except this finite wall of flesh.*
> *Then was the serpent temple form'd, image of infinite*
> *Shut up in finite revolutions; and man became an Angel;*
> *Heaven a mighty circle turning; God a tyrant crown'd.*

If thought changed the "infinite to a serpent," then it is the task of thought to change it back again, to smash the serpent temple before the last of our eternal forests are divided and devoured.

For this reason, perhaps, to assist in this process, Quetzalcoatl, airborne leviathan, flighty behemoth, his phantasmal feathers and shimmering scales woven from chaordic mirror shards of maya—"the Beast that was, and is not, and yet is"—makes his return at this time.

CHAPTER TWO

As I considered the psychophysical effects I had, apparently, unleashed by my shamanic exploration, I wondered if other, more consensual areas of inquiry might indicate a similar interconnection between "mind-stuff" and matter—if there was some public way that the daimonic reality chose to express itself in our time. Through the Internet, I discovered the crop circles, geometric patterns appearing in agricultural lands across the world, in Europe, Eastern Europe, the United States, and South America, though especially concentrated in the UK. Like most people, I had assumed they were junk—a long-running hoax designed to fool the gullible community of UFO fans and New Agers, grist for late-night exposés on trashy cable TV shows. The crop formations had developed into a far more extravagant, even awesome, phenomenon than I imagined. I spent many hours reading geometrical and biophysical studies of the formations, testimonies by hoaxers, extraterrestrial hypotheses, and personal accounts of the patterns, studying pictures of them.

On August 12, 2001, the largest crop circle ever recorded had appeared in southern England, on Milk Hill, in the county of Wiltshire, near the Avebury stone circles. The size of two English football fields, the formation imprinted thirteen spirals of circles of decreasing size, 409 total circles, radiating out symmetrically from a center point. The precisely executed design was spread over nine hundred feet of a gently rolling hillside. Despite the unevenness of the ground, the massive spiral appeared perfect when photographed from above. Among the small community who studied the formations, the Milk Hill spiral was considered an emphatic culmination of several decades of patterns that steadily increased in size and complexity.

Two days later, on August 14 (the date, I noted, of my daughter's birth), a rectangular "crop circle" was found in a wheat field beside the Chilbolton radio transmitter, used to transmit and receive signals from space, in Hampshire. When photographed from above, this glyph was shown to reveal an enigmatic face, executed in precise halftones, like a photo in a newspaper. No image in any way like it had been seen in the fields before that date.

A few days later, a second rectangular glyph appeared in the fields beside the Chilbolton transmitter. Once again flawlessly rendered in twisted and fallen crop, this image appeared to offer a direct response to a message broadcast by SETI (the Search for Extraterrestrial Intelligence) in 1973, to the star cluster M13. Less a serious attempt to contact aliens than a publicity-seeking gesture orchestrated by the popular science writer Carl Sagan, the original message from SETI included a little human stick figure, the schematics of our DNA code, the atomic building blocks of life on Earth, and the planetary configurations of our solar system. The message returned in the Chilbolton field included a smaller humanoid figure with a larger head—similar to "the visitors" described in numerous alien abduction accounts—and a slightly altered solar system. A third strand had been added to the DNA coil. The atomic elements making up the extraterrestrial life-form were expanded to include the element silicon. These formations, despite their spectacular virtuosity and startlingly suggestive content, were not reproduced or discussed anywhere in the U.S. media—where the phenomenon had long been dismissed, unreflectively, as the work of bored teenagers or drunken farmers with boards. If I hadn't sought them out on the Internet, I would never have known such images existed.

Of course I realized the subject might be deserving of ridicule, but I also felt that I couldn't know that was the case unless I investigated for myself. My exploration of shamanism and psychedelics—subjects also dismissed and ridiculed in knee-jerk fashion by the mainstream media—opened me to the possibility that there could be something happening in those fields that demanded careful consideration. At the very least, the photographs on the Internet did not suggest an amateur operation, but a highly skilled one. If the phenomenon was orchestrated by teams of human circlemakers in different countries, it would be the largest-scaled, most significant project of anonymous land art known to the world, deserving attention and acclaim on that score. If it was not entirely created by humans, the formations would clearly be of tremendous significance, compelling a paradigm shift in our understanding of the cosmos. But if any attempt to fathom it was met with derision and disregard, dismissed from public discourse, there could be no discoveries at all.

The crop circles unfolded in particular stages and sequences of patterns after attention was drawn to them in the late 1970s. Although the imprints appear in different regions around the world, the center of activity is indisputably southern England, around the Neolithic stone circles of Stonehenge and Avebury and the ancient man-made mound of Silbury Hill. This region, known as Avalon, is associated in legend with King Arthur, the Knights of the Round Table, and their obscurely vexing quest for the Holy Grail. While Britain was the birthplace of the modern scientific method, it also has a venerable esoteric tradition, from the Neolithic stargazers to Merlin and the druids; from the English alchemists of the Middle Ages and Renaissance (whose ranks include Isaac Newton) to modern-day mages such as Aleister Crowley and Dion Fortune; from the arcadian fantasies of J. R. R. Tolkien and C. S. Lewis, inspiring the folk revival of the late 1960s and Led Zeppelin, to the New Age warlocks, Fortean investigators, and pagan priestesses of the present day. The crop circles seem linked to this august heritage, especially the Neolithic culture. Patterns often evoke the "sacred geometry," geographical alignments, and particular proportions of the craggy stone monuments that continue to emanate an aura of archaic mystery.

The origin of the phenomenon is obscure. Simple crop circles may have appeared in English fields going back hundreds of years—crop circle scholars point to a seventeenth-century etching of a grinning "mowing devil" complete with horns and tail, who seems to be making a formation. Scattered reports of swirled circles date from the 1930s and the 1960s. Adults who grew up in the region recall passing random crop circles in fields as children, aware of them as a local enigma, without paying them much attention.

Contemporary interest picked up around 1980, when Terence Meaden, a climate scientist, and Colin Andrews, an electrical engineer, began to investigate these anomalies together. Despite the precise circular shape of the imprints, and the rings that sometimes appeared around them, Meaden championed the theory that they were the product of whirlwinds or vortices caused by freak weather conditions. Eventually, the precise geometrical and planned nature of the formations refuted Meaden's ideas. Andrews studied the changes in electrical and electromagnetic energy around the crop glyphs—there is often a significant shift in the electromagnetic field surrounding a new formation—and experienced a slew of psychic effects. Although he believes the vast majority of contemporary circles are made by people, Andrews still studies the phenomenon today. He considers them an intentional, and intelligently guided, project. "You get the distinct impression that there is some kind of program running here," he told me.

The early formations were simple circles or ringed circles, but other patterns soon started to appear, such as the "quintuplets," four smaller circles arrayed evenly around a single larger one. The number and complexity of the patterns grew rapidly over the next decade, from a few dozen reported circles to more than a hundred annually. It seemed that the phenomenon was responding to the attention given it. Long ribbons of a kind of alchemical or symbolic code started to appear, made of lines and ladders, circles and rings. The most famous of these was imprinted in the Alton Barnes field, near Avebury.

Like a strip of alien sign language, the 1990 Alton Barnes formation was over six hundred feet long, impressive enough to be featured on the cover of many UK newspapers. Showing up during Britain's second "Summer of Love" at the height of the Ecstasy-fueled rave scene, the Alton Barnes pictogram briefly attained icon status, teasing with the hint of some radical mind-shift or approaching galactic contact. During its brief life, the Alton Barnes formation became a pilgrimage point for people from across England and the world. Whether by accident or intention, the crop circles may have revived the area's ancient purpose: With its stone megaliths, druidic barrows, and mystic ambience, Wiltshire was most likely a pilgrimage destination in ages long past.

In 1991, the patterns evolved to a new level of complexity. Along with the continuation of circles, rings, and sign strips, formations began to display recognizable imagery. At Barbury Castle, Wiltshire, in 1991, an alchemical design appeared, in the form of a tetrahedron, connecting the alchemical symbols for salt, sulfur, and mercury, representing body, soul, and spirit.

Nineteen ninety-one was also the year that hoaxers made their mark on the phenomenon, permanently tweaking it by introducing an unavoidable and perhaps necessary element of duplicity and confusion. Two laborers in their sixties, Doug Bower and Dave Chorley, stepped forward to take credit for the crop circles. In front of the world's media, they gave a reenactment of their circle-making methods, which the press found satisfactory, but some researchers found crude and imprecise. They showed headgear made of bent wires, which they said helped them make straight lines in total darkness. They said they had created hundreds of hoaxed patterns over two decades—though not around Avebury, the epicenter of the phenomenon, and they offered no explanation for the hundreds of other formations that had appeared. Satisfied with this debunking, the reporters decamped.

The intellectual profile of the unknown circlemakers was raised considerably by Gerald Hawkins, former chair of the astronomy department at Boston University and the author of *Stonehenge Decoded,* a breakthrough book on megalithic culture. Hawkins began to study the geometry of the crop circles in 1990. He found that even seemingly simple formations contained hidden layers of intention and geometrical complexity. He analyzed a triplet of circles, in a pyramid shape, discovered on June 4, 1988, at Cheesefoot Head. He was able to draw three tangent lines that touched all three circles. These three lines formed an equilateral triangle. He drew a circle at the center of this triangle, and found that the ratio of the diameter of this central circle to the diameter of the three original circles was 4:3. He tested this out with circles of different sizes that allowed tangent lines to be drawn in the same way, and found that the ratio remained constant. The formation had yielded a geometrical theorem. Since Hawkins was well schooled in Euclidean geometry, he went looking for this theorem in the pages of Euclid and other later texts. It did not exist. The formation was displaying a Euclidean theorem that no other geometer had found.

Hawkins found three other original theorems in different crop circle patterns. From these four paradigms, he was able to deduce a fifth theorem that Euclid, and all later authors, had missed. This theorem involved concentric circles placed inside different types of triangles. Circles drawn within three isosceles triangles generated one typical formation; circles drawn within equilateral triangles generated the other. In one science journal and one magazine for math teachers, Hawkins offered a contest to see if anyone could derive the fifth general theorem from the four earlier ones. Nobody could. "One has to admire this sort of mind, let alone how it's done or why it's done," he mused.

In *Stonehenge Decoded,* Hawkins had applied rigorous astronomical and geometrical analyses to explore how Stonehenge functioned as an observatory and calculator for predicting lunar eclipses. Among other facts, he discovered a Neolithic knowledge of the precise coordination of the Earth's orbit and lunar and solar cycles—the Sun and Moon return to the same stellar alignment in relationship to the Earth every fifty-six years—that modern astronomy had not attained. He used the same techniques to study the geometry of the crop formations. He realized that the particular ratios he was finding, again and again, in ringed circles and in the smaller satellite circles, fit the series of diatonic ratios, the intervals of the musical scale. "A ratio in the diatonic scale is the step up in pitch from one note to the other. . . . The creators seemed to know of these fractions, taking care to encode them in the shapes so they could be retrieved by someone studying aerial photographs," he wrote. Even in seemingly simple formations, the circlemakers had encoded, for those who might be curious, a mastery of geometry and an interest in harmonic pitch.

Biophysicists studied the biochemical effects on plants within the formations, discovering that the "blown nodes" on bent stalks of wheat and canola were elongated toward the center of a crop circle, as if by a sudden blast of intense microwave heat. Many observers have recounted seeing "balls of light" hovering over fields in which crop circles have appeared; some have witnessed the extraordinarily fast creation of patterns—within a few seconds—seemingly orchestrated by these "balls of light." Eltjo Haselhoff, a Dutch physicist and laser engineer, collected data and advanced the hypothesis that crop circles are created by single-point sources of electromagnetic radiation. He published the results of his investigations in *Physiologia Plantarum,* a peer-reviewed scientific journal on plant physiology and biophysics. "This publication has an important consequence," Haselhoff wrote in his book *The Deepening Complexity of Crop Circles.* He argued that these publications elevated "the hypothesis that 'balls of light' are directly involved in the creation of (at least some of the) crop formations" to the status of scientific postulate. This research could only be dismissed when "someone comes forward with an alternative explanation for the circularly symmetric node lengthening, or proofs that the analysis was erroneous." He continued:

> However, such a proof will not be an article in some daily newspaper or on the Internet. The discussions about node-lengthening effects in crop circles have clearly outgrown the level of the tabloids and entered the era of scientific communication by means of scientific literature. Consequently, the only comment that can be taken seriously at this point will have to be another publication in a peer-reviewed scientific journal.

The published and peer-reviewed accounts of BLT Research and Haselhoff have not been refuted at the level of verifiable research. As with psychic phenomena, skeptics continue to dismiss these studies, without producing counterevidence of their own—and usually without examining the evidence for themselves. The skeptical perspective is parroted by the press, which does not conduct independent investigations.

In our society, the media functions as a collective nervous system or immune-response task force, inoculating the social body against any new pattern or incoming information that might threaten conventional values and biases. The "Doug and Dave" story of inventing the crop circles was accepted at face value and not interrogated by the media. Their claims had the effect of discrediting the phenomenon in the eyes of the world. "Doug and Dave"—and other hoaxers that followed them—provided a convenient excuse for ignoring and ridiculing the formations, without giving them careful attention.

In 1991, one unusual formation addressed the question of hoaxing directly. A pattern on Milk Hill appeared to be text of an unfamiliar language or hieroglyphic code. It took a team of twelve scholars several months to decipher the message, which turned out to be in post-Augustan Latin, written in an obscure script used by the Knights Templar. In this arcane medium, the message proclaimed: "OPPONO ASTOS," which, translated, states, "We oppose acts of craft and cunning."

"Doug and Dave" were followed by more sophisticated groups of hoaxers, claiming that they were responsible for the phenomenon's persistence. Some of these groups, such as "The Circlemakers" or "Team Satan," consider themselves postmodern artists, backing up their activities with dense theoretical essays on their Web site, quoting French Deconstructionists. Although the Circlemakers admit to no longer making hoaxed formations, they and other groups make sizable sums creating crop circles as logos and advertisements for corporations. The formations they have made with large teams in daylight or under spotlights at night do not achieve the aesthetic standard of the better anonymous patterns—and manifest none of the variations in node length found by biophysicists in many of the unclaimed formations.

During the 1990s, the formations in the fields continued to evolve—and this took place in relative quiet. Much of the media coverage and popular attention dispersed after the phenomenon was officially labeled a fraud. However, those who continued to follow the patterns were treated to ever-more elaborate and extraordinary configurations, using increasingly complex geometry. Southern England receives anywhere from a few dozen to a hundred configurations a year, some obvious human-made attempts, but many spectacular formations featuring seven-, eight-, and ninefold geometry, Möbius strips, toroids, DNA spirals, sunflower bursts, precise astronomical configurations revealing certain dates, pictograms, electromagnetic fields, futuristic variations on the yin-yang symbol, various complex fractals, and patterns evoking the concept of "strange attractors" from chaos science.

In the spring of 2002, I wrote about the crop circles for *Wired* magazine, tied to the release of the film *Signs,* an overhyped Mel Gibson vehicle that used the formations as a plot device. I spoke with many researchers who had been studying the formations for years, as well as skeptics. Seth Shostak, chief astronomer for SETI in California, dismissed the Arecibo Response—the potential reply to SETI's transmission that appeared beside the Chilbolton Observatory—as "good fun and a nice example of grain graffiti," but not worth taking seriously. "If aliens wanted to communicate with us, why would they use such a low-bandwidth method?" he asked. "Why not just leave an Encyclopedia Galactica on our doorstep?" He noted that their signal was aimed at the star cluster M13, and it would not reach its intended target for another 24,772 years. SETI, he said, "has no interest in investigating the phenomenon further."

The researchers studying the circles seemed reasonable, sane, utterly convinced of the reality of the phenomenon, and sincerely puzzled by it. Almost all of them recounted their own synchronistic and psychic episodes related to their research. Nancy Talbot, a member of BLT Research, studying molecular changes to plants in the patterns, said she had seen the creation of a crop circle out the window of a Dutch farmhouse. Earlier in the evening, she had complained bitterly about the difficulties of finding and documenting the formations, which had sent her roving around remote fields in Holland. In her room later that night, she realized that the dogs and cows and other animals had suddenly gone silent. Outside, she saw a spiraling beam of glowing energy hit the ground. When she went out in the morning, she saw that the energy beam had left behind a swirled circle with a crossed *T* in the middle—as if saluting her for her efforts.

The researcher who most impressed me—who inspired me to pursue the phenomenon in depth—was Michael Glickman, an English architect and industrial designer who had retired to Wiltshire to study the circles more than a decade earlier. On the phone, Glickman did not hedge his bets. He was feisty and funny, at times exasperated.

"I am an architect," he said. "As an architect, one is interested in form, design, geometry. I worked in product development. I knew about factory processes and mechanical developments. When I started to look into the crop circles in the late 1980s, it was immediately screamingly apparent to me that no human technology—no human methods—could manifest the formations."

Glickman called the glyphs "a series of profound, diverse, and complex communications of a substantial lightness and subtlety. They are using shape and geometry, number and form, to access fundamental parts of our being which have become culturally deactivated over centuries. Part of the program is reactivation—that is separate from whatever hard information the circles might be bringing." He believed the formations were linked to an accelerated transformation of human consciousness, pointing to our "dimensional shift" from the third to what he called the fifth dimension.

Glickman thought the patterns—"clearly the results of a prodigious intelligence, will, and intent"—indicated that other forms or levels of galactic intelligence were monitoring the evolution of consciousness on Earth. "We are being approached—after a standoff of millennia. What is approaching us is a federation of nonterrestrial civilizations." As signals, the patterns were coordinated with our ongoing mind-shift, providing clues as well as lessons into the new "fifth-dimensional" form of consciousness we were approaching.

Glickman said he had been able to accurately predict the geometrical progression of the phenomenon—as the symmetrical patterns went from fivefold to six- and sevenfold shapes and beyond—for several years. Although he believed that his ideas continued to fit the development of the circles, he was no longer interested in making predictions. "It is now clear to me—as clear as the effect of gravity—that those of us researching the crop circles are being fed tangible information as part of our work," Glickman said. "When you study the formations, you become some kind of message or carrier, rather than just a developer of ideas. I suppose if you said this to a professor emeritus in physics he would fall off his chair with laughter. But I would rather be on my side than his."

Patterns that could be read as calendars suggested that this process would culminate at a certain point in the near future. Glickman referred to an unusual "Grid Square" glyph that appeared in Etchilhampton, England, in August 1997. More technological in appearance than the formations that had appeared to that point, the 150-foot-long "Grid Square" was precisely executed. The grid lines suggested a map—but a map of time instead of space, Glickman conjectured. A square within a circle, it was made up of twenty-six rows of thirty squares. Glickman pointed out that the total number of squares within the grid equaled the number of weeks remaining from the date of its appearance until August of 2012. This grid also seemed to reference the Tzolkin, the calendar and Sacred Day Count used by the Classic Maya, with a matrix of 260 positions. The 780 squares on the formation could represent a Tzolkin repeated three times.

The Tzolkin, a 13-by-20 mathematical matrix that follows a logical numerical sequence, is an essential tool of Mayan culture, still used today. According to the Tzolkin, "every day is characterized by both a number and a sign," writes Carl Johan Calleman in *The Mayan Calendar and the Transformation of Consciousness,* creating 260 possibilities combined from one of thirteen "day-counts" and one of twenty "day signs." The Tzolkin, with its ever-repeating cycle, is "still regarded as the Sacred Calendar among the living Maya, reflecting a process of divine creation that proceeds without interruption," Calleman writes. In recent years, from 2003 to 2005, a series of extraordinary crop formations have referenced the Tzolkin explicitly, employing specific Mayan iconography.

"In a post-Newtonian culture, it is very difficult to put forth a conviction in something that you can't actually prove," Glickman said. "However, I am convinced that what is being predicted is a dimensional shift. My hypothesis is that the tail end of last summer, 2001, we passed through the center of a dimensional shift, which culminates in 2012. By that point, we will fully occupy an entirely different level of being." The purpose of the crop formations was twofold: first, to key us in to this change; and second, to give us specific clues as to how this shift in dimension would manifest in our world.

CHAPTER THREE

Wishes are recollections coming from the future.

RAINER MARIA RILKE

Glickman's unexpected mention of the year 2012 and the Mayan calendar caught my attention. I had originally encountered the idea of an approaching global mind-shift related to that date in the works of Terence McKenna, who proclaimed that, while tripping on mushrooms in the Colombian Amazon in the early 1970s, he had discovered the "timewave," a spiraling involution or concrescence of time that would bring history to an end on the winter solstice of 2012. During that time, he also believed he had interacted with an alien intelligence and been visited by flying saucers. "My story is a peculiar one," he admitted. "It is hard to know what to make of it." McKenna sometimes approached this bizarre idea playfully, yet at other times seemed sincerely convinced of his prophetic role.

McKenna was the most loquacious and entertaining proponent of shamanic exploration to appear on the scene since Timothy Leary. Before his death from a brain tumor in 1999, he promoted the idea of an "archaic revival," defining a new context for use of psychedelics among a global counterculture of ravers, hackers, slackers, and futurists. He was capable of speaking for ten hours straight, mesmerizing audiences with his deadpan delivery of hermetic ideas and alchemical possibilities. "We are like coral animals embedded in a technological reef of extruded psychic objects," he wrote in a 1983 essay. "All our tool making implies our belief in an ultimate tool." He saw the archetypal apparition of the UFO or flying saucer as a foreshadowing of this tool awaiting us at the end of history that would exteriorize the human soul and interiorize the body, releasing the psyche into the infinite realm of the Imagination—"a kind of Islamic paradise in which one is free to experience all the pleasures of the flesh provided one realizes that one is a projection of a holographic solid-state matrix."

One of McKenna's achievements was to popularize, and glamorize, the short-acting but revelatory DMT trip. He was fascinated with "the rose window topologies of the galacterian beehives of the dimethyltryptamine flash, that nexus of cheap talk and formal mathematics where wishes become horses and everybody got to ride." Made illegal by the U.S. government under the Controlled Substances Act of 1970, DMT is a naturally occurring chemical in the human body, found in the brain and spinal column, and produced by many plants. The DMT molecule is similar to serotonin and melatonin, tryptamines involved in the modulation of conscious states and rhythms of sleep and dream.

Dr. Rick Strassman, the researcher who conducted a 1990 study on the effects of DMT through the University of New Mexico, advanced the thesis that DMT could be the "spirit molecule," linking the physical brain to as-of-yet unknown domains of the afterlife. A practicing Buddhist, Strassman made this proposition based on an intriguing correspondence between fetal development and Buddhist scripture. According to Buddhism, the soul reincarnates seven weeks after death. The pineal gland—that singular organ in the brain that Descartes had considered the seat of the soul—develops in the embryo forty-nine days after conception. Strassman hypothesized that the pineal gland might act as a receiver for the soul, and that DMT could function as the conducting medium. "The pineal gland of evolutionarily older animals, such as lizards and amphibians, is also called the 'third eye,'" Strassman wrote. "Just like the two seeing eyes, the third eye possesses a lens, cornea, and retina." In the event of a cataclysmic shock or at death, the pineal gland might release a flood of DMT into the brain, causing the "life review" and the intense visionary phenomena reported in numerous near-death experiences.

When McKenna smoked DMT—a wrenching event accompanied by a sound similar to the ripping of cellophane—he described "bursting into a space inhabited by merry elfin, self-transforming, machine creatures . . . friendly fractal entities, looking like self-dribbling Faberge eggs on the rebound." His explorations led him to propose that the afterlife may be "more Celtic fairyland than existential nonentity." However, he also recognized the threatening nature of the experience, which forces anyone who dares it to undergo "a mini-apocalypse, a mini-entry and mapping into hyperspace." The greatest danger one encounters is the possibility of "death by astonishment."

In my own explorations of smoked DMT—the psychedelic equivalent of bungee jumping—I did not encounter any merry "elfin entelochies." I felt I was rocketing out of my body, through some intermediary tunnel of symbols and mystical geometries, vibrating and incandescent around me, then reaching a kind of vast vaulted cavernous space, some infradimensional fusion of quartz crystal and Silly Putty, mutable and fast-changing. Immense Cyclopean guardians seemed to preside over this realm, like giant stalagmites, chattering wildly, projecting toward me what I interpreted as a terrifying impartiality. During my second DMT trip—although only seven or so minutes, the subjective experience seems far longer—I found myself making alarming insect-like vocalizations, an example of the "glossalia-like linguistic phenomena" McKenna explored during tryptamine trance, and defensive gestures. Recovering my ability to speak, I turned to the supramental samurai or bardo wardens looming above me, and said, as politely as possible, "Thank you very much—I really appreciate this. And now I would like to go back to my reality."

Soon I found myself rematerializing in the musty Haight-Ashbury attic from which I had launched. Coming out of my trance, I looked down at my arm. Standing on my sleeve was an intricate metropolis, a science-fiction city of glittering spires and emerald skyscrapers in which I could almost see the movements of an infinitesimal transport grid. I looked around at my friends sitting in a circle around me, and all of them appeared radically transformed. They were themselves, but patina'd with complex multicolored tattoos, sporting celestial headdresses, prismatic auras, and elaborate comic-book-superhero armatures. The thought occurred to me that we were actually space aliens or cosmic deities who had projected ourselves backward into this time, prearranging this crude but amusing encounter in "meat-space" from our holographic homeland. "Everyone is God," I muttered, as the vision faded away. Back in conventional space-time, I hugged the woman sitting next to me, feeling I had never known anything as comforting as the solid warmth of her flesh, soft sweater, and patchouli-perfumed scent.

Nietzsche proposed, "Something might be true while being harmful and dangerous in the highest degree. Indeed, it might be a basic characteristic of existence that those who know it completely would perish, in which case the strength of a spirit should be measured according to how much of the 'truth' one could still barely endure—or to put it more clearly, to what degree one would *require* it to be thinned down, shrouded, sweetened, blunted, falsified." DMT, as well as my DPT experience and other hallucinatory forays, suggests that "truth" is not only dangerous but neither singular nor definitive—that "reality" is a dynamic and daimonic process, closer to the multi-latticed meanings implied by music or art than the confident postulates of science, which often supports conventional psychology and the antiquated mechanistic paradigm under its suspect cloak of objectivity. Like a quick bath in the quantum foam of seething "perhapsness" described by quantum physics, such trips help to undermine any cosmological or ontological framework that one might like to hold, accentuating the "curious literary quality running across the surface of existence," noted by McKenna, who thought language, or "the logos," had a crucial role in shaping reality.

McKENNA FIRST STUMBLED UPON the notion of an approaching end point to human history in 1971, when he and his younger brother Dennis, along with some friends, went to the village of La Chorrera, deep in the Colombian Amazon, hoping to find *oo-koo-he,* a DMT-containing snuff, used by the Witoto tribe—a kind of Hardy Boys investigation into the far antipodes of the psyche. The snuff was only a small part of their quest; they were interested in exploring particular aspects of the DMT trance, its relationship to language—"under the influence of DMT, language was transmuted from a thing heard to a thing seen"—and to alien forms of consciousness, as well as, perhaps, UFOs. They had great dreams. "I could feel the golden chain of adepts reaching back into the distant Hellenistic past, the Hermetic Opus, a project vaster than empires and centuries; nothing less than the redemption of fallen humanity through the respiritualization of matter," he wrote in his 1993 memoir, *True Hallucinations.*

The McKennas never found *oo-koo-he.* In fact, McKenna spent the rest of his life puzzling out exactly what they did or didn't find in the Amazon. The brothers and their friends discovered psilocybin instead, bushels of plump *Stropharia cubensis* that weren't even on their original menu. "We are closing distance with the most profound event a planetary ecology can encounter: the emergence of life from the dark chrysalis of matter," McKenna jotted after his first mushroom munch. Drawn in by the fungus, as if it were calling to them, they proceeded to ingest the *Stropharia* in vast quantities over the next weeks. The mushroom became the focus of their quest to combine the experiential techniques of shamanism with the empirical methods of science in order to investigate the invisible landscape, "the simultaneous coexistence of an alien dimension all around us," which they had first accessed through DMT.

Although McKenna was only twenty-four, he had been fascinated with fringe matters since early adolescence. "My interest in drugs, magic, and the more obscure backwaters of natural history and theology gave me the interest profile of an eccentric Florentine prince rather than a kid growing up in the heartland of the United States in the late fifties." His brother, a few years younger, shared his enthusiasms. McKenna had participated in 1960s campus politics at Berkeley, discovering DMT "at the apex of the summer of love," watched the radical counterculture collapse, traveled in the Far East, studying Sanskrit in Nepal and trafficking hash. As he had proposed it to one of his fellow travelers, DMT was "some kind of outrage" that, properly understood, could have "tremendous importance for the historical crisis everybody is in."

The McKennas dreamed of becoming the psychonautical Wright Brothers of the New Age, making charter flights through new realities, garnering Nobel Prizes for melding science and shamanism. "If the world beyond the doorway can be given consensual validation of the sort extended to the electron and the black hole—in other words, if the world beyond the doorway is found to be a necessary part of scientifically mature thinking about the world—then our own circumscribed historical struggle will be subject to whole new worlds of possibility," they wrote in *The Invisible Landscape* (1975), their book on the adventure, a precocious and inscrutable blend of scientific hypothesis and psychedelic philosophy. They proposed that the existence of these realms required scrupulous—and courageous—exploration.

It was Dennis who provided the impetus for what the McKennas would later dub the "La Chorrera experiment." When McKenna described the glossolalia he sometimes explored during DMT trips, Dennis activated the same phenomenon while under the influence of the mushrooms, emitting "a very machine-like, loud, dry buzz," threateningly insectile, which suggested to him the development "of a shamanic power of some sort." Inspired by this new ability, Dennis developed the thesis that this nonsensical sound-stream—accessible only during tryptamine trance—could be directed in some way that would result in "a molecular aggregate of hyperdimensional, superconducting matter that receives and sends messages transmitted by thought." Despite the patent absurdity of this idea, the McKennas took it seriously, and created the conditions in their rain forest shack for a test. They had read accounts of Amazonian shamanism that referred to a "violet or deep blue" magical liquid produced from the body during ayahuasca sessions, used for divination and sorcery—Terence believed he had once encountered this "psychofluid" during a confused and orgiastic night on a rooftop in Katmandu, involving LSD, datura, DMT, and UFOs. The seemingly impossible goal of their La Chorrera experiment was to produce and stabilize this extradimensional liquid in the three-dimensional world. "And end history. And go to the stars."

It didn't happen, it would seem—to the McKennas' chagrin, history refused to fold up its tent on March 4, 1971. The omega point did not arrive. But something happened, or at least McKenna continued to believe something had. For the next weeks, he found himself in telepathic rapport with some alien other that downloaded ideas and information into his mind, speaking to him like a teacher to a student. For nine days and nights he claims he did not sleep, as a deep cognitive reorganization took place. "The overwhelming impression was that something possibly from outer space or from another dimension was contacting us," he recalled. "It was doing so through the peculiar means of using every thought in our heads to lead us into telepathically induced scenarios of extravagant imaginings, or deep theoretical understandings, or in-depth scannings of strange times, places, and worlds." As McKenna received these flows, Dennis retreated further and further from consensual reality, into a temporary state of schizophrenia. While their friends worried, McKenna remained unconcerned, certain that everything was proceeding according to plan. "I used no psychoanalytical jargon in thinking about it, but I noted a reaction in myself that included the idea that he might be unfolding into a mythopoetic reality, or as I thought of it then, 'going bananas.'"

During the time in La Chorrera, McKenna noted an amassing of tiny episodes that seemed to abort, momentarily, the Newtonian constraints of space and time. Clouds formed into lenticular saucers; mysterious lights roved across the fields; the rushing waters on part of the lake suddenly appeared to stop their flow. When McKenna awoke one night and lit a candle, he saw "an intense, triple-layered corona of light," shimmering out for four feet, resembling the aura around Christ's body in a painting by Matthias Grünewald. In the aftermath of the adventure, although lacking the "consensual validation" he had hoped to establish, he continued to believe "under certain conditions the manipulative power of consciousness moves beyond the body and into the world." In particular non-ordinary states, at particular times, the mind can determine "the outcome of the normally random, micro-physical events."

The culmination, in McKenna's account, occurred on his last night in the village. Sitting by the lake in the gray predawn, he watched clouds coalesce in the distance, forming odd symmetries. One cloud seemed to swirl inward "like a tornado or waterspout," take particular form, and rush toward him. As it flew directly overhead, McKenna looked up. "It was a saucer-shaped machine rotating slowly, with unobtrusive, soft, blue and orange lights," he recalled. "It was making the whee, whee, whee sound of science fiction flying saucers."

To increase the effect of this "cosmic giggle," McKenna realized the saucer was identical to an infamous photograph of a UFO by George Adamski, widely believed to be a hoax, "a rigged up end-cap of a Hoover vacuum cleaner." The Mystery had unveiled itself to him as a cheesy stereotype, already debunked. "By appearing in a form that casts doubt on itself, it achieves a more complete cognitive dissonance than if its seeming alienness were completely convincing," he theorized. When McKenna read UFOlogist Jacques Vallee's *Invisible College* several years after the adventure, he learned that "an absurd element is invariably a part of the situation in which contact with an alien occurs." Decades later, he felt he had witnessed "a manifestation of a humorous something's omniscient control over the world of form and matter," choosing to take the shape of a saucer in this instance, but capable of taking any form.

The alien "other" that continued to speak in McKenna's mind long after the event seemed to be the mushroom itself, which represented itself as an ancient galactic intelligence interested in making a productive symbiosis with modern humanity. "It is far more likely that an alien intelligence would be barely recognizable to us than that it should overwhelm us with such similarities as humanoid form and an intimate knowledge of our gross industrial capacity," he suggested, proposing, "the *Stropharia cubensis* mushroom is a memory bank of galactic history." As McKenna related, this fungal consciousness told him that it distributed itself through the galaxies as spores, traveling on meteorites, awaiting contact with the nervous system of a higher animal capable of conscious evolution.

"The mushroom which you see is the part of my body given to sex thrills and sun bathing," the fungus informed him. "My true body is a fine network of fibers growing through the soil." These mycelial networks "may cover acres and have far more connections than the number in a human brain." Psilocybin and psilocyn were compounds produced to open the psychophysical gateway to a galactic community of mind. "You as an individual and humanity as a species are on the brink of a formation of a symbiotic relationship with my genetic material that will eventually carry humanity and earth into the galactic mainstream of the higher civilizations"—thus spake the fungus.

The other part of the message the mycelial intelligence conveyed to McKenna, as he formalized it over the next years, was a revelatory new conception of the nature of time. "Quite unexpectedly, what I now propose, based on those initial experiences, is a revision of the mathematical description of time used in physics," he wrote in *True Hallucinations.* "According to this theory, the old notion of time as pure duration, visualized as a smooth plane or straight line, is to be replaced by the idea that time is a very complex fractal phenomenon with many ups and downs of many sizes over which the probabilistic universe of becoming must flow like water over a boulder-strewn riverbed." The ancient Chinese divinatory system of the I Ching, the Taoist "Book of Changes," and the bioenergetic matrix of the DNA code seemed to be keys to unlocking a deeper pattern or underlying structure of temporality.

Far more than a divination system, the sixty-four hexagrams of the I Ching could be seen as a description or map of the processes of evolution, development, decay, and transformation as they are experienced by any organism, species, individual, or civilization. "The Logos taught me how to do something with the I Ching that perhaps no one knew how to do before," he wrote. "Perhaps the Chinese knew how to do it once and then lost it thousands of years ago. It taught me a hyper-temporal way of seeing." It was McKenna's inspiration to take the most august arrangement of these hexagrams, the King Wen sequence, and utilize mathematical postulates to transform it into a graph or wave that represented "the ingression of novelty into history." He dubbed the result "timewave zero," proposing that time was not an unvarying extension but "a medium of variables in flux," where no two instants are alike, and where new possibilities can enter the frame at certain propitious moments. The timewave revealed an "ocean of resonances," linking every moment "through a scheme of connection that knew nothing of randomness or causality."

The fractal or holographic vision of time that McKenna discovered, or that his hyperspatial ally proposed to him, contained the inherent idea that there exist different "time orders," various forms or structures of temporality. "And these kinds of time come and go in cyclical progression on many levels; situations evolve as matter responds to the conditioning of time and space." Such an understanding helped to free the individual from the burden of time: "If you know what is contained in time from its beginning to its end you are somehow no longer in time," he wrote. "Even though you still have a body and still eat and do what you do, you have discovered something that liberates you into a satisfying all-at-onceness."

The timewave also implied an imminent rupture, an end to our linear time that would be a movement into a new temporal structure. "Because we suggest a model of time whose mathematics dictate a built-in spiral structure, events keep gathering themselves into tighter and tighter spirals that lead inevitably to a final time. Like the center of a black hole, the final time is a necessary singularity, a domain or an event in which the ordinary laws of physics do not function." Such an event would mean "passing out of one set of laws that are conditioning existence and into another radically different set of laws. The universe is seen as a series of compartmentalized eras or epochs whose laws are quite different from one another, with transitions from one epoch to another occurring with unexpected suddenness."

Back in California, after La Chorrera, McKenna found himself "in the grip of a creative mania more extreme than any I had thought possible." Prodded by his ally and his own intuitions, McKenna first anchored his timewave in a presumed "novelty spike" that occurred with the dropping of the atom bomb on Hiroshima in 1945, locking in the "end of time" on November 17, 2012. Later, he conjectured that the concrescence of the "final time" could happen during a rare astronomical conjunction: the eclipse of the galactic center by the solstice sun, an event that occurred approximately once every twenty-six thousand years. Utilizing software, he discovered that this would next take place on December 21, 2012. According to McKenna, he arrived at this date without knowledge of the Mayan calendar, "and it was only after we noticed that the historical data seemed to fit best with the wave if this end date was chosen that we were informed that the end date that we had deduced was in fact the end of the Mayan Calendar." He theorized that the Eschaton, strange attractor, or merkhaba-like object awaiting us at that final time has in a sense already appeared, and what we experience as history are the shockwaves sent backward from this culmination.

"What is it that gives both a twentieth-century individual and an ancient Mesoamerican civilization the same date upon which to peg the transformation of the world?" he asked. "Is it that both used psychedelic mushrooms? Could the answer be so simple? I don't think so. Rather, I suspect that when we inspect the structure of our deep unconscious we will make the unexpected discovery that we are ordered on the same principle as the larger universe in which we arose. This notion, surprising at first, quickly comes to be seen as obvious, natural, and inevitable."

By opening their psyches to the Unknown at La Chorrera, had the McKennas retrieved some deep-buried archetypal complex, a psychophysical thought-system meshing the end-time eschatologies of various religious traditions? Or would it, in the end, turn out to be nothing more than an expression of the cosmic giggle, mocking human pretensions and alchemical dreams? McKenna could never be sure. "The notion of some kind of fantastically complicated visionary revelation that happens to put one at the very center of the action is a symptom of mental illness," he noted, realizing his story had all the trappings of the "messianic ego-inflation" that infected prophets of the past, who passed along their disease in the form of new religious doctrines and dogmas. "My theory may be clinically pathological, but unlike these religious systems, I have enough humor to realize this," he said in his defense. Although tantalized by his timewave, he saw that part of his personal drama involved "the intrinsic comedy of privileged knowledge."

CHAPTER FOUR

The dream is its own interpretation.

The Talmud

For skeptics and agnostics, the persistent belief in an approaching Apocalypse—a word that literally means "uncovering" or "revealing"—has its roots in human psychology. Since we are horrified by the idea of the death of our own individual egos, we prefer to project our end—or our supernatural transfiguration or cybernetic immortality—onto the entire human world. As Eugen Weber writes in *Apocalypses: Prophecies, Cults, and Millennial Beliefs through the Ages:*

> We yearn for some explosive, extraordinary escape from the inescapable and, none forthcoming, we put our faith in an apocalyptic rupture whereby the inevitable is solved by the unbelievable: grasshoppers, plagues, composite monsters, angels, blood in industrial quantities, and, in the end, salvation from sin and evil—meaning anxiety, travail, and pain. By defining human suffering in cosmic terms, as part of a cosmic order that contains an issue, catastrophe is dignified, endowed with meaning, and hence made bearable.

The desperation of quotidian existence is relieved by the sense of participating in the central action of the cosmic drama.

While it seems familiar and sensible, such an agnostic perspective is also based on a subjective understanding. According to this position, there is an outside stance—one of skepticism—that allows the rational mind to remove itself from the processes it observes. Secular materialism remains privileged for its presumed objectivity and rationality, which seems capable of evaluating metaphysical and mythical concepts, rather than recognizing them as constituent elements of the psyche. Skeptics hold to this perspective, even though physics has demonstrated there is no privileged position of objectivity. As John Archibald Wheeler noted, we live in a participatory universe with no place reserved for detached observation. In his determination to maintain a noncommittal stance, the agnostic is also possessed by his archetype.

The standard agnostic view does not allow for engagement with the possibility that there could be different forms of temporality—"time orders," according to the physicist F. David Peat—or that "mind-stuff," rather than matter, might be the basis of our physical world, or that transformations might take place in the kinds of sudden leaps, like quantum jumps into new contexts, that Terence McKenna describes. William Irwin Thompson proposes, "We are the climactic generation of human cultural evolution, and in the microcosm of our lives the macrocosm of the evolution of the human race is playing itself out: which is exactly what our new mythologies (such as *2001*) are trying to tell us." Such a position, for all its New Age overtones, seems at least plausible, supported by a careful attunement to fast-changing physical as well as psychic conditions.

From this alternative view, the possibility that the next few years might represent a culmination of current trends makes sense. Current trends, as noted previously, cannot continue indefinitely, although it is difficult to imagine their end point. These trends include the rapid destruction of the planetary environment, the threat of nuclear and chemical holocaust unleashed by terrorists or nation-states, the exponential increase in information, and the accelerated evolution of technology. The development of technology does seem to be approaching a concrescence point—what some futurist thinkers call the "Singularity." The Stone Age lasted many thousands of years, the Bronze Age lasted a few thousand years, the Industrial

Age took three hundred years, the Chemical Age or Plastic Age began a little more than a century ago, the Information Age began thirty years ago, the Biotechnology Age geared up in the last decade. By this calculus, it is conceivable that the Nanotechnology Age could last all of eight minutes. At that point, human intelligence might have complete control of the planetary environment, on a cellular and molecular level. This could lead to utopian creativity or dystopian insanity—perhaps both would arrive at the same moment.

Most people would immediately dismiss the tangled complex of thoughts developed by the McKennas in the aftermath of the La Chorrera experiment as a projection of their own confused psyches, amplified by the uncontrolled ingestion of large amounts of psychedelics. McKenna freely admitted this. "Any story of alien contact is going to be incredible enough by itself, but central to our story as well are the hallucinogenic mushrooms with which we were experimenting," he wrote. "The very fact that we were involved with such plants would make any story of alien contact seem highly dubious to anyone not sympathetic to the use of hallucinogens."

On the other hand, it might be that McKenna's new model of time, with all of its revelatory and apocalyptic dimensions, represented something fished from the deep waters of the unconscious, an initial phase in the emergence of an archetypal complex, connecting the end-time eschatology of the Judeo-Christian tradition to the mythological worldview of the Classic Maya. Through their efforts to recover "a lost knowledge or a lost art, or a lost consciousness" in the Amazon, utilizing the radical gnosis of psychedelic plants, the McKennas could have accessed, spontaneously, something that is, indeed, coming to presence within the collective psyche. McKenna suspected this to be the case. "Nothing is unannounced," he noted.

In his own life and the dreams of many of his patients, Carl Jung frequently experienced the phenomenon of a voice, seemingly from a source outside the individual, revealing information—similar to the telepathic intelligence that communicated with McKenna in the Amazon. The voice "always utters an authoritative declaration or command, either of common sense or of profound philosophic import," wrote Jung, who considered the voice to be "a product of the more complete personality of which the dreamer's conscious self is a part." These directives from the unconscious often manifest "an intelligence and purposiveness superior to the actual conscious insight." The downloads that McKenna received from his magic mushroom ally—about the nature of time, and the potential symbiosis between humans and the galactic intelligence of the fungi—might fall into this category.

The visions and directives produced through hallucinogens can intensify an apocalyptic perspective on reality—I found a similar tendency in the aftermath of my own trips. Psychedelics may temporarily splinter or even obliterate the ego, but soon the ego comes roaring back, seeking to create a new cosmological framework in which it is ensconced at the center of a vast drama. The effort to attain a new worldview—metanoia—can collapse into paranoia. Although McKenna codified his ideas in the timewave, the idea lacks scientific verification, since "the ingression of novelty into history" requires a subjective value judgment and cannot be mathematically quantified. Without further corroboration, McKenna's timewave would seem to belong to the realm of the eccentric and discardable, like phlogiston, or orgone energy, or one of the glittering and ineffable objects described in the fables of Jorge Luis Borges—or like the concept of the Apocalypse itself.

McKenna attained cult status in the 1980s, and his ideas were absorbed into the first flush of technological euphoria that accompanied the dot-com boom. Ultimately, his perspective was Gnostic and hermetic, informed by his psychedelic journeys far more than the Internet. He supported "the siren song of Pythagoras—that the mind is more powerful than any imaginable particle accelerator, more sensitive than any radio receiver or the largest optical telescope, more complete in its grasp of information than any computer: that the human body—its organs, its voice, its power of locomotion, and its imagination—is a more-than-sufficient means for the exploration of any place, time, or energy level in the universe." Futurist or transhumanist theorists reformulated his vision of an approaching singularity into an imminent "technological singularity," in which humans will merge with machines in order to transcend our biological limits.

In his essay "The Law of Accelerating Returns," Ray Kurzweil, author of *The Age of Spiritual Machines,* examined the exponential evolution of technology, which follows a progression defined by Moore's Law, and argues that this mathematical growth curve soon reaches a rate of change that is close to infinite. He believes this will occur in this century. "Within a few decades, machine intelligence will surpass human intelligence, leading to The Singularity—technological change so rapid and profound it represents a rupture in the fabric of human history," exults Kurzweil. "The implications include the merger of biological and nonbiological intelligence, immortal software-based humans, and ultra-high levels of intelligence that expand outward in the universe at the speed of light."

The futurist theorist John Smart, who operates the Singularity Watch Web site, shares Kurzweil's euphoria: "Technology is the next organic extension of ourselves, growing with a speed, efficiency, and resiliency that must eventually make our DNA-based technology obsolete, even as it preserves and extends all that we value most in ourselves." Unlike Kurzweil, who sees humans evolving technologies that expand out to fill up the universe, Smart sees the eventual destiny of the species in what he calls "transcension," essentially escaping this universe in the other direction, by creating immersive simulations or virtual realities that will be like new universes, drawing all information into the black hole of our information-processing and technology-generating engines.

The transhumanists believe our biological limits should be overcome through mechanical means. We are too dense, too cumbersome in our inherited meatsuits, trapped in what John Smart calls "slowspace." Immersed in virtual realities or fused with artificial intelligence agents—or some other technological genie—we will attain a speedier, snazzier state of being. Kurzweil notes: "Biological thinking is stuck at 1026 calculations per second (for all biological human brains), and that figure will not appreciably change, even with bioengineering changes to our genome. Nonbiological intelligence, on the other hand, is growing at a double exponential rate and will vastly exceed biological intelligence well before the middle of this century." By inserting nanobots into our brains or ultimately perhaps downloading our psyches into immortal silicon-based supercomputers, humans will be able to contribute our pitiful little brain-wattage and antiquated personalities to the evolution of A.I.'s higher, faster levels of functioning. Hypnotized by their futurist visions, the transhumanists neglect to note that there is no evidence, as of yet, that machines can attain consciousness.

We should even feel some compassion for the next level of synthetic consciousness we are currently gestating to succeed us. Smart writes: "Consider that once we arrive at the singularity it seems highly likely that the A.I.s will be just as much on a spiritual quest, just as concerned with living good lives and figuring out the unknown, just as angst-ridden as we are today." Even if, during some hyper-insectile phase of *Terminator*-style behavior, the A.I.s accidentally destroy the human species, he reassures us, they would no doubt want to re-create us eventually—just as we build museums and zoos to preserve cultures and animals we have made extinct.

The transhumanist vision reflects our cultural fantasies about technology and transcendence, as well as deep anxiety and misconceptions about the essence of time, consciousness, and being. Although not everyone shares the vision of an imminent "technological singularity," the notion that our technology is evolving in a teleological way toward some ever-approaching yet ever-postponed culmination is endemic to our thinking. It might be that our future lies in an entirely different direction. To see what this direction might be, it is instructive, first, to consider the essential nature of technology.

The philosopher Martin Heidegger noted that the essence of technology cannot be found in any machine or artifact; the essence of technology is the entire "enframing" of reality that is our modern or postmodern worldview. "The threat to man does not come in the first instance from the potentially lethal machines and apparatus of technology," Heidegger wrote in his essay "The Question Concerning Technology." "The actual threat has already afflicted man in his essence." Technology, he notes, is based on an ordering of reality that turns everything—including people—into a "standing reserve," a resource to be utilized for rationalized ends. The barren architecture of the vast housing projects on the edges of modern cities, where masses of humanity are warehoused as surplus labor, is a natural extension of this worldview.

That more speed, more information, or any form of quantity-based extension or synthetic transcendence of our current human reality is somehow valuable, in and of itself, needs to be questioned. An alternative perspective is offered by Eastern mysticism, which has no such vision of progress. As the Hindu guru Sri Nisargadatta Maharaj states in the spiritual classic *I Am That:* "Get hold of the main thing: That the world and the self are one and perfect. Only your attitude is faulty and needs readjustment." A faulty attitude creates a faulty world—a world of insufficiency, in which human beings are reduced to the status of things. It is a world of endless distractions, and "distractions from distraction," where individuals feel like voids that need to be filled. It is a world in which the present is devalued, and our hopes and dreams are projected on an empty future.

Heidegger notes that the origin of the word "technology" comes from the Greek word *techne,* and this word was applied not only to technology, but to art, and artistic technique, as well. "Once there was a time when the bringing-forth of the true into the beautiful was also called *techne,*" he writes. "Once that revealing which brings forth truth into the splendor of radiant appearance was also called *techne.*" He found this to be a numinous correspondence, and considered that, in art, the "saving power" capable of confronting the abyss of the technological enframing might be found.

If art contains a saving power, it is not in the atomized artworks produced by individual subjects, but in a deeper collective vision that sees the world as a work of art, one that is already, as Nisargadatta and McKenna suggest, perfect in its "satisfying all-at-onceness." Instead of envisioning an ultimately boring "technological singularity," we might be better served by considering an evolution of technique, of skillful means, aimed at this world, as it is now. Technology might find its proper place in our lives if we experienced such a shift in perspective—in a society oriented around technique, we might find that we desired far less gadgetry. We might start to prefer slowness to speed, subtlety and complexity to products aimed at standardized mind. Rather than projecting the spiritual quest and the search for the good life onto futuristic A.I.s, we could actually take the time to fulfill those goals, here and now, in the present company of our friends and lovers.

Part of the problem seems embedded in the basic concept of a concrescence or singularity, which compacts our possibilities rather than expands them. The notion of a technological singularity reflects our culture's obsessive rationality, reducing qualitative aspects of being to quantifiable factors, and imposing abstract systems over complex variables. Instead of a technological singularity, we might reorient our thinking toward a more desirable multiplicity of technique. Technique is erotic in essence; it is what

Glenn Gould or Thelonious Monk expresses through the piano—the interplay between learned skill and quantum improvisation that is the stuff of genius. Technique embraces the now-ness of our living world; technology throws us into endless insatiation.

The futurist fantasies of the transhumanists ignore the negative effects and unforeseen consequences that accompany technological progress, that have drastically increased with each new level of mechanization: Mass use of pesticides since the 1950s has negative effects on human health and animal populations around the world. The uncontrollable spread of biotechnologically altered seeds is a form of genetic pollution whose consequences may not be foreseeable as of yet. Our much-loved gadgets end up leeching heavy metals and chemicals into landfills around the planet. The toxic environment created by industrial processes, the missiles and weapons mass-produced by the military-industrial complex, suggest an imminent future that is far less dazzling.

Technology, its poisonous by-products, weapons of mass destruction, and inhumane repercussions are projections of the human psyche, expressing our current stage of development. They express not only our consciousness, but also our unconsciousness. "Look at the devilish engines of destruction!" Jung wrote. "They are invented by completely innocuous gentlemen, reasonable, respectable citizens who are everything we could wish. And when the whole thing blows up, and an indescribable hell of destruction is let loose, nobody seems to be responsible. It simply happens, and yet it is all man-made."

The Jungian perspective on the Apocalypse considers it a psychic event now taking place within the collective and individual psyche. This process of uncovering and laying bare all aspects of the psyche will have increasingly disastrous material manifestations until its full dimensions are brought up to consciousness and assimilated. Jung wrote: "The only thing that really matters now is whether man can climb up to a higher moral level, to a higher plane of consciousness, in order to be equal to the superhuman powers which the fallen angels have played into his hands. But he can make no progress until he becomes very much better acquainted with his own nature." We cannot hope to control the technological processes we have unleashed until we have mastered our own shadow material.

Rather than seeking a silicon-based solution to our civilizational cul-de-sac, McKenna believed we needed an archaic revival, restoring the visionary modes of perception known to indigenous cultures and myth-based civilizations, but integrating them with the analytical tools developed by the modern Western intellect:

> Our institutions, our epistemologies are bankrupt and exhausted; we must start anew and hope that with the help of shamanically inspired personalities, we can cultivate this ancient mystery once again. The Logos can be unleashed, and the voice that spoke to Plato and Parmenides and Heraclitus can speak again in the minds of modern people. When it does, the alienation will be ended because we will have become the alien. This is the promise that is held out; it may seem to some a nightmare vision, but all historical changes of immense magnitude have a charged emotional quality. They propel people into a completely new world.

Discussing the process of writing one of his last works, the great and difficult essay "Answer to Job," Jung noted that he experienced it as something almost inflicted upon him. "I feel its content as an unfolding of the divine consciousness in which I participate, like it or not," he complained in a letter. Unlikely as it seems, the strange story of the youthful McKennas, who received "quite unexpectedly" a new model of temporality from a telepathic mushroom in the Amazon, might represent another phase in its unfolding.

BEFORE CONSIDERING THE IDEA of Apocalypse, many of us confront two huge conceptual stumbling blocks: the book of Revelation, and the Christian tradition that produced it. Over the last centuries, official Christianity's ambience of guilt and repression and its denial of the flesh increasingly repelled the progressive-minded—throughout my early life, I never gave it a serious thought. Over time, however, I began to suspect that if I hoped to integrate spirituality into my life, I had to understand, and reclaim, the essential meaning of my own heritage.

In recent years, "spirituality" has made a huge comeback among the wealthy elites of the modern-day West, but it generally has an exotically Eastern cast, avoiding Christ and the Bible altogether. While movements related to the New Age have hordes of devotees, many of these followers cherry-pick from esoteric traditions, choosing what best fits their lifestyle: models and their stockbroker boyfriends spend thousands of dollars to attend yoga and raw-food retreats, where they practice asanas and mantras in tropical locales. Silicon Valley executives decorate their vacation homes with Hindu statues and Tibetan *thangkas*. Architects incorporate a bit of feng shui into their designs. Nightclubs are called Karma and Spirit, while bands are Nirvana and Spiritualized. Millions meditate and chant, seeking relief from anxiety and some undefined feeling of unity with the cosmos. Best-selling gurus promise infinite "abundance" through following the "laws of spiritual success." Words can be emptied of meaning through overuse, turning into their opposite. Our culture has conveniently amputated the concept of spirituality from the processes of life, using it to denote a vast range of commodifiable experiences, self-help movements, and decorative ornaments.

The figure of the sacrificed Christ represents an awkward antipode to contemporary trends, suggesting that the pursuit of meaningful spirituality requires a commitment to the collective that goes beyond the limits of what is comfortable or reasonable. Among contemporary yogis and postmodern mystics, alienation from the revolutionary figure of the biblical and Gnostic Christ may represent a necessary phase before we can return to this tradition, integrating it at a deeper level of understanding. Since we are removed from all scriptural dogma, we have the opportunity to explore our own heritage as if it were new and original.

In the Gospel of Mark, we are instructed to love God with all of our minds, as well as our hearts and our souls. In other words, we are encouraged to use our intelligence, as well as our feelings, to explore our relationship to the sacred. The Jungian model gives us a means of approaching the Judeo-Christian heritage that does not insult our cognitive faculties. From the Jungian perspective, the Apocalypse, as well as the concept of a "Second Coming of Christ," are archetypes, representing aspects of the collective psyche. Jung's follower Edward Edinger defined an archetype as "a primordial psychic pattern of the collective unconscious that is at the same time a dynamic agency with intentionality." The archetypes hover outside or offstage our human drama, awaiting their moment to "constellate" in the individual and collective psyche, catalyzing processes of transformation with far-reaching consequences.

Jung's theory of the archetypes depended upon his understanding of the reality of the psyche, the interconnection of mind and world. He defended this position against any materialist conception. "Not only does the psyche exist, but it is existence itself," Jung wrote, finding this quite obvious when one considered it: "It is an almost absurd prejudice to suppose that existence can only be physical. . . . We might well say, on the contrary, that physical existence is a mere inference, since we know of matter in so far as we perceive psychic images mediated by the senses." For Jung, mythological symbols and archetypes had definite existence as aspects of the psyche. "The fact is that certain ideas exist almost everywhere and at all times and can even spontaneously create themselves quite independently of migration and tradition," he wrote. Individuals do not invent or generate the archetypes—often they are overtaken or possessed by them, against their conscious desires.

As the repository of myths, symbols, and suppressed psychic contents, the unconscious exceeds the range of our conscious awareness. Jung believed that consciousness "must always remain the smaller circle within the greater circle of the unconscious, an island surrounded by the sea." Within the ocean of the unconscious, he perceived "an endless and self-replenishing abundance of living creatures, a wealth beyond our fathoming." Along with the psychic development of individuals, an ongoing process of evolution takes place on the level of the collective psyche. If the Apocalypse, as an archetype, is currently constellating in our world, we have the option of bringing this "dynamic agency" and primordial pattern fully into our awareness. By giving it our conscious attention, we can mediate the process, potentially avoiding its most catastrophic effects.

The book of Revelation is an awful text to contemplate, like something coughed up from the depths of a bad acid trip—disjointed, doom-riddled, insufferably self-righteous, annoyingly priggish. As the Jungian analyst Edward Edinger wrote, in *Archetype of the Apocalypse:* "It strikes the modern mind as bizarre and almost unintelligible." And yet, Edinger argued, following Jung, that the Apocalypse of St. John had to be understood, not as something specious or far-away, but as something literally taking form in our current reality. "We have evidence all around us in our daily analytic practice and in contemporary world history that this earth-shaking archetypal event is taking place here and now. It has already started. It is manifesting itself in international relations; in the breakdown of the social structures of Western civilization; in political, ethnic, and religious groupings; as well as within the psyche of individuals." He considered the Apocalypse—with its structure of separation, judgment, destruction, and reintegration at a new level of wholeness—to be a profound psychological event, representing "the momentous event of the coming of the self into conscious realization."

Jung's interpretation of the Judeo-Christian tradition, which he elucidated in his extraordinary essay "Answer to Job," is based upon his split-level model of the psyche, comprising the ego, the individual consciousness, and the self, the "more complete personality" that includes unconscious and conscious elements. In our lives, the desires of the ego and the goals of the self are often at cross-purposes. He saw the Bible as a document describing the evolution of the Western psyche, moving, in stumbling stages and violent lurches, toward an ultimate reconciliation, an integration of ego and self. The Western god-image, Yahweh, represents "the archetype of the self," containing benevolent as well as malevolent aspects. He is not only the "god of light" but also the "dark god" of punishment and retribution.

"Yahweh is not split, but is an antimony—a totality of inner opposites—and this is the indispensable condition for his tremendous dynamism," Jung realized. Not only a benevolent helper, the god-image also represents the forces that oppose us, throwing obstacles in our path, surprising and humiliating us, inciting suffering to create intensified self-consciousness. God, as a psychic fact, reveals himself in our neuroses, our allergies, our failed love affairs. "God enters through the wound," he wrote.

This idea would have appealed to Friedrich Nietzsche, who considered the modern desire to "abolish suffering" in a world of creature comforts to be an insane and ill-considered goal:

> The discipline of suffering, of *great* suffering—do you not know that only *this* discipline has created all enhancements of man so far? That tension of the soul in unhappiness which cultivates its strength, its shudders face to face with great ruin, its inventiveness and courage in enduring, persevering, interpreting, and exploiting suffering, and whatever has been granted to it of profundity, secret, mask, spirit, cunning, greatness—was not granted to it through suffering, through the discipline of great suffering?

The philosopher identified with the seemingly negative forces unleashed by the dark, instinctual god in the unconscious—the creative-destructive archetype represented by Dionysius or Shiva—desiring the deepening of consciousness through increasing anguish. He thought the modern enthusiasm for "well being" created "a state that soon makes man ridiculous and contemptible—that makes his destruction *desirable*." As though obeying Nietzsche's dictum, modern civilization thwarted its own possibilities of achieving any form of sustainable or moderate contentment, through heedless overdevelopment, intensification of ethnic and religious hatreds, and the escalations of the military-industrial complex.

Jung noted that the god-image at the beginning of the Old Testament is not exactly a benevolent father figure. The psychological portrait of Yahweh that emerges is "of a personified brutal force and of an unethical and non-spiritual mind, yet inconsistent enough to exhibit traits of kindness and generosity besides a violent power-drive. It is the picture of a sort of nature-demon and at the same time of a primitive chieftain aggrandized to a colossal size." The Western god-image underwent a steady transformation—and humanization—in the course of the biblical narrative, through his interactions with the Jews, his chosen people, as the Jews also became increasingly sensitized and conscious, through suffering, and through the tribulations induced by Yahweh. The relationship between the Jews and their god-image was a dialectical process, in which conflict incited the development of consciousness.

The story of Job—the good man chosen to endure untold suffering, subjected to personal trials by a "jealous god"—represents a turning point in the narrative, in Jung's interpretation, the point at which human consciousness reaches a level of understanding that exceeds the awareness of the god-image. "Job realizes God's inner antinomy, and in the light of this realization his knowledge attains a divine numinosity." Job's insight into Yahweh's divided nature compels the descent of the god-image into humanity, as a kind of dialectical compensation for the wrongs done to man.

This drama culminates, in the New Testament, in the incarnation of the god-image in the human world, as the "good god," Christ. Yet this long-awaited archetypal event did not complete the process of incarnation—the descent of the god-image into humanity—as Christ was not fully human, but the product of an immaculate conception. "Christ by his descent, conception, and birth is a hero and half-god in the classical sense," wrote Jung. The book of Revelation represents the logical continuation of this dialectical process, projecting this process of descent into the future, when humanity would be able to enter into direct relationship with divinity in the New Jerusalem, after transforming itself through a drastic series of trials. "The unconscious wants to flow into consciousness in order to reach the light, but at the same time it continually thwarts itself, because it would rather remain unconscious," Jung wrote. "That is to say, God wants to become man, but not quite." The Apocalypse of St. John seeks a resolution to this almost unbearable tension.

The barbaric bombast of the book of Revelation—a montage of miseries, a catalogue of cataclysm—reflects the schisms within the Western psyche; "a decisive, radical 'separation' between heaven and earth, all things upper and lower, between spirit and matter or nature," writes Edinger. St. John's intimation of the difficulty in bridging this separation elicited his frenzied, frothing rant of fear, horror, and prophetic rage. In Edinger's analysis, Revelation is seen as an alchemical text, its dire images of separation foreshadowing the "union of opposites" that is the task of contemporary humanity, confronted with the Apocalypse of its own unleashed impulses.

> The "coming of the Self" is imminent; and the process of collective "individuation" is living itself out in human history. One way or another, the world is going to be made a single whole entity. But it will be unified either in mutual mass destruction or by

> means of mutual human consciousness. If a sufficient number of individuals can have the experience of the coming of the Self as an individual, inner experience, we may just possibly be spared the worst features of its external manifestation.

The images of the golden bowls and golden cups that occur throughout Revelation are symbols of the Higher Self, even though these precious vessels release devastating plagues and destruction upon the Earth. The "Whore of Babylon," a depraved version of Venus, is described as "holding a gold cup filled with the disgusting filth of her fornication"—a clear image of opposites. It is the difficult, but unavoidable, task of the modern individual to assimilate consciously all of the contents—from darkest degradation to profoundest purpose—contained in the psyche.

The evolution of technology is a Faustian aspect of the Apocalypse drama in which we are currently enmeshed. Our one-sided fixation on mechanized progress is the result of our civilization's ingrained habit of prioritizing the "rightness" of masculine rationality, seeking to dominate nature, over the "leftness" of feminine intuition, preferring to surrender to it. Edinger writes: "Everything 'feminine' (earth, nature, body, matter) underwent a profound depreciation with the onset of our aeon. . . . The fact is that the 'depreciation of the feminine' is one of the ways by which the Western psyche has evolved; and we can only assume that it was necessary for the required sequence of events." The "depreciation of the feminine" includes, not only women, but the intuitive and shamanic forms of thought denigrated by our rigidly masculine rationality, as well as nature itself, which is treated as the soulless object of the scientist's alienated gaze. This domineering attitude continues to underlie—and belie—our mechanistic progress.

WHEN I BEGAN TO RECONSIDER the meaning of the Judeo-Christian tradition for myself, after my explorations of psychedelics and shamanism, I found it was possible for me to understand Christ's role, or his mission, through a prism that incorporated scientific and shamanic dimensions. Much of what is obscure in Christ's deeds and sayings becomes sensible if we envision him acting from the perspective of the "timeless space of a higher dimension," the transcendent domain of interconnection, defined by modern physics—similar to what Jung called the Pleroma, the atemporal realm of the archetypes. This clarifies Christ's otherwise confusing statement in the book of John, "The hour is coming, and now is," or his apocryphal saying, "I chose you before the Earth came to be." Christ offers a messianic vision of fulfilled time, of immanence rather than transcendence, as in this passage from the Gospel of Thomas:

> His disciples said to him, "When will the kingdom come?" "It will not come by watching for it. It will not be said, 'Look, here!' or 'Look, there!' Rather, the Father's kingdom is spread out upon the earth, and people don't see it."

Such a perspective contradicts the future-orientation of our world, our obsession with progress, and our habitual awaiting for some ever-postponed fulfillment.

One might think of consciousness as a kind of vibrational level or frequency range, oscillating like a wave form. My own psychedelic experiences often conveyed this impression. From this perspective, the purpose of Christ's mission could be defined as the initiation of a new threshold or intensity of consciousness for humanity. The higher level of consciousness, and intensified conscience, that Christ brought to the Earth was too much for mankind—save for a few—up until the present day. Christ realized that this would be the case: He said he did not come to bring peace, but a sword—not to unite, but to divide.

World avatars are frequency transducers who step up the voltage of mind. Christ's parables in the New Testament and in the Gospel of Thomas—found, along with other Gnostic scriptures, in a jar in the Nag Hammadi desert of Egypt in 1945—could be considered devices for storing and transmitting higher energies. The receptivity of his audience to his impacted parables and statements was in itself miraculous—as much a miracle as any of his suspensions or transmutations of seeming physical laws. Like a new cybernetic code hacking the logic of a closed system, Christ's parables break open ordinary sense to introduce paradoxes that subvert syntactical logic. His disciples listened in wonder, but understood only in part, and sometimes hardly at all. Their bafflement is apparent throughout the Gospel of Thomas, which dates from the same period as the canonical texts, and may at least equal them in authority. Repeatedly, in Thomas, Christ admonishes his disciples, "Those who have two ears better listen!"

In the Gospel of Thomas, Christ proclaims the necessity of achieving direct knowledge—gnosis—of the Divine. He declares, "If you bring forth what is within you, what you bring forth will save you. If you do not bring forth what is within you, what you do not bring forth will destroy you." His radical teaching promotes personal revelation, with no priest class or hierarchical mediation. If the Gnostic lessons presented in Thomas were inscrutable to Christ's listeners, it might be the case that they were specifically intended for a later age. It may be that we are the subjects with the capacity to understand, and it is to the advanced present-day consciousness that Christ directed his statements. We are the ones with "two ears" to listen, if we can reconcile the paradoxes of quantum physics with a renewed sensitivity to mythic structures and archetypes.

In all of Christ's words and deeds, it is as though he is taking the reality of the transcendent domain fully into consideration, and this accounts for his freedom—his liberation from the constraints of the ego—as well as his foreknowledge of fate. When Christ said, "You have to leave your father and mother to follow me," he was instructing people to awaken to self-consciousness of their own individuality. Individuation has to be achieved before the ego can be cast aside. It is only as a fully self-reflective individual consciousness that one can make the choice, out of free will, to reconcile with the Divine, through sacrifice, or supercession, of the ego.

Christ said, "He that loveth his life shall lose it; and he that hateth his life in this world shall keep it unto life eternal." In his words, his actions, and his inner being, Christ exemplified such a sacrifice. Rather than some greeting-card savior, redeeming the sins of the churchgoing faithful through his torments, he provided a model for us, demonstrating the ego-free attitude and unstinting action required from us if we are willing to incarnate the self, and complete the apocalyptic process. Unfortunately, he did not "save our souls" through the crucifixion. Instead, he showed us the path—a template for selfless action that can be internalized, and followed, if we make the free choice to evolve.

According to this hypothesis, Christ "redeems" us only when we follow his lead. We still have to save our own souls. Alas, this is no easy task. But without real sacrifice, there can be no spiritual progress.

PART THREE

LUCIFER AND AHRIMAN

Original sin, the old injustice committed by man, consists in the complaint unceasingly made by man that he has been the victim of an injustice, the victim of original sin.

FRANZ KAFKA

CHAPTER ONE

In the popular culture of our secular age, the gods, demigods, fairies, and gnomes of the old mythic realm have returned as extraterrestrials. Our mingled longing for and dread of contact with some unknown consciousness or superior alien race has been reflected in a century's worth of books, films, television, and radio plays. I grew up on *Star Trek, The Planet of the Apes, Star Wars, ET,* and *2001,* on Ursula K. Le Guin and Kurt Vonnegut and Stanislaw Lem—as an adolescent, I loved the Silver Surfer and Orson Welles's *The War of the Worlds.* The pleasure of these artifacts was in the possibilities they threw out, like so many sparks. They returned the cosmos to a capacious state of "what-if?" that our mechanistic science seemed to deny. The exploration of fictional worlds is a kind of dreaming while awake; the complex ecosystem of the cultural imagination may also have a protective function. Through such stories, we absorb ideas in sidereal fashion, perhaps readying ourselves, on some subliminal level, for future shock of various stripes, before it arrives.

After I finished my article on the crop circles, the images, and their implicit intent, continued to linger in my mind. I was perplexed by the rectangular Arecibo Response formation, dismissed by current SETI astronomer Seth Shostak as a "nice example of grain graffiti," unworthy of further investigation. I was equally confounded by the "Face" that had appeared in halftones on the date of my daughter's birth. Whether accident or synchronicity, this correspondence seemed like a personal invitation to visit what the writer Robert Anton Wilson dubbed "Chapel Perilous," that vortex where cosmological speculations, coincidences, and paranoia seem to multiply and then collapse, compelling belief or lunacy, wisdom or agnosticism.

Considering the scientific evidence, gathered by Eltjo Haselhof and others, suggesting the phenomenon had some mysterious legitimacy, as well as the many personal accounts I absorbed while doing my research, SETI's blithe dismissal of the Arecibo Response glyph, a direct response to a message beamed into space by SETI in 1974, seemed flat and unreflective. Shostak insisted that an alien civilization would not communicate in such a manner when they could simply leave an Encyclopedia Galactica on our doorstep. But how could we determine the means that an alien civilization might use to communicate? He was perhaps recalling the Fermi Paradox, which noted that any technologically evolved civilization on a nearby star system should have emitted radio waves during its development that our sensors would have picked up. The physicist Enrico Fermi asked, in the absence of these signals, "Where are they?" But the answer might lie beyond the limits of our present knowledge.

The SETI astronomer pointed out that the original Arecibo greeting was sent out to the M13 star cluster, over twenty thousand light-years away, and it therefore made no sense that it could have been answered already. It seemed equally logical to theorize that whoever—whatever—had crafted the reply knew about the original message as soon as it was sent, that they might have observed activities on our planet for a very long time. But even if one could imagine an advanced species watching the Earth, awaiting the proper moment to reveal itself to us, the Arecibo Response still made little sense. Who was meant to receive the transmission? And what were they—or we—supposed to do with it?

Small, big-headed figures with silicon added to their makeup and an extra strand of DNA, as depicted in the Arecibo Response, suggested the peculiar narrative, or evolving postmodern myth, of the Gray aliens. Over the last decades, the Grays infiltrated the global subconscious, through best-selling books such as Whitley Strieber's 1987 *Communion,* the TV miniseries *Taken,* and T-shirts, plastic figurines, cartoons, and other mass-cult detritus based on accounts of abduction. I had never paid more than a glancing attention to the UFO phenomenon or to alien abduction accounts—it seemed like some hysterical symptom of our cultural malaise, adolescent and turgid, overliteral, and deeply disreputable. The notion that three-and-a-half-foot-tall cardboard-colored aliens made nightly invasions of middle-class bedrooms across the United States and the world to insert rectal probes and take sperm samples did not seem plausible, or the type of behavior one would anticipate from a futuristic civilization.

And yet, much like the surprisingly tangible evidence on crop circles, the accumulated data on UFO sightings and alien abductions reveals jarring levels of complexity and downright weirdness that do not allow for a blanket rejection of the phenomenon. Harvard psychiatrist John Mack, author of a Pulitzer Prize–winning biography of T. E. Lawrence, dedicated the last decades of his life to studying the psychological phenomenon of adbuction by "the visitors," as Whitley Strieber called them. Considering the data gathered by a 1991 Roper poll, Mack thought it conceivable that as many as three million Americans had undergone an abduction experience. His study of abductees led him to conclude that the phenomenon had validity beyond any psychological mechanism: "There have been numerous psychological studies of these individuals; none has discovered any psychopathology in great degree that could account for the experience." In many cases, abductees "have been witnessed by their relatives to not be present during that time. They are physically gone, and families become very distressed. . . . One of the things most difficult to accept is that this can actually have a literal factual basis. . . . Abductees may wake up with unexplained cuts, scoop marks, or bleeding noses." Mack optimistically proposed that these experiences had some sort of therapeutic value.

The narrative of contact between modern culture and the UFOs has developed over a long period, beginning with mass sightings of mysterious "air ships," like souped-up blimps, in the late nineteenth century. After World War II, accounts of flying saucers became rampant. "Between 1947 and the dawn of the age of abductees in the 1970s, there were at least six major UFO sighting waves," writes Brenda Denzler in *The Lure of the Edge: Scientific Passions, Religious Beliefs, and the Pursuit of UFOs.* Each wave produced thousands of eyewitness accounts. Sometimes picked up by radar, the UFOs would execute impossible aeronautical feats, hovering, plunging, zigzagging, skipping across water, suddenly disappearing.

On July 8, 1947, the Air Force intelligence office at Roswell Army Base in Roswell, New Mexico, announced the recovery of a crashed "flying disc" in a press release published in the *San Francisco Chronicle,* among other places. "The many rumors regarding the flying disc became a reality yesterday," the release began. The military retracted the information on the following day, explaining that the disc recovered by two intelligence agents turned out to be, upon further inspection, a weather balloon. Since that time, an entire industry of conspiracy theories has developed—books and films propounding government cover-ups, secret deals made with the aliens, issuing from that peculiar incident, and a few others like it.

During the 1950s, witnesses reported seeing saucers that had landed or crashed, with small, silver-suited humanoids standing around or working on them. Sometimes, these humanoid "aliens" would wave at the bystanders. Abduction accounts began to surface in the 1960s. The first famous report—that of Barney and Betty Hill, an interracial New Hampshire couple, whose abduction memories were recovered through hypnosis and published in *Look* magazine in 1966—established the template followed by the vast body of the tens of thousands of accounts logged since then. The "salient features," according to Denzler, include "missing time, physical examination while on board the UFO, a tour of the ship, conversation with the aliens, and the use of hypnotic regression to recover lost memories."

UFOlogist Jacques Vallee links alien abductions to ancient folktales in which humans trespassed or were cajoled into the realm of the fairy folk. Putting the episodes in the same category as Patrick Harpur's "daimonic reality," he sees them belonging to "the domain of the in-between, the unproven, and the unprovable, . . . the country of paradoxes, strangely furnished with material 'proofs,' sometimes seemingly unimpeachable, but always ultimately insufficient. . . . This absolutely confusing (and manifestly misleading) aspect . . . may well be the phenomenon's most basic characteristic." The visitors usually appear at night when the abductee is sleeping, often paralyzing them and then floating them out of their bed and onto a ship, where rapid, confusing, painful, and often repugnant events transpire.

Once selected, abductees tend to be picked up and tormented by the Grays again and again—and hypnosis often reveals that these contacts go back to early childhood. The visitors communicate telepathically, their tiny mouth slits and large black eyes never moving. They seem lacking in affect—although some abductees find them afraid or sad or amused at certain moments—and are puzzled yet fascinated by human emotional reactions. Their behavior is consistently bizarre and unpleasant, as if their actions represent a kind of mangled syntax, their true intentions concealed or distorted in some way. To take one of many examples, at the end of an abduction, the visitors exhorted one abductee, over and over again, to "eat only cow things." In another account, a male Gray paraded in front of its victim wearing her high-heeled shoes. Another abductee described a group of "small Grays" (they come in different sizes) gathered around a Christmas gift they had found in her car, opening and clumsily rewrapping it. Their hectic movements and the seemingly senseless operations they perform give the visitors an odd, fugitive quality, somehow out of sync, like figures from an old silent movie.

For the abductees, the most prevalent emotional response is one of extreme terror and violation—although some abductees, in what might be an extradimensional version of Stockholm Syndrome, come to believe in and trust their visitors, overcoming their initial reactions of horror. They convince themselves they are in league with the visitors—or were (or are) Grays in another life. They accept the claims sometimes made by the visitors, that they are here to salvage humanity and the planet from our destructive mania. Abductees often report rapes and procedures where small BB-sized implants are painfully deposited under their skin, deep up their nose or their rectum. In some cases, these implants have been retrieved and analyzed in laboratories—but they are of indeterminate origin and inconclusive proof of anything otherworldly.

In 1981, the abductions were declared an "invisible epidemic" by researcher Budd Hopkins, author of *Missing Time*. In the 1980s, Hopkins and other researchers noted the prevalence of reports describing the removal of eggs or sperm, and the compiled accounts began to suggest that the Grays were engaged in a massive "hybrid" human-alien breeding program. In dozens of reports, women are abducted, gynecological procedures performed, and then, back in their normal lives, they test positive for pregnancy. A few weeks later, their mysterious pregnancy disappears. Under hypnosis, they would recollect an intervening abduction and the removal of a tiny fetus. In future abductions—as revealed under hypnosis—the women would be shown developing fetuses, babies, or children and told that these were their hybrid offspring. A sordid ambience of accusation and guilt clings to these memories. In several accounts, abductees trying to escape from the tortures or experiments the Grays had designed for them were told by their captors: "Don't you remember? You agreed to this." As his captors inserted a needle into his brain, Strieber shouted, "You have no right to do this." The visitors answered calmly, "We do have a right."

The abductions have the ambience of intensely lucid nightmares, and some researchers suspect they are hypnagogic, chaotic, or nonlinear events that the experiencers reorganize into a more logical narrative afterward. To a certain extent, hypnotherapists may help shape the abduction narratives through subtle cues. Yet the similarity of encounters reported on different continents, the identical details picked up again and again, in thousands of reports from credible and often reluctant subjects around the world, suggest, at the very least, that something is happening that cannot be reduced to current categories of psychology, or fit into accepted frameworks of meaning. As John Mack noted, "What characterizes the abduction experience is that it is physically real and it enters the physical world, but it is also transpersonal and subjective. It crosses that barrier between the hard-edged physical world and the spirit/transpersonal world."

Although perfectly willing to concede his experiences could represent something other than alien contact, Whitley Strieber wrote: "If it is an experience of something else, then I warn you: This 'something else' is a power within us, maybe some central power of the soul, and we had best try to understand it before it overcomes objective efforts to control it."

In *Close Encounters of the Fourth Kind* (1995), an intelligent and honestly astonished account of the abduction phenomenon, *New Yorker* contributor C. D. B. Bryan crafted a physical portrait of the visitors, exploring their many anomalies. "The aliens' bodies are flat, paunchless. Their chests are not bifurcated; they have no nipples. Nor does the chest swell or diminish with breathing," he wrote. Culling from reports and his own research, he found, "The lower part of their anatomy does not contain any stomach pouch or genitals; it just comes to an end. . . . The Small Gray's body appears frail, with thin limbs and no musculature or bone structure." Some researchers assume they are less like biological organisms than machines, powered in some way we cannot comprehend, as they do not seem to eat, drink, or excrete. Nor do they have a slot for inserting batteries. Incongruous details abound: In Britain, the Grays are associated with the odor of cinnamon; in the United States, their smell is of ammonia, almonds, and lemon.

Bryan's book offered no coherent thesis to explain the phenomenon. More alarming and pointed in his conclusions than Bryan was David Jacobs in *The Threat: Revealing the Secret Alien Agenda* (1999). A hypnotherapist and professor of history at Temple University, Jacobs believes that he has, after years of work, distilled a completely logical and entirely horrifying picture of what the Grays are doing and planning—and he is disconsolate over it. He describes the breeding program, including haunting details from abductee accounts.

Captured humans are often brought to play with the children of the visitors, who are described as melancholy and lethargic. The Gray children play with blocks, similar to the blocks used by human children. But the alien blocks do not have letters or numbers on them—instead, they emit different emotions when they are turned. Since they seem to be telepathic, the visitors have no need to learn spelling or counting. The toys seem to indicate, instead, that they are trying to learn how to feel. Could it be that this yearning for affect is one reason the visitors seek human contact? Does it indicate something of their intent?

"I can discern a visible agenda of contact in what is happening," Strieber wrote in *Communion*. "Over the past forty or so years their involvement with us has not only been deepening, it has been spreading rapidly through the society. At least, this is how things appear. The truth may be that it is not their involvement that is increasing, but our perception that is becoming sharper." Even the difficulties of retrieving memories of these experiences could be part of a process in which the visitors are slowly acclimatizing us to their existence, Strieber speculates.

During the encounters uncovered in Jacobs's hypnotherapy sessions, abductees are often shown images, like propaganda films, of an apocalyptic event—nuclear war or sudden climate change—followed by clips of hybrid human-aliens walking arm in arm across a transformed Earth, the sun shining down on them peacefully. The Grays state that their breeding program will repopulate the Earth after the approaching cataclysm that makes the planet uninhabitable for our type of life. The alien agenda, Jacobs believes, has three stages—"gradual, accelerated, and sudden." We are currently in the accelerated phase. Under hypnosis, abductees report being trained to operate the Grays' saucers, and to help herd masses of people, like frightened sheep, into them, when the moment is right for the "sudden" phase.

Jacobs hypothesizes that the visitors' frequently nonsensical and bizarre behavior is a way of covering their direct intent. Like cunning cartoon villains, the visitors have used our own propensity for disbelief to render us defenseless to their agenda: the incipient takeover of the Earth. One abductee reports, "After The Change, there will be only one form of government: The insectile aliens will be in complete control. There will be no necessity to continue national governments. There will be 'one system' and 'one goal.'" Jacobs ends on a note of dread: "We now know the alarming dimensions of the alien agenda and its goals. . . . I do not think about the future with much hope. When I was a child, I had a future with much hope. . . . Now I fear for the future of my own children."

I found something wearying—not just foggy but almost smutty—about studying the abduction accounts. Almost from the first moment of pursuing it, it was as if a veil was falling down over my mind and my senses. Whether projections of our own mind or literal entities or both, the Grays call to us from a feverish twilit world of shades of grayness without clear definition. The path to understanding what may or may not be known about them by the government leads to an opaque barrier of reports of unverifiable authenticity, military and CIA panels with names like "Project Grudge," "The Robertson Panel," "Project Blue Book," and "Majestic 12," a plausible yet unreal history of covert operations, secret underground bases, cattle mutilations, alien crashes, possibly paranoiac accounts of former military personnel, and disinformation campaigns. The endless mass-market books on the subject include, inevitably, black-and-white photographs of disc-shaped objects and blurry streaks that look entirely unconvincing and somehow antiquated—a kind of 1930s idea of what a futuristic technology might look like.

But what if there was a literal truth to David Jacobs's narrative? Was it possible that the Grays, as horrible as they sounded—as disreputable, somehow, as the entire enterprise seemed—were actually orchestrating an imminent evolutionary shift for the hapless human species?

Reading scores of abduction accounts, I felt a pitiful sense of helplessness against this telepathic, sorcerous, affectless enemy—"the bugs," as many abductees call them. I thought about my disappointment with the human race, seemingly hurtling toward ecological collapse and nuclear disaster, unable to control our worst impulses. Was this all part of a process, to create the forced conditions for a transformation that would, indeed, be apocalyptic at a very deep level? And if this might be the case, then what would be our best response to "The Threat," as Jacobs called it? Should we try to resist the visitors? Should we surrender to their morbid mastery? But then, why was there something so laboriously theatrical, tacky, and fraudulent about all of it?

In June 2002, I went with my partner and our baby to the opening of the Documenta11 exhibition, in the West German town of Kassel. As a journalist writing about art, I had always hoped to visit this exhibit, which takes place every five years. I associated Documenta with the hard-core cool of the conceptual art of the early 1970s, with the German conceptual artist Joseph Beuys, known for his neo-shamanic self-mythologizing and iconic displays of iron plates, felt, and fat. Unfortunately, by the time I finally managed to attend, my mind was filled with other, stranger matters.

We stayed at a hotel in the Wilhelmshöhe Bergpark, across the road from the city's baroque castle. The castle had beautiful gardens, old gnarled trees, and a towering stone monument of Hercules clad in a lion skin at the end of a long reflective pool and fountain. The Brothers Grimm had lived near the park, and their house was a local attraction. During the day, we toured the exhibition halls spread across the city, in old factory buildings, breweries, and railway stations. Organized by Okwui Enwezor, a Nigerian curator, the exhibit was starkly political. It featured numerous Third World artists, and minority artists from the West. Many works addressed the destructive excesses of globalization, allegorically or directly. One film documented the bleak, monotonous lives of South African diamond miners, in bunkers and tunnels deep under the Earth. Sculptures mocked the modernist visions of utopia, parodied colonialism's slave-driven delusions of grandeur. The exhibit was angry, inspiring—a post-Marxist assault on global imperialism.

In the hypersophisticated ambience of Documenta11's numerous receptions and parties, standing amidst espresso-drinking aesthetes and stylish art dealers chattering in various Romance languages about museum shows and beach resorts and the latest art world gossip, I found myself thinking incessantly about the abduction saga, the postmodern myth of the visitors. Were these glamorous and well-heeled aesthetes soon to be fodder for an orchestrated alien takeover, doomed to explain Neo-Conceptualism and Post-Pop to short, affectless, hyperdimensional invaders?

One night, after a long day of art-going under the pouring rain in Kassel, I had a vivid dream about the visitors. In the dream, I went with two friends to meet one of the "Gray Alien" commanders in an Upper West Side lobby. The alien resembled a Chinese woman. She wore a red silk dress, had large almond-shaped eyes, and four fingers on each hand. She spoke as if we were going to make some kind of deal.

"It's going to be great for you when we take over your planet," she told me. "We can't wait to help you. We want to show you around the galaxy." She called for her assistants. They were hunchbacks with bulbous features, resembling medieval trolls. They put my two friends on their backs and gave them piggyback rides. The alien commander pointed upward, where cheap tinsel stars and planets were stuck on the lobby's domed ceiling. She acted as if this were an impressive sight, and my friends did seem impressed. I was disappointed: Was this all they had to show us?

Confused, I left the lobby and went, alone, to a crowded, seedy nightclub where a long-haired weirdo came up to me with his girlfriend. They were "hybrid" human-aliens. The man laughed and put one of his four fingers deep into my mouth. Immediately, in the dream, I turned around and put my finger just as far into his girlfriend's mouth. Then we all laughed about this almost obscene exchange.

I awoke from the dream and recalled the details before reaching for my notebook—over the last years, while exploring shamanism, I had developed disciplined habits of dream recollection. Wide awake, I reflected on the dream's particular seamy, swampy ambience. Before I started to write it down, my partner, in deep sleep, suddenly sat up and leaned toward me. She opened my mouth with one hand. She brought her other hand to my face, and put one of her fingers into my mouth.

Startled, I woke her immediately. But she remembered nothing of it—or what she had been dreaming.

Later, I learned that the area around Kassel—an ancient area similar in some ways to the stone-circle-studded landscape of Wessex in England—is the center of German crop circle activity. Several new formations appeared in local fields on the weekend that we were there.

The doors of Chapel Perilous swung open to welcome me inside.

CHAPTER TWO

Only if we grant power to something can it have power over us. It becomes a serving and sustaining potency when we again are able to place it into the realm where it belongs, instead of submitting to it.

JEAN GEBSER, *The Ever-Present Origin*

Writing in the early decades of the twentieth century, the German Jewish critic Walter Benjamin noted that modernity was causing an emptying-out of experience, as well as destroying the aura that had previously belonged to precious artworks and natural objects, giving them their unique presence. "Less and less frequently do we encounter people with the ability to tell a tale properly. More and more often there is embarrassment all around when the wish to hear a story is expressed," he wrote in his essay "The Storyteller." "It is as if something that seemed inalienable to us, the securest among our possessions, were taken from us: the ability to exchange experiences."

Modernity unleashed a succession of shock effects, changing the nature of perception, as well as the individual's relationship to his own personal history. "Experience has fallen in value. And it looks as if it is continuing to fall into bottomlessness." The old culture of contemplation gave way to the mass absorption in distractions. The wise counsel embodied in the storyteller's art was supplanted by the endless parade of statistics and information in the daily newspaper. To enter the modern world, we forfeited our capacity for intimate exchanges requiring slowness and reciprocity and the play of the imagination. "To perceive the aura of an object we look at means to invest it with the ability to look at us in return," Benjamin wrote. The value of perception and the meaning of personal history were degraded and denigrated to institute a mass society focused on abstractions, impelled by a "sense of the universal equality in all things."

The loss of the art of storytelling corresponds to a change in our experience of time. "The value of information does not survive the moment in which it was new. It lives only at that moment; it has to surrender to it completely and explain itself to it without losing any time," Benjamin wrote. "A story is different. It does not expend itself. It preserves and concentrates its strength and is capable of releasing it even after a long time." He compared stories to those "seeds of grain which have lain for centuries in the chambers of the pyramids shut up air-tight and have retained their germinative power to this day." Information, on the other hand, is always accompanied by explanation—"no event any longer comes to us without being shot through with explanation. In other words, by now almost nothing that happens benefits storytelling; almost everything benefits information." It is almost our tragic fate as modern people to long for meaning and receive only explanation.

The earliest form of the story is the fairy tale, "the first tutor of mankind," according to Benjamin. "Whenever good counsel was at a premium, the fairytale had it, and where the need was greatest, its aid was nearest. This need was the need created by the myth. The fairytale tells us of the earliest arrangements that mankind made to shake off the nightmare which the myth had placed upon its chest." In the fairy tale, we learn to overcome our terror of the unknown, to trust in natural powers and the instinct of animals, to have faith in the secret workings of destiny. "The wisest thing—so the fairytale taught mankind in olden times, and teaches children to this day—is to meet the forces of the mythical world with cunning and with high spirits." In the fairy tale, we discover the power of names—how fear is dispelled, how the "other" loses its power over us, when we learn the secret name of what frightens us, as in the story of Rumpelstiltskin.

One classic form of the fairy tale, Benjamin tells us, is "the man who sets out to learn what fear is." Whitley Strieber's abduction memoir, *Communion: A True Story,* provides the postmodern inverse of this archaic quest for knowledge—a much uglier tale, but one that also suggests a moral: If you do not go out to find fear, fear finds you—or, what might be worse, it does not.

Communion was a best-seller—one of those endless amusements absorbed without reflection by the distracted masses—and it imparts the basic form of the abduction narrative, a story with an essential lack or aporia at its center—not a story at all, in fact, but a bundle of clues and fragments that lead nowhere, that do not cohere. "The visitors marched into the middle of the life of an indifferent skeptic without a moment's hesitation," notes Strieber, a writer of "imaginative thrillers." He was home at his secluded cabin in upstate New York one night, when the abduction occurred—although he doesn't remember the events until days later. Persistent pain—the physical traces left by the traumatic events—finally awakens his dulled recall. "I believe that the combination of the infected finger, the rectal pain, and the aching head were what finally brought my memories into focus."

Indifference and skepticism are two potent forces in the modern mentality. We have elevated them to the status of values. They are part of the way we have learned to inure ourselves from shock—what Benjamin calls "the price for which the sensation of the modern age may be had: the disintegration of the aura in the experience of shock." They furnish us with a certain shabby level of comfort. Franz Kafka wrote: "There is an infinite amount of hope—but not for us." To be an "indifferent skeptic" is to have reached the end of a certain evolutionary line—for a passionate skeptic, or even an indifferent believer, there might still be hope. For an indifferent skeptic, all that remains is the piling up of fact and statistic, to be sorted into categories of explanation. Strieber's memoir is a document of alienation, as well as aliens. Like the technological "enframing" defined by Martin Heidegger, the trauma inflicted upon him by the visitors is one that has already afflicted him in his essence.

What he begins to recollect is an experience—the last in a series of experiences—of such paralytic impotence and violation that his mind, heart, and will recoil from accepting it. Against his powers, the aliens remove him from his house and float him up to a spaceship, into a "messy round room," "actually dirty" as well as stuffy, dry, and confining, where "some clothing was thrown on the floor." Unable to move, he finds himself "in a mental state that separated myself from myself so completely that

I had no way to filter my emotions or most immediate reactions. . . . My mind had become a prison." He is shown an "extremely shiny, hair-thin needle mounted on a black surface," and told that the visitors intend to insert this into his brain. The operation is performed with "a bang and a flash." Almost weeping, he sinks down "into a cradle of tiny arms."

There is a female among the visitors with whom Strieber feels a slight kinship. "She was as small as the others, and appeared almost bored or indifferent." In the spaceship, he asks her if he can smell one of the visitors. "Oh OK, I can do that," says one of the males absently, holding his hand to Strieber's nose. "There was a slight scent of cardboard to it, as if the sleeve of the coverall that was partly pressed against my face were made of some substance like paper," he recalls. "The hand itself had a faint but distinctly organic sourness to it, but it was unmistakably the smell of something alive. There was a subtle overtone that seemed a little like cinnamon."

He next finds himself in "a small operating theater. I was in the center of it on a table, and three tiers of benches were populated with a few huddled figures, some with round, as opposed to slanted, eyes." He is passed along to another room, his fear increasing whenever they touch him. "Their hands were soft, even soothing, but there were so many of them that it felt a little as if I were being passed along by rows of insects." He is shown a foot-long mechanical device, "gray and scaly, with a network of wires on the end." This instrument is inserted into his rectum. "Apparently its purpose was to take samples, possibly of fecal matter, but at the same time I had the impression that I was being raped, and for the first time I felt anger." An incision is made on his forefinger. He is returned to his bed, awakening with no immediate recollection of the episode, but a deepening malaise.

Hypnotherapy restores the full dimensions of these events to his awareness—and it also introduces him to a suppressed pattern of past experiences of lost time and alien encounters, hours and days and weeks, stretching back through his entire life. He recalls one night, a few months before the abduction, when a light passed his window, and a small, hooded figure appeared in his room, "something totally unknown to me, glaring at me from right beside my bed in the dead of night. . . . I relived fear so raw, profound, and large that I would not have thought it possible that such an emotion could exist." The visitor takes a silver ruler and touches Strieber's forehead. He is shown pictures of the world blowing up. "It's a picture of like a whole big blast, and there's a dark red fire in the middle of it and there's white smoke all around." The visitor tells him, "That's your home. That's your home. You know why this will happen." He finds corroboration for this account from friends at his house, who also recall a light and other disturbances.

This recollection leads him back to earlier ones, from childhood, from a period in 1968 when he had dropped out of college and experienced "four to six weeks" of missing time, associated "with a perfectly terrible memory of eating what I have always thought was a rotten pomegranate, which was so bitter that it almost split my head apart." He spits up this substance in this encounter, but in later abductions, the visitors find ways to make him swallow what they are forcing on him. He is disturbed by the "structural coherence of the thing. First I am fed and it comes back up. Then I am fed again and this time drops are used to prevent the material from returning. . . . In short, my hallucinatory friends seem to have learned something about getting me to digest whatever it is they are trying to feed me." Strieber wonders if he has the capacity to live with the notion "that my whole life might have proceeded according to a hidden agenda." In one abduction, he was told, "You are our chosen one," but this declaration was accompanied by a sterile humor on the part of his captors, who did not seem to take it seriously.

"The goal does not seem to be the sort of clear and open exchange that we might expect," Strieber writes. "Whatever may be surfacing, it wants far more than that. It seems to me that it seeks the very depths of the soul; it seeks communion." The title of his book was given to him as a directive. He was planning to call it *Body Terror* until his wife talked in her sleep one night. Suddenly speaking in a "strange basso profundo voice" that was not her own, she said, "The book must not frighten people. You should call it *Communion,* because that's what it's about."

In Strieber's memoir and other abduction accounts, it is as if the whole inhuman mythical world, banished beyond the margins of the thinkable by the modern consciousness, returns to stare at us with detached longing and assault us with implacable fury. It is as if the muteness of created things, denied their aura and their place, demands recognition and redemption. Refused this hope of redemption, the mythic forces tell us quite clearly that they will happily assist in the extinction of our world. So similar, yet different, from the old fairy tales of fairies and gnomes, lurking on the archaic borderlands of our reality, the manifestation of the visitors seems to be something bodied forth by our throttled imaginative powers. With their mood of guilt and accusation, with their infernal and enigmatic and explicitly sexual punishments, with their peculiarly bureaucratic and murky quality, the abduction accounts suggest the particular fabulistic territory defined by Franz Kafka, known as the Kafkaesque.

The closest literary parallel to the abduction experience is found in Kafka's fables—in "The Metamorphosis," the story of the man who awakens as a bug one morning; in the inescapable accusation and existential guilt that permeates *The Trial*; in the infernal contraption that inscribes the condemned man's sentence over and over into his flesh in "The Penal Colony"; in the jaunty, dispossessed figures of the assistants and guardians who multiply in his tales, and whom Benjamin connects with the *gandharvas* from Indian mythology, "celestial creatures, beings in an unfinished state." The horror of the visitors is matched by the horrific passivity of modern man, trapped in technological encumbrances and bureaucratic systems designed by modern man to keep himself distracted, comfortable, and asleep. Strieber undergoes a battery of psychological tests and learns he is "sane," but what does that mean? Like the accusers in Kafka, his visitors do not seek to redeem or reform or change him in any way—they just are, like the somber reality of a crime scene.

Ransacking his mind and library for possible explanations, Strieber wonders if "we were creating drab, postindustrial gods in place of the glorious beings of the past." He conjectures the intruders might represent a more primitive species that had attained a form of hive intelligence in a different manner from us. "What if intelligence was not the culmination of evolution but something that could emerge from the evolutionary matrix at many different points, just as wings and claws and eyes do." He considers that humanity might be a larval form, transmuting into something entirely other after death, and the visitors were the dead returning in unknown form. "Perhaps the dead had been having their own technological revolution, and were learning to break through the limits of their bourne." He reiterates theories of Carl Jung, Sigmund Freud, and P. D. Ouspensky, rehearses ideas from Navajo, Greek, and Egyptian myth, from Zen, quantum physics, and Hinduism. None of these provide satisfaction or consolation. Accumulation of explanations is like the accumulation of things, turning increasingly cold and desolate, punishing us for our lack of presence, our insufficiency, our stubborn refusal to meet reality on its own terms. "Warmth is ebbing from things. The objects of daily use gently but insistently repel us. Day by day, in overcoming the sum of secret resistances . . . that they put in our way, we have an immense labor to perform," Benjamin noted.

The weight of the dead upon the living is a common theme in Kafka, linked to the draining and dehumanizing pressure exerted by all authorities. In an essay on Kafka, Benjamin wrote: "Uncleanness is so much the attributes of officials that one could almost regard them as enormous parasites. . . . In the same way the fathers in Kafka's strange families batten on their sons, lying on top of them like giant parasites. They not only prey upon their strength, but gnaw away at the sons' right to exist. The fathers punish, but they are at the same time the accusers. The sin of which they accuse their sons seems to be a kind of original sin."

This original sin, for Friedrich Nietzsche, was *ressentiment,* bad conscience, which developed due to our amputation from our original condition as warriors, hunters, and "beasts of prey," in civilization. Civilization domesticated humanity, made us docile, *"a household pet,"* although our inner nature continues to rebel against this deformation. "Man, full of emptiness and torn apart with homesickness for the desert has had to create from within himself an adventure, a torture-chamber, an unsafe and hazardous wilderness—this fool, this prisoner consumed with longing and despair, became the inventor of 'bad conscience.'" Modern nihilism revealed "man's sickness of man, of himself," but this "forcible breach from his animal past" suggested, and pressed toward, new conditions of existence, "as though something were being announced through him, were being prepared, as though man were not an end but just a path, an episode, a bridge, a great promise. . . ."

Strieber seems to represent a particular type of modern man who has recalled the need to seek for a soul—he notes that he worked for fifteen years at the Gurdjieff Foundation—yet in whom that desire, in itself, has somehow been blandified, routinized, losing its savor. It has backfired or sputtered out. The search for a soul becomes another detached pursuit—like the search for a new sensation or new possession one cannot quite identify. The quest stops short of the radical otherness of enlightenment or illumination, which robs the ego of its most precious possession: its ego-hood. As the Hindu guru Sri Ramana Maharshi put it, "Liberation exists—and you will never be liberated." The "coming of the self" described by Edward Edinger is an apocalypse for the ego, the "you" that wants to hang along for the ride. It may be that the only way to survive the Apocalypse is to undergo it, first, within your own being.

Strieber's memoir warns us that explanations are meaningless if we lack the inner resources—the cunning and high spirits—necessary to meet the forces of the mythical world, as they well up within us. "The liberating magic which the fairy tale has at its disposal does not bring nature into play in a mythical way, but points to its complicity with liberated man," Benjamin tells us. It is through the continuing activity of self-liberation that we escape the parasites that would prey upon us, overthrowing, at the same time, our delusory burden of original sin.

Goblins are clever: Our fairy tales tell us this. These entities—"unfinished," half-created, or soulless—set age-old traps for humanity. The Grays or the visitors are perhaps best understood as an updated form of goblin or gnome. According to the Austrian visionary Rudolf Steiner, an extraordinary cartographer of the supersensible worlds, what gnomes and goblins are made of is "entirely sense, and it is a sense which is at the same time *intellect*." Lower creatures such as these "have no bony skeleton, no bony support; the gnomes bind everything that exists by way of gravity and fashion their bodies from this volatile, invisible force, bodies which are, moreover, in constant danger of disintegrating, of losing their substance," he taught, in *Harmony of the Creative Word: The Human Being and the Elemental, Animal, Plant and Mineral Kingdoms,* a series of lectures given in Germany in 1923. Despite its interstellar pretensions, nothing about the visitor narrative suggests an extraterrestrial origin. The Grays seem to possess no inner life, no history, nothing beyond their vindictive and macabre relationship to the human species. Researchers of UFOs have noted that many, if not most, observations seem to occur along earthquake fault lines, and the caverns that abductees sometimes describe being taken to seem to be deep underground. The "flying saucer myth" seems a construct of these entities, a theatrical staging, to compel our belief in them. They do not come down from empyrean reaches, but out of the mythological depths of the Earth—from a lower dimensional reality, perhaps, that is saturated with ancient gloom and dark yearnings.

According to Steiner, gnomes constantly move along veins of minerals and metals underground, and they "laugh us to scorn on account of our groping, struggling intellect with which we manage to grasp one thing or another, whereas they have no need at all to think things out." These "root spirits" are also connected to the waste processes of the Earth, to toads and amphibians. Working beneath the surface of the planet, in the Earth's mineral layers, the gnomes take a similar role in our dream lives. We are terrestrial beings, and therefore the gnomes are also at work "beneath the Earth" of our psychic life, in our subconscious depths. "Anyone who gets to the stage of experiencing his dreams in full consciousness on falling asleep is well acquainted with the gnomes," Steiner noted. For somebody unprepared, the experience would be an alarming one: "at the moment of falling asleep, he would behold a whole host of goblins coming towards him. . . . The form in which they would appear would actually be reflections, images of all the qualities in the individual concerned that work as forces of destruction. He would perceive all the destructive forces within him, all that continually destroys." If we were able to remain conscious as we dreamed, what we would perceive would be "a kind of entombment by gnomes" in the astral world.

One of the few modern intellectuals to appreciate Steiner's work, William Irwin Thompson characterizes the "Steinerian vision" as one that "looks at the human as so completely embedded in the animal, vegetal, and mineral evolution of the solar system that it becomes nonsense to separate a fictive 'matter' from mind, and a mere three dimensions from ten. . . . All of the seemingly mystical perceptions of Steiner have a biological relevance that fits a new kind of science, and a new kind of culture." For Steiner, the universe is a harmonic orchestration in which our minds are embedded, not as alienated organs, but as integrated aspects of the whole. Within this whole, other forms of mind also exist, a vast ecology of consciousness, that can be registered by us, if we work to attain what he called "supersensible perception." Throughout his innumerable lectures and books, Steiner describes "lower creatures" such as the goblins and other "root spirits," as well as great cosmic entities, angelic hierarchies, and higher spiritual beings.

In earlier phases of development, humanity had a deep affinity, an unconscious attunement, to the Earth and the cosmos. This affinity is reflected in our oldest stories and mythologies, which reveal cosmological secrets about the movements of the planets and the stars, as well as evolutionary processes in the history of the Earth. Thompson considers fairy tales and myths to be "forms of cultural storage for the natural history of life." He proposes that the ancient saga telling of fairies defeating goblins and banishing them beneath the Earth is a representation, as myth, of the original "pollution crisis," more than 500 million years ago, when cyanobacteria developed photosynthesis, creating an oxygen-rich environment that was poisonous for other life forms, radically transforming the biosphere to make it sustainable for larger mammals. Translating from myth to biology, Thompson proposes that "the goblins are the anaerobic bacteria who live on and in our wastes and garbage, and the fairies, those airy creatures of light, are the cyanobacteria that were the first to invent photosynthesis to feed on light to give off the oxygen that would become the new atmosphere of an illuminated world." Anaerobic cells contain no nucleus—like the Grays, whose bodies are undifferentiated and without internal organs.

In the saga of the visitors, we are witnessing a return of the repressed, the mythic world, surging into the postmodern consciousness in a form that strangely fits our fixation on technology, our space fantasies and genetic obsessions and dingy bureaucracies, and our terror of the unknown. The Grays exist on the boundary of the sensible, seeking entry into our realm. According to mythic thought, time is not linear but cyclical, or, more accurately, it turns in a spiral. It appears that we are experiencing a reenactment of the ancient narrative described by Thompson—of the battle between forces of light and darkness, or of degradation and evolution—but taking place on a different level, a higher octave of being and knowing. Burrowing away at the gaps in our subconscious and fragmented knowledge systems, gnawing on our bad conscience, the visitors have worked like industrious ants to create certain conditions, and to imprint a particular narrative into our popular culture, absorbed by the masses in a state of distraction. They force-feed their abductees, over decades, until they learn to swallow what they want them to eat.

The Grays are analogous to bacteria, but operating on the level of the psyche. Earthly bacteria derive energy from breaking down complex organic compounds that they take in from the environment. The visitors seem to be entities that sustain themselves from the negative emotions, such as fear and anxiety, emanated by the human nervous system and energy body. Just as earthly bacteria play a crucial role in the global ecosystem, we may eventually understand that our infraterrestrial intruders occupy a necessary niche in the ecosystem of consciousness, as it develops or decays.

Are the visitors "real" or "imaginary"? They are both, and they are neither. Like quantum phenomena, they do not exist or not exist, they also do not exist and not exist, nor do they both not exist and not not exist. According to Dzogchen, a tradition of teachings and practices linked to Tibetan Buddhism, ultimately there are no entities. Any entity only possesses relative reality, including ourselves. Overcoming dualism is essential to Dzogchen, as Chogyal Namkai Norbu writes in *Dzogchen: The Self-Perfected State:*

> Duality is the real root of our suffering and of all our conflicts. All our concepts and beliefs, no matter how profound they may seem, are like nets which trap us in dualism. When we discover our limits we have to try to overcome them, untying ourselves from whatever type of religious, political, or social conviction may condition us. We have to abandon such concepts as "enlightenment," "the nature of the mind," and so on, until we no longer neglect to integrate our knowledge with our actual existence.

Entities who manifest in other forms of consciousness, such as the Grays, are, at the same time, separate from us and aspects of our psyche. We are the ground for their manifestation, and it is only by attaining a nondual perspective that we can understand them.

Reality, according to Eastern mysticism, is maya, illusion—a tonal tapestry or spectrum of vibrations called, in the Mayan tradition, *tlalticpac,* the dream-world of Earth. There is, according to Dzogchen, neither being nor nonbeing. However, there may be an infinite number of relatively real entities, possessing varying forms of consciousness. Overcoming dualism, we can recognize such beings as fractal shards or autonomous archetypes of our own psyche, as well as self-directed entities undergoing their own forms of evolution.

From such a nondualistic perspective, it makes sense that some encounters with the Grays are neutral or even positive. Betty Andreasson, a woman whose accounts of abduction were published in the 1960s, described a classic mystical vision in which the Grays showed her a giant golden phoenix that was consumed by a flame. In her descriptions, and many others, the intruders reveal an ambivalent polarity. Monsters produced by "the sleep of reason," they have a necessary part to play in the evolution of human consciousness—the process of psychic transmutation that alchemists called the "Great Work."

Connected to our technological development, the Grays embody a malignant, supersensible element lurking beneath our fascination with mechanization, revealing the irrational basis of our constricted rationality. They also have lessons to teach us. As the critic Lewis Mumford noted, "Our capacity to go beyond the machine rests on our power to assimilate the machine. Until we have absorbed the lessons of objectivity, impersonality, neutrality, the lessons of the mechanical realm, we cannot go further in our development toward the more richly organic, the more profoundly human." Like transhumanist zombies, the Grays embody the reductive perspective that sees everything—matter, genes, human souls—as resources to be used for purposes of control and domination. In this way, the visitors serve as a warning, as well as an innoculation against a nightmarish fate we can recognize, and reject, in the time that remains to us.

As Prospero says of Caliban in *The Tempest:* "This thing of darkness I acknowledge mine." According to occult tradition, humanity has a responsibility to all of the elemental beings, those other forms of consciousness that express themselves through the natural world. We are supposed to learn to work with the elementals and, also, alleviate their suffering—it is clear from the abduction accounts that the visitors are suffering. Like dusty insects attracted to flame, the Grays yearn for our qualities of soul-warmth;

despite their cunning and technological acumen, these qualities remain beyond them. They are intelligent and sentient, hence aware of their exiled status. Unable to escape their de-souled condition, they desire to draw humans into their lower world, sustaining their half-lives on our subtle energies. They appear to be utilizing their dream-world technologies in a serious and desperate attempt to find a way out of their cul-de-sac.

The incomprehensible murkiness of the phenomenon suggests that it is occurring, for the most part, at a different plane of consciousness, not the daylight awareness of the waking state—the only form of consciousness that the modern mind believes real—but in a twilight or dream-like subliminal state. This quantum indeterminacy extends to manifestations that would appear to be purely material—such as the Roswell incident of the crashed "flying disc," or the various government reports and semiofficial statements that reveal nothing, and perhaps conceal nothing as well. On books and Internet sites, there appear testimonies of former officials in the military or government who step forward, admitting to taking part in an organized cover-up of alien activity, without offering convincing evidence. Could high-level political and military figures be colluding with the visitors, even to the point of signing Faustian pacts, yet themselves flickering in and out of awareness of what they are doing? The conspiratorial route descends into deep levels of tabloid darkness, generating waves of fear and paranoia. The obsession with finding the literal truth of the abduction phenomenon may be, in itself, a red herring, obscuring the significance of the visitors as psychic archetype.

In one of Carlos Castaneda's books—which are of appropriately sketchy provenance—Castaneda starts to encounter his "double" in dreams. At first, he assumes he is dreaming the double. Eventually, the sorcerer Don Juan tells him he is wrong: "The double is dreaming you." Perhaps we are the dream of the visitors, as they are our nightmare. With their invasive probings and technological acumen and genetic obsessions, the visitors seem to be a kind of field effect or echo of our fixation on materialist technologies and hierarchies of dominance. While not the Other we would hope to encounter, the Grays may be, at this stage, the other we deserve.

In Mayan and Aztec traditions, the calendrical end date, or birth date, in 2012 "signified an open door to the cosmic center, in a sense the arrival of the cosmic source into local space-time, or, in other words, a renewal by returning into the central fire," the independent Mayan scholar John Major Jenkins wrote in *Maya Cosmogenesis 2012*. "The local and Galactic planes would then be aligned, opening the way, as the Aztecs said about the end of the Fifth Sun, for the *tzittzimime* (celestial demons) to pour down out of the sky to devour mankind." The visitors appear to be the *tzittzimime,* an archetype from Aztec myth taking postmodern form as we approach the completion of this cycle. Now that they have colonized our movies and dreams, a literal manifestation of them cannot be ruled out. Against the wishes of the conscious mind, the psyche appears to be unleashing tremendous forces of creative destruction to attain self-knowledge and noospheric activation—to force its own evolution.

Or—or perhaps and—it is all just a fairy tale.

CHAPTER THREE

Quantum measurements inject our consciousness into the arena of the so-called objective world. There is no paradox in the delayed-choice experiment if we give up the idea that there is a fixed and independent material world even when we are not observing it. Ultimately, it boils down to what you, the observer, want to see.

AMIT GOSWAMI, *The Self-Aware Universe*

Do places exist on the Earth where the veils separating matter from myth, psyche from spirit, are thin or partly lifted? Glastonbury, England, would seem to be such a spot. The air feels a little different, intangibly charged, as if ancient mysteries were imprinted in the local ether, awaiting the right touch to return to our world. The medieval town also attracts more than its fair share of long-haired louts and lost drunks, as well as spaced-out kids with mangy dogs who come for the annual music festival and forget to leave. They squat in front of the rows of New Age shops hawking mawkish trinkets—crystals, magik wands, tarot cards—insulting passersby, swigging stout, and begging for coins.

Except for its New Age adornments and ruffian factor—also a high local incidence of UFO sightings—Glastonbury and the region around it are remarkably serene and well preserved. Gardening is a popular local pastime; front yards are adorned with shimmering blooms and fragrant blossoms nourished by the abundant rains. I stayed at a bed-and-breakfast three miles from the town center, and enjoyed my daily walk past emerald-green farmers' fields and cow pastures, keeping my eyes on the Tor, the ancient hump-shaped hill and pilgrimage site that looms over the town, topped by a fifteenth-century tower. "There is such magic in the first glimpse of that strange hill that none who have the eye of vision can look upon it unmoved," wrote Dion Fortune, an early-twentieth-century mystic, founder of the Fraternity of the Inner Light, who lived beneath it. She called Glastonbury "a holy place and pilgrim-way from time immemorial, and to this day it sends its ancient call into the heart of the race it guards, and still we answer to that inner voice."

Man-made ridges circle the Tor, where cows and sheep often graze, in a spiral pattern. The hill was inscribed with a labyrinth at some distant, long-forgotten point in the past, perhaps thousands of years ago. The path that winds around it may have been used in archaic ceremonies and initiation rites. "There can be but little doubt that the priests of the ancient sun-worship had here their holy place," Fortune wrote, in her lilting antiquarian style. According to legend, this curiously anthropomorphic mound is a place where the human realm and the spirit world brush each other. "Gwynn ap Nudd, the lord of Annwyn or Hades, was allotted an invisible palace inside the Tor, to which he repaired with his Wild Huntsmen," wrote the local scholar Geoffrey Ashe in *King Arthur's Avalon*. He considered the Tor the probable site of the spiral castle described in the Grail legends. According to local tradition, Morgan Le Faye, King Arthur's faery half-sister, lived at the foot of the Tor, by a natural underground spring, now a local garden sanctuary called the Chalice Well, providing a counterpart of feminine yin energy to the Tor's masculine yang. The water from the Chalice Well, reddish and iron-tinged, is believed by New Age pilgrims to have mysterious healing properties.

"Behind all the legends, prophecies and revelations at Glastonbury can be discerned one single theme: that the will of God will finally prevail, and humanity will rediscover its natural condition within an earthly paradise," wrote John Michell in *New Light on the Ancient Mystery of Glastonbury*. An independent scholar-anthropologist with a large cult following, Michell pursues trails of myth and geomancy leading from ancient paganism and neolithic lore to the establishment, according to legend, of the world's first Christian Church at the site of Glastonbury Abbey by St. Joseph of Arimathea, meshing with medieval fables of King Arthur and the Grail.

For archaeologist Philip Rahtz, such "alternative archaeology" has turned the so-called "Sacred Isle of Avalon" into "the Mecca of all irrationality," attracting "an incredible variety of hippies, weirdoes, drop-outs, and psychos, of every conceivable belief and in every stage of dress and undress, flowing hair and uncleanliness."

The morning after I arrived, I took a bus tour to visit a few crop circles, some of the newest and best of the season. The tour guide on our bus, David Kingston, a polite gray-haired man in his sixties, dated his involvement with the phenomenon to the 1970s, when he worked in British military intelligence. In 1976, he was part of a group investigating UFO activity in Wiltshire. One night, stationed atop a hill, he watched six or seven "spheres of light" hover in the air, then merge into one long sphere. The long sphere ascended to the height of the stars, then blinked out. "At daylight, we found a straightforward circle, thirty feet in diameter, in wheat," he recalled. He went on to give a short history of the phenomenon since that point. He was especially interested in shifts in the magnetic fields around formations, and the discovery that particular glyphs emitted high-pitched sounds. "Quite often, we don't hear anything by ear. We pick it up by machine, and analyze the sounds later."

The first crop circle we visited had just appeared the day before. Our guide passed around the bus a photograph of the formation, taken by airplane. On one level, the formation showed a tree bearing a large number of round fruits, its roots spreading out in a fan shape beneath it. However, the image could also be read as a mushroom—not just any fungus, but the Fly Agaric or Amanita muscaria, the red-and-white-spotted visionary sacrament of Siberian shamanism. When the image was flipped upside down, the underside of the "tree" appeared as another mushroom—the roots turned into the small ridges or gills of the classic *Psilocybe cubensis,* the "magic mushroom" used in South American shamanism, also prevalent throughout England and Ireland.

In *Breaking Open the Head,* published later that year, I discussed a scholarly dispute over these two species of fungus, both considered candidates for the original Soma, the visionary intoxicant or entheogen described in the Rig Veda as a milky liquid pressed from a plant. Although its nature is long forgotten, Soma was the archaic catalyst of Hindu cosmology. The banker and mycologist Gordon Wasson—rediscoverer of the mushroom cult of the Mazatec in Mexico in the early 1950s—argued that amanita was the mushroom used by the Hindus. Terence McKenna countered that Soma was most likely a species of psilocybin-containing fungus. McKenna theorized that psilocybin, not Amanita muscaria, sparked human language development, and was the lost dispenser of "vegetable gnosis" at the heart of most world religions. In one hieratic icon, the double—or triple—reading of the crop glyph suggested all of these Gnostic possibilities.

I also noted the thesis that early Christianity concealed a secret cult devoted to Amanita muscaria, developed in the 1970s by Princeton art historian John Allegro, in *The Sacred Mushroom and the Cross,* a controversial work. This idea was recently revived by the Italian scholar Giorgio Samorini, who assembled images of frescoes from early Christian churches in support of Allegro's thesis. Some of these pictures, dating back to the second or third century AD, showed a plant that would be recognized as a Tree of Life by churchgoers, or as an amanita by initiates into the Mystery. An entire lexicon of hermetic imagery and psychedelic scholarship had been artfully evoked and skillfully compressed into a single glyph imprinted in Wiltshire wheat by unknown perpetrators.

There also seemed the suggestion of a figure in the formation, hanging upside down from the tree. As the glyph related to shamanism, this figure suggested the Norse god Odin, or Votan, a prototypical shamanic figure. Odin hung from Yggdrasil, the World Tree, for nine days, in order to achieve knowledge of his own fate and the fate of his fellow Norse gods. During that time, he learned that the gods were going to perish during Ragnarok, a final struggle with their titanic foes. For Odin, it was a lesson in humility, forcing acceptance of the inescapable shape of fate. The World Tree is a universal shamanic symbol, representing the various levels of being traversed by the shaman, from the underworld to the realms of the gods.

The Tree of Life crop circle was in Adam's Grave, a few miles from the stone circles of Avebury and the peculiarly tit-shaped man-made mound of Silbury Hill. We parked the bus and walked out into it—the Glastonbury conference had made deals with local farmers, paying them in advance for access to their fields. Farmers in the area run the gamut in their response to the formations—some try to exploit the phenomenon, others cut out the patterns as soon as they appear, or bar visitors access to the land. Still others superstitiously tolerate them. We followed arrows into a field not too far from the road. The gold-colored grain came up to my waist, tickling my arms. It was a beautiful, sunny day.

The formation was over three hundred feet long, and it had been executed with exactitude. There was a little pip at the center of each fruit dangling in the tree's branches (or each spot on the amanita). In the field, this effect was produced by taking the central clump of standing wheat stalks and twirling them around—it was difficult to see how hoaxers would have accomplished this, so elegantly, in darkness. On the ground, the crop had been laid down in careful patterns, in some places designed to create a three-dimensional effect. Each of the thin—perhaps foot-wide—lines of fallen wheat making the ridges that appeared as the roots of the tree, or the gills of the mushroom, had been toppled in the opposite direction from the ones next to it. The crop appeared to have been knocked down by a compressed force of some kind, in one fast and precise movement. There did seem to be an energetic charge within the formation—I felt, almost, as if I were tripping. I had the impression of an invisible eyeball suspended overhead, as Terence McKenna noted in the Amazon, "something in the sky, calmly omniscient but closely observing us." Many of the people in our tour group quickly sank to the ground. Falling silent, they assumed meditation postures, or lay flat on the fallen wheat. This reaction is typical, almost non-volitional, in new formations.

As I wandered the pattern, crushing the fallen stalks underfoot with each crackling step, I met a young bearded man with an ascetic, intelligent face, who seemed to be examining the lay of the crop with a certain casual expertise. We started talking, and I learned he was one of the presenters at the 2002 Glastonbury Crop Circle Conference, which I was attending over the weekend. His name was Allan Brown. He was just driving from his home in Brighton to Glastonbury, and he happened to see this formation from the side of the road. Coincidentally, he had been listening to a recording of McKenna while he was driving. He had not yet seen a picture of the formation from above. I told him my double mushroom interpretation.

"I find the crop circles to be a very psychedelic phenomenon," Allan said. We discussed our admiration for McKenna's work. He told me he was staying with Michael Glickman, the crop circle researcher who had talked about the "dimensional shift."

We walked over to a middle-aged woman in an embroidered blouse who had taken out a dowsing pendulum, a pointed rock crystal dangling on a chain, and was testing various spots inside the crop circle. Dowsing is traditionally used to find sources of water underground, and its utility for that purpose is generally recognized, but it has also become popular among followers of the New Age as a way to trace patterns of energy and discover "ley lines," mysterious currents of geomantic energy. The historian Ronald Hutton dismisses this interest in energy patterns and geomantic currents as "a modern mythology." He notes, "During the last twenty years, thousands of hitherto unsuspected prehistoric monuments have been rediscovered by means of geophysical surveys and aerial photography," yet "not one has been found by all the psychics and dowsers who abound in 'alternative' archaeology." I asked if I could try the pendulum for a moment. Holding it still, I noted that the crystal was definitely pulling at a strong angle, perhaps indicating an alteration in the magnetic field beneath the formation.

We left the World Tree and drove to a second glyph that had appeared a few weeks earlier, on July 4, 2002. If anything, it was more extraordinary—certainly more beautiful—than the tree formation. This crop circle was only a short distance from Stonehenge, and for the first time, from a distance, I saw that ancient monument, its ragged obelisks encompassed by tourist throngs, as we exited the bus, filing across another farmer's land. The pattern showed a vine with leaves along it, forming six symmetrical loops in an oval shape. Over 450 feet long, this crop circle made impressive use of the landscape, skirting the edges of three Neolithic barrows. Earth-covered hillocks made of stones, originally tombs or initiation chambers or both, these barrows dot the lands around Stonehenge and Avebury. If this crop circle was crafted by humans, a global positioning system must have been used, or some other land-surveying technique at a high degree of accuracy.

Returning to Glastonbury at dusk, I hiked to the top of the Tor to watch the sunset, a harmonic orchestration of silvery slivered clouds and fading purple hues. From the Tor, I had a panoramic overview of the region's green rolling hills and fields. Much of the area was originally underwater, in marshy peat bogs, dredged over the course of centuries yet retaining its fertile lushness.

When I tried to sleep that night, I was kept awake by hypnagogic visions that continued until dawn. I observed a misty parade of creatures—gaudy, gauzy entities that could have marched out of the paintings of Hieronymus Bosch. They waved banners and staves and strange spiky things. I awoke many times, peering out my curtained window over moonlit fields, and when I tried to sink back into sleep, the flickering festival started up again behind my closed eyelids.

Ever since my DPT experience and other shamanic forays, I had accessed an occasional visionary capacity beyond my ability to control. From time to time, interior images or scenes appeared spontaneously, projected by my third eye, without any psychedelic catalyst. Buddhist philosopher Ken Wilber, much given to defining the evolution of consciousness and ingression of "spirit" in endless levels and ladders and graphs, noted that "psychic facts" became evident in the "psychic worldspace" at a certain level of development, which he called "vision-logic." Wilber proposed that these "psychic facts" were just as apparent as the material things in the physical world, once you were able to see them. Something like this had happened to me, if I understood it properly. The other option, of course, was that I was sliding down a slippery slope toward an unusual form of madness.

To my surprise, the thinker who gave me the deepest insight into my new situation, rescuing me from an acute occult crisis, was Rudolf Steiner, best known for creating Waldorf Schools and biodynamic agriculture, a forerunner to today's organic farming. Despite these achievements, a fusty ambience of Victorian dustiness surrounds Steiner, whose works are rarely read outside the Anthroposophic movement he founded over a century ago. Steiner is vaguely associated, in the mainstream mind, with fringe movements and mystics such as Emanuel Swedenborg and Madame Blavatsky—his image is of a vaguely sentimental sanctimony and old-fashioned Christian rectitude. Some confuse his movement with the Amish. This dour reputation is tragically undeserved and absurdly off the mark.

When I discovered Steiner, I was amazed by the exalted quality of his thinking—opening onto vast new realms of visionary possibility—and his deep humanity. Against the materialist tendencies of his age, he proposed a "spiritual science" as rigorously grounded as any of the natural or behavioral sciences, but based upon "supersensible" perceptions, attained through an evolution of our cognitive capacities. "Spiritual science attempts to speak about non-sensory things in the same way that the natural sciences speak about sense-perceptible things," he wrote. He argued that it was a prejudice of materialists to believe that the form of empirical cognition they practiced was the only legitimate means of approaching reality. "No one can ever deny others the right to ignore the supersensible, but there is never any legitimate reason for people to declare themselves authorities, not only on what they themselves are capable of knowing, but also on what they suppose cannot be known by any other human being."

He argued that the development of the modern consciousness, with its individuated and alienated ego, concealed a higher purpose. Although our civilization cut itself off from spiritual knowledge, we have the free choice, as individuals, to seek it out, through our own efforts. Our empirical and analytical mind-set provides a firm basis for exploring supersensible realms—if we reorient ourselves toward this different goal.

Steiner prepared the philosophical ground for exploring what he called "higher worlds," carefully describing the stages in acquiring visionary modes of perception, as well as the dangers one might encounter along the way. One can learn to attune one's self to the threshold of the sensible, using the psyche like an antenna for subtler perceptions and secret revelations. This effort of esoteric training requires a development of the whole person, inseparable from moral progress. The use of psychedelics seemed a faster but less conscientious means of attaining supersensible insights, perhaps through stimulus of production of DMT in the pineal gland.

Once he laid the groundwork, Steiner proceeded to explore cosmic domains beyond anything I had encountered before, or conceived as possible, elaborating a vast hierophany. Studying him, I felt I had found a comprehensive alternative to modern rationalism—and one that offered tremendous hope and shining inspiration. He reclaimed the occult traditions that the modern world had forfeited, within an ethical framework. His perspective incorporated the empirical methods of science, and transcended it.

Terence McKenna later suspected that the episode he and his brother had in the Amazon had occurred due to their immersion in visionary realms, coupled with a willing suspension of disbelief. "One reaches through to the continents and oceans of the imagination, worlds able to sustain anyone who will but play, and then lets the play deepen and deepen until it is a reality that few would even dare to entertain." He proposed "that the human imagination is the holographic organ of the human body, and that we don't 'imagine' anything. We simply see things so far away that there is no possibility of validating or invalidating their existence." Walter Benjamin, similarly, thought the imagination was like a fan that could be opened infinitely, unfurling vast Proustian vistas in the smallest moments of perception. Steiner systemizes this understanding, proposing imagination, intuition, and inspiration as three modes of visionary cognition that can be developed through intense discipline.

Where Western philosophers such as Immanuel Kant believed there are "things in themselves" separate from our perceptions of them, Steiner noted that such a division can't be maintained. To separate thoughts from the "things in themselves," to alienate matter from mind, is already, always and only, a thought—therefore it lacks intrinsic validity. Thinking is an aspect of reality—as much a part of the world as any physical object or process—and cannot be amputated from it. "Doesn't the world bring forth thinking in human heads with the same necessity that it brings forth blossoms on the plant?" he asked. Carl Jung—who also realized the reality of the psyche—thought the mythic archetypes contained in the psyche had autonomy and agency beyond the individual. Similarly, Steiner proposed that thinking, in itself, was neither objective nor subjective, but universal. "In thinking we have that element given us which welds our separate individuality into one whole with the cosmos. In so far as we sense and feel (and also perceive), we are single beings; in so far as we think, we are the all-one being that pervades everything." In our feelings and perceptions we are separate, but our thinking connects us to the larger community of mind—what Teilhard de Chardin dubbed the noosphere.

If we conceive of thinking as a part of the world, inseparable from natural processes, then there can be no limits to our cognition. Thinking is not a passive tabulation of objective facts, nor can science lead to a definitive "final theory" about the universe. Thinking becomes, instead, a creative and participatory act that transforms reality. In Steiner's science of the imagination, philosophy becomes the art of reconciling "percepts"—those entities and things we perceive outside of ourselves—and "concepts"—the elements of our cognition. There can be no conceivable end point or limit to this work of reconciliation. It is an infinite task—an infinite play. "All real philosophers have been artists in the realm of concepts," he wrote. I found Steiner's perspective to be shockingly liberating. He offered a non-dualistic revisioning of reality.

Steiner was compelled to develop his philosophy out of his circumstances and unique gifts. He was born in Kraljevec, on the border between Hungary and Croatia, in 1861, of Austrian parents. His father worked for the Austrian Southern Railway, and was promoted to the job of station-master at Pottschach in Lower Austria, where Steiner grew up in modest circumstances. As a child, he was a natural clairvoyant—he could perceive the beings, energies, and fluctuating patterns of the supersensible realms, as well as the spirits of the dead. In his memoir, he recalled how a relative from a distant town appeared to him one afternoon, asking him to take care of her where she was going. He later learned she had died that day. Such episodes are common in indigenous cultures, where the shaman is known to act as the intercessor for the dead—in our modern world, they are seen as aberrations of mental illness.

Since those around him did not possess his clairvoyant powers and didn't understand them, Steiner learned to be quiet about his gifts. At the same time, he kept trying to understand the barriers that prevented people, not only from attaining supersensible perception, but even from being able to conceive that anything like it might exist. He felt a great sense of relief when he encountered geometry in school, which introduced him to a realm of pure mental forms resembling some of his inner visions.

Realizing that he needed a training in philosophy if he was going to convey the hermetic knowledge that was freely available to him but hidden from others, he completed his doctorate at the Vienna Institute of Technology, then took a position editing Johann Wolfgang von Goethe's science papers, moving to Weimar in Germany. In the work of the great Romantic poet and scientist, Steiner found a theory of knowledge that supported his supersensible investigations. While living in Weimar, he visited Friedrich Nietzsche, who had already lost his senses and was confined to his bed. Steiner described seeing the soul forces, depth of spiritual longing, and supersensible beings that had animated Nietzsche's tragic quest for knowledge. He later wrote a book about the philosopher, calling him a "fighter for human freedom."

At the age of thirty-six, Steiner moved to Berlin to work as editor of the *Magazine for Literature,* a literary journal. In Berlin, he became part of a bohemian cultural milieu that included the writers Frans Wedekind and the literary bon vivant Otto Erich Hartleben. He found that the writers and artists in his circle were so obsessed with aesthetics that they cut themselves off from the living processes of the world. "Even very fine people—even those of distinguished character—had an interest in literature (or painting or sculpture) so deeply ingrained that the purely human element completely withdrew into the background." Steiner perceived his connection to these artists extending back to past incarnations. Their character flaws were the "results of previous earthly lives, which could not be fully resolved in the cultural environment of the present life. It became clear that future earthly lives were needed to transform those imperfections."

With admirable patience, he waited until he was forty years old before he began to speak in public about his visionary knowledge, sharing the astonishing cosmology he had developed through painstaking investigation of his inner world. Steiner claimed to have access to the "Akashic Record," a spiritual imprint of the evolution of humanity, from distant past to far-flung future. According to occult philosophy, the Akashic Record becomes available to clairvoyants at a high level of attainment; in the United States, the twentieth-century mystic Dr. Edgar Cayce was famous for his ability to access this "library" while in trance states, recounting past incarnations of his patients and prescribing cures for them. After Steiner went public with his supersensible knowledge, he published dozens of books and gave over three thousand lectures during the last twenty-five years of his life, establishing anthroposophy as a "white occult" movement to oppose the negative occult and nihilistic trends he saw around him, which would culminate in the rise of Hitler's Nazi Party, its diabolical delusions of Aryan purity leading to global war and the Final Solution.

I would not have been receptive to Steiner's cosmological frame if it weren't for my own shamanic journeys, which seemed to open doorways in my psyche that never completely closed. In the vast ecology of mind he describes, our consciousness is inseparable from other forms of sentience—such as the goblins, or postmodern Grays, who swarm us as we pass between sleep and dream. Steiner split up the unitary Christian devil into different forces that constantly work upon the human psyche, seeking to divert us into deviant paths. Lucifer, the "light-bringer," draws us upward into realms of imagination, beauty, fantasy, and egoistic pride, pulling us away from the Earth. The other power is Ahriman, dark and malevolent Earth spirit of the Zoroastrian faith, also known as Satan, Mephistopheles, and Moloch. Ahriman drags us downward—into the mineral world, materiality, mechanization, and death. Our age of materialism represents the temporary ascendance of Ahriman, who strives to make the world into a machine. Studying Steiner, I recognized Lucifer and Ahriman as "psychic facts," part of the wallpaper of my new "psychic worldspace."

The occult hangover of my DPT trip seemed, intuitively, a kind of psychic combat, ending with the integration of my shady nemesis or dream daimon. Afterward, I felt significantly sharper, though this sharpness was double-edged. I had to recognize and learn to resist the tendency to feel contempt for the lumbering crudity of our human world—contempt for my own shortcomings, as well. For months I felt as if an invisible snicker or sneer was suspended over my head. In Steiner's work, I encountered detailed descriptions of the Luciferic forces, the "sceptered angels . . . spirits reprobate" crowding the pages of Milton's *Paradise Lost*. According to Steiner, these entities were not evil, but deviant. They were far more advanced than humanity in many ways—too advanced to take physical incarnation in our world—yet they required us as the ground for their further development. Luciferic spirits seek to "take refuge in a kind of substitute physical body," by making an alliance with a human being who has passed a certain threshold of psychic development. "We are the stage for the Luciferic evolution," Steiner notes. "While we simply take the human earthly body in order to develop ourselves, these Luciferic beings take us and develop themselves in us." He cited Romantic poets and Baroque composers as examples of men who had made temporary alliances with Luciferic forces. I sensed he was also describing what had happened to me.

Lucifer is mocking and outraged, caustic and impolite—like some debauched, Dionysian rock star of the late sixties, he wants the world, and he wants it now. The Luciferic tone is perfectly embodied in the works of Aleister Crowley, the self-proclaimed "Great Beast" of modern magic, who apparently entered into a daimonic alliance. As he recounts the story, Crowley received a channeled text, *The Book of the Law,* in Cairo, in April 1904, after a number of synchronicities and magical manifestations, and a night spent with his mistress in the Great Pyramid's central hall. For one hour at a time, over three days, a voice spoke in his head, dictating the work—according to Crowley, its validity was proved "by the use of cipher or cryptogram" relating "some events which had yet to take place, such that no human being could possibly be aware of them." The voice identified itself as Aiwass, a disembodied intelligence—"a messenger from the forces ruling this Earth at present." The transmission reads like a histrionic work of late Romantic poetry—and much like the rhyming reams of mediocre verse produced by Crowley himself. Aiwass's *The Book of the Law* samples Swinburne, Nietzsche, bits of Blake, expressing a streak of maniac cruelty:

> We have nothing with the outcast and the unfit: let them die in their misery. For they feel not. Compassion is the vice of kings: stamp down the wretched and the weak: this is the law of the strong: this is our law and the joy of the world. Think not, o king, upon that lie: That Thou Must Die: verily thou shalt not die, but live.

Utterly convinced by this episode, Crowley believed himself chosen to be the messiah and prophet of the New Aeon. He called himself "the Master Therion, whose years of arduous research have led him to enlightenment."

Crowley spent the rest of his life following *The Book of the Law,* including its call to immoderate excess.

> To worship me take wine and strange drugs whereof I will tell my prophet, & be drunk thereof! They shall not harm ye at all. It is a lie, this folly against self. The exposure of innocence is a lie. Be strong, o man! lust, enjoy all things of sense and rapture: fear not that any God shall deny thee for this.

In Jungian terms, one might say that an archetype, an immaterial cluster of psychic energy—"spontaneous phenomenon" with "a certain autonomy," according to Carl Jung—attempted to constellate in Aleister Crowley: the archetype of the "Great Beast 666" from the book of Revelation. The "Great Work" Crowley believed he would accomplish did not come to fruition in his lifetime. Seeking to overcome the Victorian repression of sexuality and the body, Crowley, "the prince-priest the Beast," indulged in techniques designed to shock and offend. Crowley borrowed ideas from the Eastern disciplines of Tantra, esoteric practices in which the desires and passions of the sensory world are used to attain liberation. He debased these ideas into a system of "sex magick," in which sex acts, including homosexual intercourse, are a means of initiation into endless hierarchical stages with pompous titles. Instead of an achieved Tantrism, his life's work was an extended tantrum. If Crowley made a daimonic alliance, it intensified his ability to produce psychophysical phenomena, as described in the many accounts that embellished his legend. But the alliance magnified his character defects, leading to his ruin.

The glamorous ego-inflation of Lucifer is countered by the mechanistic deflation of Ahriman, who reduces humans to the status of things, a "standing reserve" to be used for inhuman ends. The techno-futurist fantasies of transhumanists in which artificial intelligence takes over, science fiction such as *RoboCop* and *Terminator,* the sociobiological perspective of books such as *Guns, Germs, and Steel* and *The Selfish Gene,* the glacial technological predation of the alien abduction accounts—such artifacts reveal Ahriman's one-dimensional viewpoint, which reverberates through our culture.

Steiner did not believe it was possible to separate ourselves from these influences—like Jung, he considered the "devil" a psychic reality that had to be assimilated into consciousness, rather than rejected or repressed. In Steiner's view, humanity walks a balance beam between these opposing forces, and Christ's sacrifice gave us a model for maintaining the necessary equanimity and deep humility that our position requires. Since Moloch rules the roost in our current world, Steiner believed that we needed to recover the Luciferic influence to counteract him—he titled his anthroposophic journal *Lucifer Gnosis.* I suspect that Steiner would have seen psychedelics as

Luciferic catalysts, helping us access lost modalities of supersensible perception, but possessing the danger of deluding us with hubris, leading us away from earthly responsibilities.

In certain hermetic traditions, the Holy Grail is described as "a jewel that fell from Lucifer's crown"—one of the myriad meanings circling around this ancient palimpsest. The jewel in Lucifer's crown refers to the third eye—pineal gland or ajna chakra—located in the middle of the forehead. The opening of the third eye brings about spiritual vision, supersensory perception—unless you are the Hindu god Shiva, who destroys the manifest universe when he opens his third eye, at the end of a kalpa. For those who are unprepared, like the mind-blown hippie masses in the background of the film *Altamont,* this opening leads to disorderly hallucinations and an unmooring from reality. In supersensible domains, according to Steiner, the individual psyche shapes what it beholds, magnifying neurotic disorders into phantasmagoric horrors. To approach the supersensible world in a healthy way requires inner strength, patience, and calmness. "Our reasoning must be perfectly lucid, even sober, at all times," Steiner warned.

THE STEINERIAN VISION seemed particularly appropriate to Glastonbury; the Holy Grail is one of many legends blanketing the region like morning fog. According to legend, St. Joseph of Arimathea, the disciple of Christ who took Christ's body down from the Cross, was instructed by an angel to take his followers to England, bringing with him two "cruets" containing the blood and sweat of Jesus. After a long and arduous journey, they landed in Glastonbury. They were given permission by the local king to found a community in the area. They built their simple huts of wattle and mud and a modest wooden church on the spot where the powerful Glastonbury Abbey was later founded.

It is impossible to determine whether or not this story has any validity. However, there is evidence supporting it and none contradicting it. When he and his disciples first docked, St. Joseph is said to have planted his staff, which came from the branch of a thorn tree, near the top of Wearyall Hill, the second highest promontory in the area. His staff took root, and a single specimen of a type of thorn tree native to the Middle East still grows there today. Another tree of the same species spreads its gnarled charm in the garden at the abbey. The abbey thorn tree blossoms once a year, around Christmas, and a sprig of its flowers is placed on the queen's breakfast table every Christmas morning.

Whatever its origins, the Abbey of Glastonbury was the center of English Christianity for many centuries, until it was smashed up and burned to the ground by King Henry VIII in 1539, its last abbot executed on the Tor. The abbey's ruins display a number of stone structures that must have been formidable Gothic edifices—perhaps even more beautiful as jagged vestiges. According to tradition, the monks of the abbey were able to hide many of the abbey's treasures and manuscripts in secret troves around Glastonbury—still awaiting rediscovery, according to local lore. Beneath the wrecked abbey, secret underground tunnels are said to exist, snaking to neighboring valleys and villages, built by the monks to escape Henry's wrath.

The "mysto steam" rising from Glastonbury kept affecting me. On a break from the conference in Glastonbury's town hall, I visited the abbey for the first time. Sitting near the center of the old church, before a stone slab commemorating the spot where King Arthur and Guinevere's tomb was discovered in the twelfth century—an event generally dismissed by mainstream historians as an early prototype of the modern publicity stunt—it was easy to close my eyes and catch a momentary, flickering vision of a white-robed ceremony or coronation, presided over by faint hooded figures. Was I sensitive to some energetic trace or psychophysical imprint left in the area, or was I prey to suggestibility that brought these images, unbidden, into my mind-screen?

The Crop Circle Conference consisted of an enjoyable series of talks given by a range of eccentric characters, including local celebrities and best-selling scholars on the alternative archaeology circuit. It was held in the town hall, in a large auditorium decorated with wildly colorful murals that jumbled crop circle symbols with Egyptian, Mayan, and outer-space iconography. While some talks focused on the formations, others explored loosely connected areas such as the Qabalah, ley lines, the conspiratorial secrets of the Bilderburg Group and the Trilateral Commission, and the methods of divination used in ancient Greece. My new friend Allan Brown spoke about some of the ways in which the crop circles seemed to be reawakening knowledge of the ancient Neolithic landcape and culture. Following clues provided by some of the patterns, he thought he had found the outlines of huge ancient goddess figures indicated by natural and man-made landmarks across Wiltshire. "The whole landscape was a living ritualistic expression, designed or encoded during the Neolithic Golden Age," he theorized.

A star of the conference was Michael Glickman, a charming man in his sixties with a short salt-and-pepper beard and large, expressive hands. Suffering from multiple sclerosis, Glickman walked with two canes, yet his handicap did not keep him from the center of the action. When I had interviewed him from New York, Glickman said that researchers into crop circles split into two camps—seekers after mechanism, or seekers after meaning. "I consider myself to be a seeker after meaning," he said. Because of his willingness to go out on a limb, Glickman was a favorite target of skeptics and hoaxers.

In his lecture, which he delivered in a tuxedo—a pointedly grandiloquent gesture, indicating his own sense of possessing elite knowledge, against the ignorance of the masses—Glickman theorized that the crop circles were giving hints of an accelerated process of psychic transformation. In our daily lives, we were also getting cues that we were on the cusp of this new world reality. These cues were being given "so gracefully, so respectfully and so gently" that you had to attune your mind to them, or you would miss them. Glickman listed a number of "signs of the dimensional shift." These included the experience of time speeding up; an increase in synchronicities and coincidences; acceleration of premonitions and intuitions; and the occasional occurrence of "little light events without a source," demonstrating "Newtonian mechanical impossibility" or violations of probability—he showed a slide, taken by a friend of his, of a herd of sheep forming a perfect circle in a Scottish meadow, for no discernible reason. He also thought the "karmic elastic band is snapping back faster and faster," with rewards and punishments meted out almost instantly.

"Thoughts seem to be effecting reality more than before," Glickman conjectured. "The universe is increasingly obedient, compliant, even dumb."

The formations depicting different kinds of fractals displayed an essential truth of the universe: "Everything is everything. Everything contains everything. Everything is part of everything else." He showed a slide of one enormous crop circle that had flattened hundreds of feet of crop in all directions, but left one ignominious thistle bush standing, exactly at the center of one of its spokes. "This whole enormous feat was organized around this one modest plant," he said. "The circlemakers are showing us their reverence for other life forms."

The gigantic thirteen-armed spiral formation of 409 circles that appeared on Milk Hill in August 2001 represented "the hinge point of the dimensional shift. It was like the pounding of an enormous fist on an oak table, or the seal on the bottom of a big parchment degree saying, 'That's It!' The end of Book One. Everything that comes from now on is another volume." He said that thirteen was "the number of transformation." Not only were there thirteen spirals, but according to gematria, an occult method of revealing hidden numbers, the number 409 became 4 + 0 + 9 = 13.

Glickman had less definitive comments about the Face and Arecibo Reply glyphs that had appeared last summer. He thought the Face meant: "'We are watching you. We are seeing you.' The authors of the crop circle phenomenon know us better than we know ourselves." The fact that the image was executed in halftones, like an image from a newspaper, seemed significant. "Perhaps they are saying that reality only exists where pure light and pure dark merge to form an image." He had scathing comments about the original Arecibo message sent by SETI: "This is how we represent ourselves to the stars? Quantity and chemistry? Size and number? What about Shakespeare? What about Gothic cathedrals?" By giving us exactly what we said we wanted, the Arecibo Response was mocking us.

"Is there anyone out there as stupid as us?" Glickman asked the audience.

AT THE CONFERENCE, I bought a black T-shirt featuring one of the most elegant formations, of circles overlapping to form crescent shapes, abstractly suggesting a winged angel. I stuck the T-shirt in my shoulder bag. When I returned to my room that night, I was disappointed to find the T-shirt was gone—unusual for me, as I tend to be good at holding on to my possessions. I couldn't remember opening my bag, and I had no idea how the shirt could have fallen out.

The final talk at the conference required a separate fee. The speaker was Dolores Cannon, a hypnotherapist from Arkansas, author of books on Nostradamus and on the extraterrestrial visitors she called "The Custodians." Her books were international hits, especially popular in East Europe and Russia. A grandmotherly woman in her sixties with white hair and glasses, Cannon sported a lime-green polyester pantsuit.

She asserted that The Custodians were responsible for all the formations. "They have overseen the evolution of humanity since the planet cooled down enough to support life," she said. Despite the terrifying and repulsive dimensions of most abduction accounts, Cannon insisted that "ET" was a benevolent species who had attained a high level of spiritual and technological perfection. "ET doesn't need to eat or sleep the way you and I do. They are so much closer to The Light," Cannon said. "On their ships, they have special rooms where they go to bathe in The Light. That's all they need, that's how spiritually advanced they are."

The Custodians were responsible for the abductions, but these events had been misconstrued by many people. According to Cannon, ET was only interested in our welfare. "Recently things haven't been going too well for the human race and ET is here to help reset our evolutionary destiny by making genetic changes to ensure the human race continues." They were deeply concerned that we might cause an unintentional environmental catastrophe. According to Cannon, if there was a catastrophe, they would take us away to another world that they were preparing for our continued evolution.

Cannon spoke in a lulling monotone. Her flat vocal style seemed a product of her hypnotherapy training, and it had a numbing effect on her listeners—at least it did on me. The town hall was packed for her lecture, drawing a different crowd than the conference, including many older people and local hippies. The area around Glastonbury is a center for UFO encounters—in the pubs, I had spoken with locals who all described seeing strange craft or "balls of light." It made sense that they would come out for her talk—they were struggling for answers. The audience seemed rapt in attention as Cannon recounted stories from her regression therapy. She said that sometimes her sessions would be interrupted by ET, who would communicate with her directly, speaking through the bodies of her hypnotized patients.

Generally, I am not particularly sensitive to the types of "subtle energies" that psychics and "light workers" pick up—they tend to be too subtle for me. However, as Cannon spoke, I became increasingly aware of something unusual in the room, like vortices swirling above the heads of the audience. It seemed I could sense the entities she was discussing, hovering in the air above us. My intuition was that those entities were probing and testing the lulled awareness of the listeners, looking for entry points, seeking to fasten on to their psyche, like mind-parasites.

A strange thought occurred to me: Dolores Cannon, as benign and matronly as she seemed in her lime-green ensemble, was allied, consciously or not, with the visitors. As I listened to her, I felt I had entered the frame of a David Cronenberg movie, staring at the weird and improbable banality of polyester-pantsuited malevolence. I shivered—as I had, months after my DPT trip, when I happened to encounter a quote from St. Paul: "For we wrestle not against flesh and blood, but against principalities, against powers, against the rulers of the darkness of this world, against spiritual wickedness in high places."

In Tibetan Buddhism, magical practices involve creating "tulpas," known in Western occultism as thought-forms. Tulpas, thought-forms, are imaginal entities that can be given energy and artificial life through rituals, meditations, and other workings. The poet W. B. Yeats started a magical order in which he tried to revive certain Celtic deities through visualization and ceremonies. One of these, the "white jester," apparently gained enough independent vitality to become visible to a few of Yeats's friends, who were not aware of what he was doing. John Michell, chronicler of Glastonbury legends, considered old fables in which the stone circles, built around 3000 BC, were constructed by giants summoned up by tribal warlocks. "Those who are knowledgeable in the arts of raising spirits and creating thought-forms have written that it is often easier to produce phantoms than to dissolve them. In the course of time they become more solid and may even bleed when wounded. The magicians of the age of Taurus were adept in forming giants and monsters, but like all technologies theirs had unwanted side effects. Not all the spirits they raised were properly laid to rest, and some lingered about the countryside to establish a breed of monsters." It occurred to me that hypnotic regression might be the return of this unsavory magical practice—the ensouling of thought-forms—hidden under a cloak of psychology. With or without conscious intent, the psychologist acts as sorcerer, turning psychic energies into entities, through his attention and articulation.

After one hypnotherapy session, Whitley Strieber found a "vivid image" of one of the visitors emerging in his mind, as perfectly rendered as a photograph, staying there for several days. He was able to make the entity move in any way he pleased, taking the opportunity to describe it in great detail in his book, even having a friend paint what he saw. "We don't know a thing about conjuring and magic," Strieber wrote. "We've dismissed it all, we who love science too much. It could be that very real physical entities can emerge out of the unconscious." Through her work, Cannon—like the unwilling, unwitting Strieber—was helping the visitors to gain solidity, bringing them nearer to our realm. Like a villain from the Harry Potter series, she was a sorcerer allied with underworld forces—at any rate, that was my exciting and unnerving conclusion.

When it was time for questions, I decided to test these murky waters. "It is quite a story you tell," I said. "But how can you be so certain that these entities are benevolent? How do you know they aren't just using you to promote their own agenda? And who do you think are the 'unclean spirits like frogs' described in the book of Revelation?"

"Oh, no," Cannon replied in her affectless monotone. "ET is here to help us and to look after us. They would never do anything to harm us. Where would you get that idea?"

FRIENDS OF MINE from New York came to visit me at the end of the conference. Angelica and Lee Ann, traveling with Angelica's two kids, had rented a car to visit me and see some crop circles. Lee Ann was married to Angelica's younger brother, Tony, one of my closest friends, and pregnant with their first child. We drove to the Crop Circle Cafe, on the road to Avebury, a gathering point for international tourists and UFO enthusiasts pursuing the phenomenon. The cafe featured topological surveys of the region on the wall, dotted with pins indicating the precise locations of recent formations, and photographs of the season's circles. At the cafe, we learned about the latest formation, which had just been found that morning. We took down the directions and drove out to it.

Crop circles can be hard to find. Often they are set way back on farmers' lands, on gently sloping hills making visibility difficult. From a distance, along a level surface, a crop formation appears as little more than a slight shading in the color of the field. We walked a long way, and finally found a formation of large rings that was a blatant human creation. The grain had simply been smooshed down by a stick. If this was all we were going to find, it was truly disappointing.

As we walked back toward the car along one of the tractor lines, we cut through the wheat and tried another path. Up ahead, atop a rise, we noticed irregular patches in the rows of crop that seemed to indicate something different. We entered a brand-new pattern with its now-familiar palpable energetic charge—a sense of a hovering presence over the scene, inducing hyperreality. As Kirtana, Angelica's thirteen-year-old daughter, entered the perimeter of this formation, she burst into tears—a common reaction in new crop circles. From the ground, the pattern was difficult to read. We thought it might be a flower, with various rings radiating out from a center. We spent some time sitting in the formation, which had a traditional "nest" of swirled-together wheat at its center.

A few days later, a photo of the pattern was posted on the Internet. The image showed three rectangular frames, their edges suggesting braided rope, each enclosing the next. Appearing on Monday, July 22, 2002, this crop circle was "the first in the history of the phenomenon to utilize an unambiguous rope or cord motif," wrote Allan Brown.

> It is interesting to note that knots are an artifact of our 3 dimensional material reality, as topologically a knot will unravel itself when rendered in any higher dimensional space. . . . I am reminded of an umbilical cord whenever I look at this formation.

> Perhaps the formation is indicating a birthing into higher dimensional space, in which the paradoxical polarities of duality are unraveled and made whole again.

The image showed a union of circle and square—traditional representations, in sacred geometry, of Heaven and Earth. Karen Douglas, a researcher and organizer of the Glastonbury conference, theorized that "the rope motif signified the binding connection between these two realms."

We returned to the car and Lee Ann opened the passenger door. The black T-shirt I had bought and lost several days ago, seventy miles away, was lying there, draped across the backseat, witnessed by Lee Ann and Angelica. The T-shirt had disappeared before they arrived in Glastonbury. There seemed no conceivable way I could have been responsible for the shirt ending up in the car, unless I had forgotten a whole series of actions. I recalled my night-long "fairy-world" visions after my first ascent of the Tor. Living in a realm adjacent to our own, hopscotching through our world like quantum tricksters, fairies in fairy tales are often mischievous beings, playing with humans by stealing objects or moving things around, sometimes returning what they borrowed. Were they teasing me? Was my mind playing tricks on me, or were the old fables more than fictive? Were the flickering fairies another archetype, like the gobliny Grays, elusively reasserting their archaic privileges in our postmodern present tense?

CHAPTER FOUR

Advancing the work of Fritjof Capra in *The Tao of Physics,* the physicist Amit Goswami proposes that the paradoxes of quantum mechanics—nonlocality, action-at-a-distance, quantum uncertainty, and so on—can only be resolved through the hypothesis that consciousness, not matter, is the fundamental reality of the universe. Instead of a dualistic split between mind and matter, or subject and object, Goswami puts forth a philosophy that he calls "monistic idealism." According to this position, "the consciousness of the subject in a subject-object experience is the same consciousness that is the ground of all being. Therefore, consciousness is unitive. There is only one subject-consciousness, and we are that consciousness," he writes in his 1993 book, *The Self-Aware Universe: How Consciousness Creates the Material World*. Goswami posits a "brain-mind" system in which the physical brain functions as a measuring and recording device, following the rules of classical physics, while a "quantum component of the brain-mind . . . is the vehicle for conscious choice and for creativity."

It is the activity of consciousness, determining the "quantum collapse" of a wave-form into a particle, that brings the world into being. "Consciousness is the agency that collapses the wave of a quantum object, which exists in potentia, making it an immanent particle in the world of manifestation," Goswami writes. Our subjective awareness arises as a result of a "tangled hierarchy" in the brain, a closed loop of self-reference similar to the famous "liar's paradox" (the man from Crete who insists, "All Cretans are liars"). Since quantum collapse can only occur through a physical brain, the ego is "an assumed identity that the free-willing consciousness dons in the interest of having a reference point." Esoteric disciplines and techniques of meditation teach us to observe our subjectivity, our ego-hood, with its continuous babble of thoughts and worries, from an outside perspective, a "witness consciousness." By doing this, we jump out of our individually conditioned viewpoint, the self-referential circuit, and take the transcendent perspective.

The ego—which Goswami calls the "Classical self"—develops habitual patterns of thought and behavior in response to conditioning. Over time, this conditioning creates a probabilistic bias in favor of past patterns of response. "Once a task has been learned, then for any situation involving it, the likelihood that the corresponding memory will trigger a conditioned response approaches 100 percent." In the same way that classical Newtonian physics is now recognized as a subset of quantum physics, "behaviorism is recovered as a special case of the more general quantum picture." But beyond our habitual patterns of reaction, we always preserve the potential for spontaneous insight and unconditioned action. "When we act in our conditioned modality, the ego, then our thoughts . . . seem algorithmic, continuous, and predictable, which give them the appearance of objects of Newtonian vintage. But there is also creative thought, a discontinuous transition in thinking, a shift of meaning from conditioned to something new of value," writes Goswami, proposing that "creative thought" is "the product of a quantum leap in thinking." Similarly, the physicist F. David Peat suggested that "synchronicities, epiphanies, peak, and mystical experiences" reveal the mind "operating, for a moment, in its true order . . . reaching past the source of mind and matter into creativity itself."

The quantum aspect of consciousness also breaks through the crust of habit in the form of intuition, which is not an irrational process, but an arational one, often containing "a superior analysis or insight or knowledge which consciousness has not been able to produce," wrote Carl Jung. "You do not *make* an intuition. On the contrary, it always comes to you; you *have* a hunch, and you only catch it if you are clever or quick enough." Rudolf Steiner noted, "Intuition is for thinking what observation is for the percept. Intuition and observation are the sources of our knowledge." Most creative breakthroughs, whether in the arts or hard sciences, are the result of intuitive leaps, or quantum jumps into new contexts.

Like quantum objects, thoughts seem to obey the uncertainty principle. Goswami notes, "You can never simultaneously keep track of both the content of a thought and where the thought is going, the direction of thought." He proposed that "mental substance"—thought—was made of the same intangible elements that build up the "macro objects" of the physical world, but "mental substance is always subtle; it does not form gross conglomerates." Steiner also conceived of thoughts as possessing a substance. Lacking the terminology of quantum physics, he expressed the idea in a mystical framework, perceiving "thoughts as living, independent beings. What we grasp as a thought in the manifest world is like a shadow of a thought being that is active in the land of spirits." The "land of spirits" could be considered the transcendent domain of potentia.

In *The Self-Aware Universe,* Goswami shifted from a materialist orientation to one based upon the essentiality of "mind-stuff." "I had vainly been seeking a description of consciousness within science; instead, what I and others have to look for is a description of science within consciousness," he realized. "We must develop a science compatible with consciousness, our primary experience." In a later work, *The Physics of the Soul* (2001), he puts forth a scientific hypothesis, based on his understanding of the quantum nature of consciousness, for the mechanisms of reincarnation, and the existence of the "subtle bodies" and chakra system described by many esoteric traditions.

Neither current Christian doctrine nor the secular thought of the modern West supports the idea of reincarnation, which is a basic element in many spiritual traditions, including Hinduism and Buddhism. Ian Stevenson, former professor of psychiatry at the University of Virginia, spent forty years studying young children who spontaneously recall past lives, compiling the data from over 2,600 case studies in his massive book, *Reincarnation and Biology: A Contribution to the Etiology of Birthmarks and Birth Defects.* He has found numerous incidences in which children recall specific details of their last life, and has documented, in some instances, startling physiological evidence—birthmarks and birth defects—that seem to carry over from their former existence. In a number of cases, especially ones in which the past life ended in violent death, Stevenson has found, and photographed, highly unusual birthmarks on the children that perfectly match the placement of the mortal wounds that ended their remembered past life.

"Most of the children speak about the previous life with an intensity, even with strong emotion, that surprises the adults around them," writes Stevenson, who has documented past-life memories among the Tlingit in Alaska, the Druze in Southern Lebanon, across India, and other regions. "Many of them do not at first distinguish past from present, and they may use the present tense in referring to the previous life. They may say, for example, 'I *have* a wife and two sons. I *live* in Agra.' Although some children make fifty or more different statements, others make only a few but repeat these many times, often tediously." The recollection usually commences when the child is very young—between two and four—and ends within the next five years. Stevenson has accompanied some of these children to meet the families, corroborating their specific memories. "The birthmarks and birth defects in these cases do not lend themselves easily to explanations other than reincarnation," he notes. In the cases he has studied, the average period between death and rebirth is fifteen months.

Goswami hypothesizes that conditioned and habitual patterns of thought, feeling, and action create what he terms the "quantum monad," an aggregate of probabilistic patterns, and it is this aggregate that returns in life after life. Demonstrated quantum effects such as "action at a distance" and "non-local correspondence" prove that quantum objects possess a "memory"—although that memory is not recorded physically, like a photograph or a tape, but only activated when consciousness causes the wave-collapse. "Objects obey quantum laws—they spread in possibility following the equation discovered by Erwin Schrödinger—but the equation is not codified in the objects. Likewise, appropriate nonlinear equations govern the dynamical response of bodies that have gone through the conditioning of quantum memory, although this memory is not recorded in them. Whereas classical memory is recorded in objects like a tape, quantum memory is truly the analog of what the ancients call Akashic memory, memory written in akasha, emptiness—nowhere." According to Eastern traditions, it is the accumulation of karma—the word literally means "action"—that induces reincarnation, and Goswami's thesis supports this idea. "Certain incarnate individuals are correlated via quantum nonlocality," he writes. "They have privileged access to each other's lives through nonlocal information transfer." He defines the "quantum monad" as an intermediate level of individuality between the ego and the "quantum self," which is equivalent to the transcendent consciousness.

The distinct patterns of conditioned thought, feeling, and action can also be considered the "subtle bodies" defined by mysticism. The "mental body" represents the individual pattern of thought, and the "vital body" is the individual pattern of energy and feeling. The individual "quantum monad" develops vital and mental memories of past contexts over its successive lives. Beyond these habitual tendencies, worn like grooves in an old record, Goswami posits a "supramental intellect" or "theme body," "the body of archetypal themes that shapes the movement of the physical, mental, and vital" bodies—what Jung called the self. According to Steiner, the theme body does not express itself through thoughts, but "acts through deeds, processes, and events." The self "leads the soul through its life destiny and evokes its capacities, tendencies, and talents."

The subtle bodies and esoteric organs defined by different esoteric traditions—chi, prana, the nadis, Carlos Castaneda's "assemblage point," the chakras, and so on—do not possess physical substance, but have a real existence as quantum phenomena, "the possibility waves of an undulating infinite medium." Consciousness determines their manifestation. In Goswami's scheme, consciousness is fundamental, hence these subtle energy streams are integral aspects of our existence. He suspects the chakras may provide "blueprints" through which the vital body maps itself onto the physical, representing itself in the health and function of our organs.

He presents physical evolution as a process through which the preexistent transcendent consciousness remembers itself—self-organizing through increasingly complex levels of light and matter, life and mind—until it regains its original and unconditional state of freedom. Through the gradual process of natural selection, as well as the sudden jumps into new contexts abundantly suggested by the fossil record, organic life developed in phases of "punctuated equilibria," from amoeba to dinosaur to mammal, until the human species emerged into prominence, its capacity for language and culture and thought creating the tools to undertake the "hominization" of the planet. At first gradually and then with increasing speed, humanity assimilated and learned to utilize the stored resources of energy within the biosphere to construct civilization. At that point, the organic evolution of the biosphere went into entropic decline, and the cutting edge of future evolution moved to the mind—Teilhard de Chardin's noosphere, the thought-envelope around the Earth.

Goswami argues that major changes in complexity, such as the transition from reptiles to birds or primates to humans, cannot be explained by Darwinian mechanisms of natural selection. "Instead, they show the quantum leap of a creative consciousness choosing among many simultaneous potential variations." The transcendent consciousness "employs matter, as we employ a computer, to make software representations, which we call life in the living cell and conglomerates, of vital functions." At the stage of human evolution, "software representations of the mind" can be programmed into the brain's hardware.

According to linguist Noam Chomsky, the development of human language could not have occurred as a straightforward step from animal communication. "There seems to be no substance to the view that human language is simply a more complex instance of something to be found elsewhere in the animal world," he writes. "This poses a problem for the biologist, since, if true, it is an example of true 'emergence'—the appearance of a qualitatively different phenomenon at a specific stage of complexity of organization." Humans possess an innate or inherent capacity for language development, for grammar and syntax, that no current model for the acquisition of traits seems to fit. "In fact, the processes by which the human mind achieved its present stage of complexity and its particular form of innate organization are a total mystery. . . . It is perfectly safe to attribute this development to 'natural selection,' so long as we realize that there is no substance to this assertion, that it amounts to nothing more than a belief that there is some naturalistic explanation for these phenomena." The sudden emergence of language as a cognitive structure seems to support Goswami's thesis of the "quantum leap of a creative consciousness," operating from a transcendent domain. Such a perspective leaves open the possibility of future "quantum leaps" of cognitive complexity.

"Any attempt to dismiss a phenomenon that is not understood merely by explaining it as hallucination becomes irrelevant when a coherent scientific theory can be applied," Goswami notes. He believes that his model of evolution points to a future phase of human development when we will be able to map the "supramental intellect" or "theme body" onto the physical body, as with the vital and mental bodies—this would be a massive phase-shift in human potential, to a condition beyond the constraints of mortality, as it is now experienced. If possible, it would be a state foreshadowed by the "resurrection body" of Christ, the *sambogakaya* (body of pure luminosity) of Buddhism, and similar ideas from various sacred traditions. He draws inspiration from Sri Aurobindo, a twentieth-century Indian philosopher who theorized an evolution into a "supramental" condition, a point at which the "upward causation" of evolution would meet the "downward causation" of a self-empowered and self-directed consciousness, able to transform matter, the body, and the world out of a direct engagement with the "inner creativity" of the quantum self.

"But what after all, behind appearance, is the seeming mystery?" Aurobindo asked, "We can see that it is the Consciousness which had lost itself, returning to itself, emerging out of its giant self-forgetfulness, slowly, painfully, as a life that is would-be sentient, to be more than sentient, to be again divinely self-conscious, free, infinite, immortal."

GOSWAMI'S "SCIENCE WITHIN CONSCIOUSNESS" fits the philosophy of Rudolf Steiner, who also saw mind as the basis of reality, describing the workings of our distinct "subtle bodies" and the chakra system, as well as supersensible entities evolving on other planes. Steiner said that the specific mission of his life on Earth was to bring the knowledge of reincarnation back to the West. In his cosmology, not only do individualities return again and again, but the Earth itself reincarnates, and this is the fourth incarnation of the Earth. I doubt that Steiner knew of the Hopi cosmology when he formulated this, but it correlates with their idea of the Fourth World, on the verge of phase-shifting into the Fifth—similar, also, to the waning Fifth Sun of the Aztecs.

In their remote mountain kingdom, Tibetan Buddhists developed, over centuries, a highly evolved spiritual science of reincarnation. They recognize the return of certain individualities, *tulkus,* who are reinstated in their previous role as lamas, or lineage-holders. The current Dalai Lama, for instance, was identified, as a young child, as the reincarnation of the previous one, through oracles and foretellings. He passed a series of tests in which he was asked to choose from a variety of objects the ones that belonged to his predecessor. Modern Westerners generally consider this a symbolic process, a culture-specific practice rooted in tradition, rather than an empirically verifiable method. Steiner believed that reincarnation has continued over the course of human development, not only in Eastern cultures that believe in the transmigration of souls, but, unbeknownst to us, in the West as well. He wrote a series of books, titled *Karmic Relationships,* in which he used his supersensible faculties to follow certain individualities—Goethe, Garibaldi, Voltaire, Eliphas Levi—as they evolved over successive incarnations. The supramental intellect, or theme body in Goswami's model, chooses the hereditary factors and cultural circumstances that will allow it to unfold particular talents and character traits.

"We return to Earth again and again, whenever the fruit of one physical life has ripened in the land of spirits," Steiner wrote in *An Outline of Esoteric Science.* "Yet this repetition does not go on without beginning or end. At one point we left different forms of existence for ones that run their course as described here, and in future we will leave these and move on to others." His esoteric philosophy was thoroughly evolutionary. He proposed that everything in the cosmos perpetually transforms—not only human beings and the planets and the higher "spiritual beings" who, he believed, advanced themselves further through sacrifice, but even basic laws of the cosmos are in flux. Of his difficulties in translating scenes from the Akashic Record into language, he wrote: "One must be completely clear about the fact that the evolutionary forms of the distant past as well as of the future are so entirely different from those of today that our present appellations can only serve as makeshifts, and really lose all meaning in relation to these remote epochs." Time and space, matter and mind, body and soul take on different characteristics in each new phase of our development.

A number of contemporary scientists are currently exploring the hypothesis that everything, even the seemingly immutable laws governing mineral processes or the cosmological constants underlying space and time, change and evolve. According to the biologist Rupert Sheldrake, "the assumption that the laws of nature are eternal" is a vestige of the Christian belief system that informed the early postulates of modern science in the seventeenth century. "Perhaps the laws of nature have actually evolved along with nature itself, and perhaps they are still evolving? Or perhaps they are not laws at all, or more like habits?" Sheldrake writes in *The Presence of the Past* (1988), proposing the existence of "morphogenetic fields," functioning like quantum memory, shaping patterns of formation and development on every level, from atom to crystal to cell to organism to social organization and beyond.

In *The Life of the Cosmos* (1997), physicist Lee Smolin suggests "the idea that the laws of nature are immutable and absolute . . . might be as much the result of contingent and historical circumstances as they are reflections of some eternal, transcendent logic." He offers a hypothesis of "cosmological natural selection" in which black holes open onto new regions of the universe, where the cosmological constants are slightly altered. "In each birth of a new universe the parameters change by a small random step," Smolin theorizes, analogous to the "small random change" between the genes of a child and its parents.

Such postulates resemble the macrocosmic mindscape elaborated by Steiner, who saw the universe as a staging ground for infinite transformations and permutations of consciousness, taking multidimensional forms. In his vision of the solar system, "the planets are not hunks of stuff out there but nodes of vibration that resonate in multiple dimensions that enfold themselves into one another in patterns of complex recursiveness in which Sun, Moon, and Saturn are also modalities of Earth," wrote William Irwin Thompson, who suggested that such a view was not a violation of rationality, but an expansion of our cosmological framing. Like Goswami, Steiner offers a nondualistic vision, perceiving the human being as integrated, embedded, within a universe that is psychophysical in its essence, where there is no "out there" opposed to "in here," where our thoughts and imaginings are extensions of the natural processes of the world.

Studying the occult, I absorbed the ideas of Dion Fortune, G. I. Gurdjieff, Julius Evola, Aleister Crowley, Steiner, and so on, finding similarities among them but also significant differences that confused me. Over time,

I realized that apprehending the pattern of occult thought was more important than adhering to the details of any particular system. The occult vision offers a way of conceiving reality that is closer to art—or the theme-and-variations of music—than science. "What if at the higher levels of meaning consciousness is like a hyperspace in which each point is equidistant from the other and where 'the center is everywhere and the circumference is nowhere'?" Thompson proposed, in *Passages about Earth*. "The mythologies of the occult seem like baroque music: there is an overall similar quality of sound and movement, but, upon examination, each piece of music is unique; Vivaldi and Scarlatti are similar and different." The movement away from the linear and dualistic thought of "one-dimensional man" opens on multivalent meanings and mythopoetical modalities, requiring equanimity and a firm grounding in reality—a difficult balance to strike.

In his lectures, Steiner proposed that humans were "inverted plants"; that the Buddha reincarnated on Mars after he left Earth; that birds, butterflies, and bats were "cosmic thought, cosmic memory, cosmic dream"; that the beehive represents a higher-dimensional consciousness and is a gift from Venus; that the blood spilled by Christ transmuted the inner nature of the Earth, altering human evolution; that we would have conscious control over the rhythms and pulsations of the heart in our next phase of evolution. Yet he established enduring institutions such as Waldorf Schools, the Anthroposophic movement, biodynamic agriculture, techniques of holistic and homeopathic medicine, and Camphill Villages that develop skills in the mentally retarded, all of which continue to benefit the world, eighty years after his death. He said that the purpose of human existence was to "transform the Earth," insisting that the only way to oppose nihilism and degraded institutions is to build what is good and pragmatic for human evolution—and he made good on his word.

"If we can simply distinguish between the different successive stages of evolution, it is possible to see primeval events within the earthly events of the present," Steiner wrote. Describing macrocosmic stages of evolution, he denoted past and future incarnations of the Earth according to the names of the planets—as if the various planets in their current embodiments represent husks of past stages of consciousness, or symbolic indications of future levels that we are still incubating. In his philosophy, the planets in our solar system are like resonant frequencies attuned to different levels of mind, alternate realms in which supersensible entities unlike ourselves are working out their fate. "The interrelationships of spiritual beings inhabiting celestial bodies are a cause of these bodies' movement," Steiner wrote. "Causes that are soul-like and spiritual in character move these celestial bodies into position and set them in motion in ways that allow spiritual circumstances to be played out in the physical realm."

According to Steiner, rudimentary humanity developed during the "Saturn" and "Sun" incarnations of the Earth—in those early phases we were plant-like, deeply unconscious—followed by the "Old Moon" incarnation, where we attained a dream-like awareness. Steiner describes the "Old Moon" as the "planet of longing," where we apprehended our future forms in dream-like flickers, but could not yet embody them. Consequently, the Earth is "the planet of fulfillment," where we reached the stage of individual self-consciousness. From the current Earth, our evolutionary stream flows toward the Jupiter phase—identical, perhaps, with the Fifth World of the Hopi. According to Steiner, we will, eventually, become self-created entities of cosmic wisdom, as he described in one of his lectures:

> The Jupiter Beings are unlike the men of Earth. When a man of Earth wants to grow wise, he must undergo inner development, he must struggle, battle inwardly and overcome; through periods that are filled with active development the human being struggles to acquire an unpretentious form of wisdom. Not so the Jupiter Beings. They are not "born" as earthly beings are born, they form themselves out of the Cosmos. Just as you can see a cloud taking shape, so do the Jupiter Beings form themselves in the etheric and astral worlds, out of the Cosmos. Neither do they die. They interpenetrate one another, do not, as it were compete with each other for space. These Beings are, so to speak, wisdom that has become real and actual. Wisdom is innate in them; they cannot be other than wise. Just as we have circulating blood, so have the Jupiter Beings wisdom. It is their very nature.

Each successive interval represents an upgrade into higher consciousness, with a concomitant shift in our physical body as well as our various subtle energy systems—but also, at every evolutionary threshold, there are many who are not ready to advance. Restrained by past karma—conditioned patterns of thought, feeling, and action—they undergo an alternate process of development, passing through lower worlds.

In Steiner's view, everything possesses awareness, at its own level, and continually develops new depths, or degrades into flightier forms, of consciousness. "To supersensible perception, there is no such thing as 'unconsciousness,' only various degrees of consciousness. Everything in the world is conscious," he wrote. In the approaching Jupiter state, plants will attain a higher level of consciousness—shifting from their current slumbering form which he equates with "dreamless sleep," to a level of sentience similar to what humans now experience in dreams. Minerals, as well as animals, will also attain a new level of consciousness, as humanity splits into several "human kingdoms" diverging into different evolutionary streams. The organ of reproduction will move from the genitals to the throat chakra—we will sing, or enunciate, other beings like ourselves into existence.

From previous, less tangible incarnations of the Earth, we inherited the current "cosmos of wisdom," symphonically orchestrated by higher echelons of spiritual hierarchies. It is the ultimate purpose of human development, according to Steiner, to transform this cosmos of wisdom into a freely determined "cosmos of love." In the present phase of Earth evolution, humanity received the perfected physical body, and, through our historical development, attained self-consciousness, identifying ourselves as singular beings denoted through our usage of the "I." "The actual essential nature of the I is independent of anything external; therefore, nothing outside of it can call it by its name," he wrote. "Religious denominations that have consciously maintained their connection to the supersensible call the term I the 'ineffable name of God.'" Because we only attained this individuated ego-based awareness in our recent development, the "I" is still weaker than the other bodies we possess—which he calls the physical body, the astral body, and the ether body. Cravings and desires are constantly pouring into us through the astral body, and the goal of our present phase of evolution is to master those cravings, and the astral body itself, through the strengthening of the I, transmuting lower passions into higher energies. "Fundamentally, all of our cultural activity and spiritual endeavors consist of work that aims for this mastery by the I. All human beings who are alive at present are involved in this work, whether or not they are conscious of it."

As we transform the astral body through painstaking conscious labor, we slowly create a fifth body, which Steiner calls the spirit self. "The spirit self constitutes a higher element of our human makeup, one that is present in it in embryonic form, as it were, and emerges more and more in the course of working on ourselves." In the Jupiter phase, or Fifth World, the spirit self will experience its full unfolding. In future stages, we will also learn to transform the ethereal body (the "vital body" of energy and feeling), and eventually the physical body itself. What Steiner describes of our future state seems essentially identical to Aurobindo's vision of the "supramental" condition—what Goswami termed the "theme body."

But what could it mean to transform the astral body, and how would this occur? Our dream-life, perhaps, provides the clue. The perception of our bodies that we have in dreams—as though watching a dream-being who is ourselves but outside of our control—could be considered a representation of the astral body. According to Steiner, when we sleep, the I and the astral body detach from the physical body and the ether body, and visit other worlds. "While we humans, as physical beings, are part of the Earth, our astral bodies are part of a world that embraces additional heavenly bodies. During sleep, therefore, we enter a world that encompasses other worlds in addition to our Earth." In my own life, in my personal pursuit of gnosis, I discovered a profound transformation in my dream-life—entering lucid dreams, receiving direct messages and prophetic hints—that seemed an extension of my waking efforts. While my dreams became more vibrant, reality became more dream-like.

In indigenous shamanism, dreaming is considered a talent and a tool, as well as a skill that can be cultivated. The story of Black Elk, a medicine man of the Lakota Sioux, chronicled in *Black Elk Speaks,* reveals the essential place ascribed to dreaming and dream-visions in Native American cultures. As a child, Black Elk received a series of powerful visions as he lay stricken with illness. When he related these dreams to the tribal elders, they recognized their importance and staged them as a spectacle for the tribe.

Black Elk spent the rest of his life trying to understand, and fulfill, the prophetic message of his dreams. "Without a vision, the people perish," he said, echoing the Gospel of John.

Among contemporary scientists, it is increasingly fashionable to dismiss dreams as meaningless. In the last decades, even mainstream psychology has turned away from the extensive analysis of dream data pioneered by Freud, toward a biology-based model of mental disorders, which are thought to have a genetic basis. The recent thrust of cognitive science is the attempt to establish a brain-based theory of consciousness, confirming the materialist and mechanistic perspective. Typical of this effort is *The Dream Drugstore: Chemically Altered States of Consciousness* by J. Allan Hobson, a Harvard professor of psychiatry. Hobson writes:

> Is it possible to fuse the subjectivity of consciousness with the objectivity of brain activity? If the brain—or its informational content—becomes aware of the outside world and of itself as the instrument of that awareness, it seems possible that the awareness itself is brain activity. We leave out the usual qualifier "nothing but" (brain activity) to avoid giving offense but that, of course, is what we mean.

Hobson's thesis is that all altered states of consciousness—including out-of-body experiences, lucid dreams, and alien abductions—can be explained by "organic etiology," physiological changes to the brain. Because the brain works overtime to coordinate all incoming data into a coherent framework, it can be easily bamboozled and discombobulated. Overloaded, it will make patterns where none exist. While it is sensible that all changes in consciousness would have some measurable component—such as the changes in brain waves induced by meditation, or an increase in DMT production during visionary states—that does not prove consciousness is brain-based. Scientists such as Hobson could be confusing a correlation for a cause.

My own dream experiences—such as the alien-finger incident—have been mirrored by numerous stories gathered from people I know, suggesting a model of dreaming essentially identical to the esoteric or shamanic perspective. One acquaintance—an Episcopal priest—told me he had been actively recruited, while living on the West Coast, by a charismatic and forceful leader of the Liberal Catholic Church, a small sect. In the next months, he began to have frequent dreams in which he was in a living room with different members of this church, who were trying their best to convince him to join. One night, he awoke in the middle of the night to hear his partner, sleeping beside him, suddenly say out loud, "The Eucharist of the Liberal Catholic Church is an appropriate vehicle for the dispensation of Divine Grace." He woke up his friend, who did not remember the dream but was upset to have been used as a messenger.

Another friend of mine, a journalist in San Francisco, went, on a lark, to an initial consultation at the infamous Church of Scientology. He found, over the next months, that many of his dreams took place in Scientology clinics. In these dreams, histrionic cult members pressured him to join the cult. Through his dreams, he felt as if a disproportionate amount of psychic force was being exerted on him.

I befriended a woman who has frequent lucid and prophetic dreams. She also suffers from autoimmune deficiencies and other physical maladies that often accompany the reception of shamanic gifts. My friend met a woman suffering from a repetitive nightmare in which she was pursued and attacked by a frightening male figure. She suggested they sleep in the same bed. During that night, they had variations on the same dream, in which the scary man entered the bedroom to menace the woman once again. In the dream they shared, my friend yelled at the man to get out of the room, to stop bothering them. Shocked, the dream-villain departed. The woman's nightmares ended after this.

Shamanic cultures use dreams for prophetic and healing purposes. They accept, from the evidence of personal experience and accrued wisdom, that dreams can be communications from other orders of sentience, from occult entities or energies, from spirits of the dead as well as the living. Beyond this shamanic use of dreams, there are esoteric techniques of "dream yoga," utilized by Tibetan adepts to establish continuity of consciousness during sleep. By following this practice, "all experience and phenomena are understood to be dream," writes Tenzin Wangyal Rinpoche in *The Tibetan Yogas of Dream and Sleep.* "This should not be just an intellectual understanding, but a vivid and lucid experience. . . . Genuine integration of this point produces a profound change in the individual's response to the world. Grasping and aversion is greatly diminished, and the emotional tangles that once seemed so compelling are experienced as the tug of dream stories, and no more."

According to Julius Evola's *The Hermetic Tradition,* conscious mastery over dream-states and the achievement of continuity of consciousness was the secret goal of Western alchemy—their obscure terminology of planets, metals, and "dosages" referred to the work of psychic intensification required to reach this state. Achieving the "separation" meant learning to remain lucidly aware during dream sleep and dreamless sleep and, in those states, to explore other realms or spiritual dimensions at will. To make this separation into a permanent condition was considered a "labor of Hercules," extraordinarily difficult, and the route to immortality. Once this state was attained, bodily death was no longer a frightening mystery, but merely a transition to "other conditions of existence, having no resemblance to the Earth." Evola quoted the mysterious alchemist and eighteenth-century Italian count, Cagliostro:

> I belong neither to any country nor to any particular place; my spiritual being lives its eternal existence outside time and space. When I immerse myself in thought I go back through the Ages. When I extend my spirit to a world existing far from anything you perceive, I can change myself into whatever I wish. Participating consciously in absolute being, I regulate my action according to my surroundings. My country is wherever I happen to set foot in the moment. . . . I am that which is . . . free and master of life. There are beings who no longer possess guardian angels: I am one of those.

According to Evola, for one who has completed the "Work," sleep is no longer a pause for obliteration, but the staging ground for a "cosmic awakening." As one transforms the astral body through intensified self-awareness, one gains increasing mastery of the quantum flickers of dream-time and dream-space. Steiner, according to various accounts of his life, had reached this stage of development, consciously exploring different supersensible realms each night in his sleep.

The mad poet Ezra Pound once described artists as "the antennae of the race." Shamans, mystics, and alchemists also deserve that appellation—the unknown territory they mapped can belong to us, in the same way that the once-obscure and shocking distortions of modern art are now commonplace. They direct us toward an inner landscape—invisible, intangible to normal perception, psychic, spiritual, supersensible—approached through the discipline of non-ordinary states. The alchemists' dictum—"Explore the interior of the Earth!"—may indicate a human future woven from the fine filaments and silvery strands of the liberated imagination.

PART FOUR

THE LOOM OF MAYA

Why couldn't the world that concerns us—be a fiction? And if somebody asked, "but to be a fiction there surely belongs an author?"—couldn't one answer simply: why? Doesn't this "belongs" perhaps belong to the fiction, too?

FRIEDRICH NIETZSCHE

CHAPTER ONE

Are you and I perchance caught up in a dream from which we have not yet awakened?

CHUANG TZU

Our conventional notion of history, like our conventional understanding of space and time, supports a linear, evolutionary, and causally deterministic view of events. Our historians, anthropologists, and economists begin with a deep-buried assumption that civilizations are meant to progress in one way, as ours did: from non-technological and "primitive" to technological and therefore "advanced." If a civilization collapses or disappears, it means that environmental or cultural factors kept it from achieving our level of science and rationality, which is seen as a necessary advance over previous forms of social organization.

This is the standard perspective from which we view the civilization of the Classic Maya, which flourished in the jungle-covered Yucatan Peninsula from AD 250 until it vanished, suddenly and completely, in the ninth century, abandoning extraordinary city-complexes of temples and observatories that disappeared beneath the foliage until they were discovered by the Spanish invaders and painstakingly exhumed by archaeologists in the last two centuries. In his 2005 book, *Collapse: How Societies Choose to Fail or Succeed,* UCLA geography professor Jared Diamond has identified "increasingly familiar strands in the Classic Maya collapse," including "population growth outstripping available resources," "increased fighting as more and more people fought over fewer resources," and the short-term blindness of its leaders:

> We have to wonder why the kings and nobles failed to recognize and solve these seemingly obvious problems. . . . Their attention was evidently focused on their short-term concerns of enriching themselves, waging wars, erecting monuments, competing with each other, and extracting enough food from the peasants to support all those activities. Like most leaders throughout human history, the Maya kings and nobles did not heed long-term problems, insofar as they perceived them.

Diamond directed his dissection of the fall of the Classic Maya—based on analysis of bone fragments from skeletons and archaeological evidence—toward a particular polemical purpose: "The United States is also at the peak of its power, and it is also suffering from many environmental problems," he noted in *Harper's* magazine.

A deep irony is implicit in the sudden disappearance of the Classic Maya—for never in the history of our world was there a civilization as obsessed with time, and vast cycles of time, as the Maya. "The Mayan civilization was based on a 'chronovision,' a total absorption of the individual and collective life in the rhythms of nature, mapped into a mathematical system that had several cyclic counts running simultaneously," writes J. T. Fraser in *Time: The Familiar Stranger.*

The Classic Mayan accomplishments in math, astronomy, and calendrical accuracy are considered among their most impressive. The Maya employed three main calendrical systems: the Tzolkin, a sacred calendar of 260 days, where each day combines a number and a day-sign; the Haab, a civil calendar of 365 days, consisting of eighteen months of twenty days each, plus five extra days, considered unfavorable; and the Long Count, a cycle of approximately 5,125 years in total, based on 360-day cycles called the tun. Every fifty-two years, the number and day-sign of the Tzolkin match up with the same date on the civil calendar of the Haab, inaugurating a new cycle. The Aztecs, who inherited their calendar from the Maya, threw a "New Fire" ceremony on these occasions, putting out all the fires throughout the land for one night, rekindling them the next day, and forgiving debts.

The complex structure of the Mayan calendar is at least 2,500 years old, based on earliest evidence for use of the Long Count. "The Sacred Calendar is the main entry to the thinking of the advanced civilizations that existed in the Western world before the arrival of the Europeans," writes Carl Johan Calleman in *The Mayan Calendar and the Transformation of Consciousness.* Throughout the Yucatan, indigenous Maya continue to use the system today. When the revolutionary Zapatista movement, based in Chiapas, organizes protests, they choose significant days in the sacred calendar, representing particular forms of energy and intention.

While we use a base-ten system of counting, the Maya use a base-twenty, or vigesimal (as opposed to decimal), system that includes zero (0–19), which was modified slightly, but significantly, for computing calendrical dates. A day is a "kin"; twenty kins make a "uinal"; for mathematical calculations, twenty uinals make a "tun," equaling four hundred, but for calculating dates, eighteen uinals equal a tun, making an approximate "solar year" of 360 days. After this, the pattern of dating becomes regular once again: twenty tuns make a "katun" and twenty katuns create a "baktun." There are 144,000 days in a baktun—close to 394 years. According to the latest calculations and corrections, the Mayan Long Count of thirteen baktuns began on August 11, 3114 BC (in *The Mayan Factor,* independent Mayan scholar José Argüelles presents this date as August 13, 3113 BC, which is aesthetically satisfying, but misses the fact that there was no year zero in the Gregorian calendar, which flipped from 1 BC to AD 1, thus putting the origin date back one year). The Great Cycle, or the Mesoamerican Fifth World, ends on December 21, AD 2012. We are currently in the final phase of the thirteenth baktun, which began in AD 1618. After the baktun in factors of twenty, there is the piktun, calabtun, and kinchiltun, and the alautun. The alautun represents a span of time that is slightly over 63 million years. Beyond that is, apparently, the hablatun, equivalent to 1.26 billion years.

"One is amazed at the mastery over tremendous numbers implied in the various terms for higher units which have survived," wrote J. Eric S. Thompson, author of *Maya Hieroglyphic Writing.* "Surely no other people on a comparable level of material culture have had such a concept of vast numbers and a vocabulary for handling them." On their stone stelae and friezes, the Maya recorded dates as distant as 400 million years in the past. "These very long periods seem to have been used to weld mythical and historical events into a continuity of cycles," suggests Fraser.

A basic feature of Mesoamerican religious thought was "the idea of cyclical creations and destructions," wrote Yale anthropologist Michael Coe in *The Maya,* offering the standard interpretation, extrapolated from sculptures and inscriptions.

> The Aztecs, for instance, thought that the universe had passed through four such ages, and that we were now in the fifth, to be destroyed by earthquakes. The Maya thought along the same lines, in terms of eras of great length, like the Hindu kalpas. There is a suggestion that each of these measured 13 baktuns, or something less than 5,200 years, and that Armageddon would overtake the degenerate peoples of the world and all creation on the final day of the thirteenth; the Great Cycle would then begin again.

The Classic Maya calculated the Long Count—what Argüelles named the "Great Cycle"—and obsessively memorialized and encoded this 5,125.36-year period, ending in the 2012 alignment, in their friezes and sculptures, correlating this date with the end of one World Age, or Sun, and the beginning of the next. Precisely what this future state represented to them, however, is unclear—perhaps unknown, or perhaps unfinished. In fact, it is even a matter of dispute whether the Maya believed there would be another Great Cycle beyond this one, or if the conclusion of the current epoch represented the end of cycles, and the shift into a different cosmological and temporal framework altogether.

Mayan mythology chronicles a progressive evolution and development of humanity through the cycle of previous worlds. According to the Mayan creation myth, *Popol Vuh,* the gods made several attempts to create humanity, first using mud, then wood, and finally corn. Each iteration was better than the last, though still flawed, and intentionally denied the knowledge and omniscience of the gods—and ultimately destroyed by cataclysm. If humanity is ascending a spiral toward conscious creativity and self-knowledge, if we are reaching our maturity as a species, it may be that we will determine the meaning of this imminent transition for ourselves. The Maya may have intentionally left this future cycle a blank slate for us to occupy with our own deeds and intentions—or it may be that the ultimate meaning of this cyclical completion lay beyond the boundaries of their knowledge and worldview. Their shamanic science may have been capable of indicating the threshold before us, without revealing—purposefully or not—what lies beyond.

THE PYRAMIDS, glyphs, and artworks left behind by this civilization suggest an extraordinary refinement and subtlety of thought—vastly different from the later Aztecs, who, Octavio Paz noted, "confiscated a singularly profound and complex vision of the universe to convert it into an instrument of domination." The most celebrated monuments of the Maya include the Temple of Kukulkan (the name of Quetzalcoatl in the Toltec tradition) at Chichen Itza. This nine-step pyramid is four-sided and has ninety-one steps on each side, with the platform on top equaling 365, the number of days in a solar year. Twice a year, on the spring and fall equinox, the shadow of a serpent appears and undulates along the staircase of the temple for more than three hours, creating a pattern of isosceles triangles similar to that found on local rattlesnakes. Puns on all levels, games played with stone and light, in which architecture and sculpture reflect stellar alignments and yearly cycles, fascinated the Mayan mind. Acoustics were also of interest to them. Archaeologists have recently discovered that clapping at certain spots around the Temple of Kukulkan produces a sound identical to the chirp of the quetzal bird. "Among the ancient Maya, it seems, artists and scribes belonged to the very highest stratum of a rigidly ranked society," wrote Michael Coe. As far as we know, they built their imposing structures with the most rudimentary means, lacking even the wheel. Instead of mining metal for coins, they used cacao beans as currency.

Even a cursory examination of Mayan imagery and myth suggests that this civilization was based on different principles, with a completely alien mind-set, from anything we know today. Whereas our society is fascinated by material accumulation, linear evolution, and technology, the center of the Mayan worldview was a vision of vast cycles or cosmic spirals of time, embodied and expressed by a seething pantheon of extravagant deities, hero-twins, and cosmic monsters. Trained in shamanic practices and initiatory disciplines, members of the nobility were thought to be able to contact these entities directly. In many sculptures, Mayan leaders emerge from the jaws of writhing vision serpents or grinning demigods that act as portals conveying them between different worlds or planes of reality. These nonhuman entities required appeasement—sometimes through blood—and could be directly evoked in shamanic trance states. "For the ancient Maya, human beings released their *ch'ulel* [soul-stuff] from their bodies when they let their blood. Through bloodletting, they 'conjured' (*tzak*) the *way* and the *ch'u,* the 'companion spirits and gods,'" wrote David Freidel, Linda Schele, and Joy Parker in *Maya Cosmos.*

The Mayan wizard-king was considered the chief possessor of *itz,* the cosmic sap or magical liquid of shamans and alchemists—akin or identical to the violet-tinged "translinguistic fluid" sought by the McKennas—which can be used to heal or kill. It was the task of the rulers to maintain the balance of cosmic forces through their magical powers. The king "was the embodiment of the World Tree of the center," and his task was one of "world centering." "A king who controlled creation and the power of the Otherworld was by definition a successful king." In the elaborate temple cities of the Maya, the central courts were "the settings for great dance pageants that enabled the lords and their people to travel to the Otherworld to greet the supernatural beings who gave power and legitimacy to the human community," wrote Friedel and Schele.

A long tradition of alternative archaeology challenges the conventional historical model of the Maya and other premodern civilizations. These scholars—a group of wide-ranging quality including Graham Hancock, Zecharia Sitchin, Erich von Däniken, Maurice Cotterrel, Robert Bauval, Adrian Gilbert, among others—fixate on the advanced aesthetics, paradoxes, intellectual puzzles, and anomalies that shadow the standardized accounts of the Mayans' warring dynasties and ecological collapse. Fanciful interpretations have been offered of the most famous icons of Mayan art, such as the sarcophagus cover of the tomb of K'inich J'anaab Pakal, known as "Pacal the Great," the seventh-century king of Palenque. For von Däniken, Lord Pacal was clearly an "ancient astronaut," operating the controls of some futuristic craft: "This strange being wears a helmet from which twin tubes run backwards. In front of his nose is an oxygen apparatus. The figure is manipulating some kind of controls with both hands." For Sitchin, Pacal the Great was part of a vast cosmic plot, in which advanced, inhumane star-beings made their way to Earth for pirate plunder, cloning slaves, and mining for gold.

In *Fingerprints of the Gods,* Graham Hancock, a former correspondent for *The Economist* and the London *Sunday Times* and one of the better alternative archaeologists, traversed the globe, exploring hints and clues from Egyptian, Mayan, Olmec, and Hopi artifacts and myths, among others, to make an argument for the existence of a prehistoric civilization that had attained an extraordinary level of science and culture before it was destroyed by a cataclysm. Hancock was persuaded by evidence suggesting that the civilizations that produced Machu Picchu in Peru, as well as the Sphinx in Egypt, were many thousands of years older than conventional accounts would allow. He followed the arguments of renegade Egyptologist John Anthony West, who had observed that the earliest monuments of Egypt were their most sophisticated achievements and argued that "Egyptian civilization was not a 'development,' it was a legacy." Hancock thought that these civilizations might have descended from a common predecessor—where Plato wrote of "Atlantis," Mayan legends tell of "Aztlán," their ancient home—perhaps located in a once-temperate Antarctica.

Studying the Maya, he was amazed by the "computer-like circuitry" of their calendars, which maintained the exact synodic orbits and stellar conjunctions of Mercury, Venus, Mars, and Saturn; calculated the length of the Earth's orbit around the Sun and the Moon's orbit around the Earth with extraordinary precision; and kept track of eclipses. Hancock proposed that the calendar was a bequest from a technologically advanced civilization of prehistory, and its ultimate purpose was to predict "some terrible cosmic or geological catastrophe" in the year 2012, ending the world. He also thought that the image of Lord Pacal on his sarcophagus, with its "side panels, rivets, tubes and other gadgets," suggested a "technological device."

In the conventional archaeological view of this artifact, Pacal is seen to be superimposed on an elaborately decorated "foliated cross" that represents the classic "World Tree" of shamanism, reaching down to the underworld and the land of the dead, and upward to transcendent realms. He is suspended along the "Kuxan Suum," defined in *Maya Cosmos* as "the navel of the world . . . a life-sustaining cord traversing the layers of the cosmos, connecting humanity to the gods, the source of life." According to Freidel and Schele, contemporary Maya believe this cord "was cut by the Spanish invaders."

IF IT WASN'T FOR José Argüelles—visionary Mayan scholar and self-proclaimed prophet, the instigator of the 1987 "Harmonic Convergence"—I would never have considered the possibility that the chronovision of the Classic Maya might be relevant to our current reality. Months after the fall of the World Trade Centers, I was sent his latest book, *Time & the Technosphere,* by his publishers. Apparently, he dashed it off quickly after 9-11, hoping that, in the aftermath of the disaster, more people would be receptive to his extremist alternative worldview. *Time & the Technosphere* blends a wild compendium of theories, mathematical postulates, confusing graphs, deadpan ravings, and occult predictions. While some sections of it made straightforward sense, other parts read like some zany cartoon revelation, Japanese manga montage as messianic text, spattered with unfamiliar terms like "radiosonics," "codon cube cosmology," and "psi bank plates."

A onetime professor of art history, Argüelles admitted in his introduction that, after the 1987 Harmonic Convergence, "I no longer sought any conventional means to communicate. I was now operating prophetically, on behalf of the noosphere," and indeed, his book was a mantic thought-machine—pitched into the stratosphere, far beyond the pale of the mainstream discourse. Although I didn't know what to make of it, somehow I couldn't dismiss it. His hectoring style was off-putting, but his prophetic angle compelled me, reflecting many of my own strangest and most secret musings. His self-imposed exile from convention freed him to articulate ideas and make bold leaps of vision that few other thinkers could even imagine, let alone dare to take seriously.

Time & the Technosphere argued that the 2012 end date—or birth date—of the Mayan calendar signifies the "conscious activation of the noosphere," the thought-envelope around the Earth defined by Teilhard de

Chardin. Argüelles proposed that, like the McKennas' timewave, the Classic Mayans' model of time was recursive and fractal, with patterns repeating on ever-more-concentric spirals of decreasing duration as we approached the imminent "end of time"—at least, of one form of time. He insisted that he had inherited "the mantle of the prophetic stream of the Chilam Balam, the jaguar priests, the wizard knowers of the 'night script.'" He liberally cited the Quran, the book of Revelation, and *Tao Te Ching,* as well as the work of Russian scientists such as V. I. Vernadsky, a geologist who studied the biosphere, defining it as "the region on Earth for the transformation of cosmic energies."

According to Argüelles, the 5,125 years (5,200 tun) of the "Long Count," from 3113 BC to AD 2012, measures the entire span of what we call history, from its beginning to its end—from the building of Stonehenge and the Pyramids, the creation of the solar clock and written language (all around 3000 BC) to the "end of history" in our planetary birth into higher consciousness. The Long Count codified this fractal cycle as a necessary phase, transitioning from the original organic balance of the biosphere, progressing through increasingly artificial, destructive, and destabilizing layers of mechanized civilization that he dubbed the "technosphere," and finally, after the imminent collapse of our technological support systems, attaining the "pristine conditions" of the noosphere—a state in which humanity, as telepathic collective, would be directly attuned to the crystalline precision of the Gaian Mind.

Argüelles discussed the theories of the geologist Vernadsky, one of the Soviet Union's leading scientists before his death in 1945, who, like Chardin, foresaw the transition into the noosphere: "Man by his work and his conscious attitude toward life is remaking a terrestrial envelope, the geological domain of life," Vernadsky realized. "He is transforming it into a new geological state, the noosphere." In this transition, the destruction of the World Trade Centers was an "inevitable event." Following their fall, Argüelles conjectured, "the whole technospheric system will be coming down slowly over the next few years, like a giant circus tent that has lost its central prop." He thought it was necessary for the technosphere to collapse, soon, before its parasitical slurps consumed the life-support systems of the planet.

While the mainstream analyses of 9-11 published in popular newspapers and magazines could only conceive of it in material terms, as an event with global strategic consequences, Argüelles considered this planetary emergency to be "a profound theological moment," leading to a world war "between the religion of choice and the religion of submission." I couldn't escape the feeling that there was something to this idea—a disquieting thought suppressed by all of the frantic mobilizations and public declarations in the wake of the attack. The essence of Islam is submission to the will of God. Such a prospect is anathema to the West's individualistic and ego-centered culture, based on defiance of limits or outside authorities. "I have taken seriously the claim that the Quran is the final revealed text for humanity," he wrote.

He proposed that "prophecy," its reception by particular "human channels" and its dispersion through religious and esoteric traditions, was a carefully timed release of information, guiding us toward the 2012 transition to the noosphere. He suggested the 9-11 attack was part of the prophecy presented in the book of Revelations, as well as the Quran. But he interpreted the significance of these foretellings in a much different fashion from Christian or Islamic Fundamentalists: "The Apocalypse is about 'the end of time.' But the end of whose time? The new paradigm that we need to explain 9-11 should be a paradigm that is all about time." He thought that the Mayan conception of time, and his peculiarly modified version of their sacred calendar, provided the essential new paradigm.

I had read one other book by him a few years earlier, but it left little impression on me. I found it intriguing and fun, but fantastically implausible. *The Mayan Factor* (1984) inspired the Harmonic Convergence, a global gathering and peace-meditation held at sacred sites around the world, on August 17, 1987, the end of the cycle of "Nine Hells" that began on April 21, 1519—the date that Cortés arrived in Veracruz—and ushering in the cycle of "Thirteen Heavens," according to his interpretation of Mayan prophecy. Trying to understand my fascination with *Time & the Technosphere*—which seemed so nutty yet somehow encompassed the post 9-11 unreality better than any other text I had found—I returned to *The Mayan Factor*, and on this second reading, it resonated with me.

Inner events can be as dramatic as outer ones, though they are much harder to convey. I was sitting in the Balthazar Cafe on Spring Street early one morning—a lofty high-ceilinged space with giant mirrors and red leather booths, a Manhattan simulacrum of some fin-de-siècle Parisian sanctuary—when I found myself entering the frame of what Argüelles was describing. He was exploring the Mayan conception of time, energy, and consciousness, as he understood it. He argued that, despite their lack of material technology, the Maya actually possessed a more advanced science than we do: "What distinguishes Mayan science from present-day science is that it is a system operating within a galactic frame. A science operating within a genuinely galactic frame of reference cannot be separated from what we call myth, art, or religion." With their synthesizing and holistic worldview, based on mind as the foundation of the universe, inseparable from time and space, "not only do the Maya challenge our science, but they play with our myths."

The basic goal of Mayan civilization, underlying their obsession with astronomical orbits and vast cycles of time, was synchronicity, synchronization, or what Argüelles calls "harmonic resonance." Their "exquisitely proportioned" number system was not primarily a counting code, but "a means for recording harmonic calibrations that relate not just to space-time positionings, but to resonant qualities of being and experience." Hidden within the rubble of their civilization, the Maya intentionally left behind a trove of secret knowledge, hidden teachings on the nature of time and being. This information, in his model, was not just some quantity of statistic or fact, but a new pattern that had to be sensed and felt as well as logically grasped. To receive a new pattern, one must be open to it. "The essence of information . . . is not its content but its resonance," he wrote. "This is why feeling or sensing things is so important. To sense the resonance of incoming information co-creates a resonant field." As I understood what he was saying, sitting amidst the chattering Balthazar flock of stockbrokers and fashion plates, I felt the weaving of that field, like a pattern of magnetic energy, between myself and these unusual yet strangely satisfying ideas.

SOMEDAY, if a history of postmodern metaphysics is written—if books are still written then—a side plot may pit the mandarin cerebrality of William Irwin Thompson against the cruder will-to-cosmic-destiny represented by

Argüelles. Men of the same generation, minds similarly scourged by the critical chiasmus of the 1960s, both recognized, as Thompson put it, that mysticism "seems impractical in technological culture because it is the dialectical negation of that culture and the affirmation of the next culture." Seeking the escape hatch from one-dimensional doom, they plunged into Eastern disciplines—Thompson into yoga, Argüelles exploring Tibetan meditation—and emerged with intellectual syntheses of West and East. Although he considered Argüelles an exemplar of an unhinged paranoid strain, Thompson was also critical of the literalist notion that extraterrestrials would appear on Earth through the scenarios of von Däniken, Dolores Cannon, or *Close Encounters of the Third Kind*—or even through the reception of some transmission of radio waves, as SETI imagined.

"The meeting we are expecting in front of us in linear time has already occurred, is now occurring, and will continue to occur," Thompson proposed. Gods and advanced aliens "do not talk to us, they play with us through our history. . . . Our subjective-objective distinctions about reality are incorrect. As in the world view of the Hopi Indians, Matter, Energy, and Consciousness form a continuum." In his thought, Thompson encompassed this realm, but Argüelles penetrated it, risking visions that unshackle the mind's grammar to embody a nascent cosmology.

Believing he had gleaned the secret purpose of the Maya, Argüelles saw their wizard-kings as representatives or avatars of a galactic civilization that is "post-technological." Powerful potentates like Lord Pacal of Palenque, he conjectured, were agents who had incarnated on the Earth at the precisely perfect juncture, to leave behind, in the ruins of Palenque, with its secret chambers and hermetic puzzles, noospheric news linking time and consciousness via synchronicity, for modern humanity to rediscover as we reached the end of the cycle:

> As everyone knows, there is no intelligence in coercion or forcing another into action or realization. And if the name of the galactic game is superior, intelligent harmonization, it must be played so that the local intelligence is taught or shown how it works in such a manner that it comes to its own conclusions. In other words, the

galactic code of honor is to manifest and demonstrate harmony by whatever means possible.

Once the Mayan avatars completed their task, they decamped. "Their purpose was to codify and establish a system of knowledge, a science, and having codified it in stone and text, to move along."

If the "Galactic Maya" had space-time-tripped to Earth from Hunab Ku, the Galactic Center, this did not happen through cumbersome rockets, but through "harmonic resonance," patterns of genetic information intentionally transmitted from star to star, encoded in light—a "wave-harmonic means of transmission, communication, and passing from one condition of being to another." The "so-called sun worship" of the Maya and ancient Egypt was actually "the recognition and acknowledgement that higher knowledge and wisdom is literally being transmitted through the Sun, or more precisely, through the cycles of the binary sunspot movements."

Argüelles proposed that the Tzolkin, their 260-day ceremonial calendar, was the basis of Mayan esoteric technology, linking them to vast cosmic cycles and evolutionary patterns. The seemingly simple 13-by-20 matrix of the Tzolkin functioned as the basis of Mayan science and space-time travel, allowing them to receive and transmit information—as well as themselves—between star systems. He called the Tzolkin the "Loom of Maya," and suggested they utilized it in their science of divining harmony and creating resonant patterns—a science that was also an art of being in the right place at the right time.

"Like galactic ants, the Maya and their civilization would be the synchronizers of momentary need—represented by planetary or solar intelligence—with universal purpose, full conscious entry into the galactic community." Argüelles compared this mission with the goals of our present civilization: "Who can say what the goals of our civilization are? Do these goals even have a relation to the planet, much less to the solar system?" It was the apparent simplicity and intricate subtleties of this Tzolkin, rather than any "Encyclopedia Galactica," that the Maya had bequeathed to us to help us enter the community of galactic intelligence.

Reading over this in the clattering cafe, I suddenly found myself attuning with this Galactic Mayan perspective on time, synchronicity, and consciousness. I felt, in that moment, as though a presence—watchful, serene, a tad impersonal, almost wistful, and ancient beyond my conception of age—had descended into me, peering out at the book and the restaurant through my eyes. It was a thrilling presentiment, an atemporal apprehension. For Argüelles, this is how the Maya are revealed to us—not as alien entities, but as a resonant attunement with a higher frequency of mind. "The Maya were returning, but not as we might think of them," he wrote. "Ultimately their being, like ours, transcends bodily form. And precisely for that reason, their return can occur within us, through us, now." For one swooning, Proustian moment, I came unstuck in time. Vast vistas presented themselves—not a flood of personal memories, but of impersonal cycles, mystical arrivals, cosmic unveilings.

Argüelles's radical theory about the Maya subverts and overturns our received ideas about higher intelligence, how it would manifest and how it might function. Instead of a mechanistic race for faster and faster microprocessors to create some kind of artificial intelligence that can merge with or even replace us, the paradigm shift he proposes returns us to the "zero time" of meditation or psychedelic insight, to the mystical understanding that the world and the Self are already perfect, to a conscious deepening of technique rather than any mechanistic extension of technology. If the "Galactic Maya" hypothesis has validity, it reveals a reality vastly different from our current perspective on it—unfurling, out of modern linear thought, a multidimensional array of infinite possibility.

Argüelles points out that the name of the Maya is already suggestive. As Fritjof Capra noted, "maya" is a central concept in Hindu and Buddhist traditions, where it signifies the "magic creative power" of the gods. "Maya" also came to mean the world of illusion, which is mistaken for reality by those who are under its spell. "Not surprisingly, we find that Maya is the name of the mother of the Buddha," Argüelles notes. "In Egyptian philosophy we find the term Mayet, meaning universal world order." The term also appears in Greek mythology where "the seven Pleiades, daughters of Atlas and Pleoine, number among them one called Maia, also known as the brightest star of the constellation Pleiades." The Greek Maia is the mother of Hermes, or Mercury, trickster god of magic, thievery, and writing.

Could such correspondences conceal a hint? Could the reality we currently experience be a kind of puzzle or, as recent movies have it, a matrix that conceals other orders of time, other levels of being, that we will be able to access when, as the Rig Veda puts it, "without effort, one world passes into another one"? My explorations of the rapid-transport of DMT and other psychedelics seemed to suggest this—they were like the sudden raisings-and-shuttings of the blinds on opaque other realities.

The solution of the puzzle would be hidden in the structure of time—and the possibility that we could attain a deeper realization of it. "Does it really exist, this destroyer, Time?" the poet Rainer Maria Rilke asked. According to modern physics, it doesn't exist in the way our ordinary brains and nervous systems process it. There is, also, a transcendent domain, spaceless and timeless, of which matter, space, time, and consciousness are interrelated expressions, and in which everything we perceive as separate is intimately intermeshed.

Argüelles proposes it is not a coincidence that the number of the "elect" in the book of Revelation, 144,000, is the same as the number of days in a "baktun." *Time & the Technosphere* is filled with such numerical conjunctions and synchronicitics, some of them more tantalizing than others. He also noted that a flaming, falling tower with people dropping from it was depicted on one of the traditional tarot cards, which some believe draw their symbolism from the branching paths along the "Tree of Life" in the Qabalah. In the tarot, "The Tower" represents the collapse of a false belief system—a necessary first step before attaining a new understanding. I had also noted that odd conjunction of symbolism in the wake of 9-11—as if the crash of those skyscrapers was a tarot card come to life.

Of course, most intelligentsia in the secular culture find it ridiculous to think about a literal fulfillment of biblical prophecy as something now taking place. However, I could not shake the sense that 9-11 possessed a metaphysical significance that overwhelmed the number of the dead, the scale of destruction, the geopolitical consequences. I went back to the text of Revelation and found a passage in Revelation 11 describing the destruction of the "two witnesses . . . clothed in sackcloth," also called "the two olive trees, and the two candlesticks standing before the God of the earth." The candlesticks are destroyed, their dead bodies lying "in the street of the great city." I found it possible that this section prefigured, in symbolically cloaked form, the fall of the towers. Artworks by children were put up around the city after the attack, and in some of the paintings, the towers were represented as candles. If this archetype had in fact alighted, who else but children, still operating freely from unconscious depths, could access what was denied by our adult mind-set, disavowed by modern beliefs?

According to Carl Jung, "One does not become enlightened by imagining figures of light, but by making the darkness conscious." Had this biblical foretelling constellated in our human reality, making its impact in shattering steel and glass? And if so, who would dare to interpret the deep-darkening world that those candles illuminated? I found an uncanny resonance between the 9-11 event and its aftermath and the narrative of Revelation 11—however, I do not mean to enforce belief in the idea, only to note it as part of the pattern of clues and intuitive hints that kept amassing as I pursued my investigations. Such ideas would only make sense if, as Argüelles suggests, the Maya have been playing with us through our history, peeping out from odd prophetic corners, giggling at us from behind the seams of our linear matrix, in preparation for the unveiling ahead.

CHAPTER TWO

We know time.

DEAN MORIARTY, *On the Road*

More than five hundred years after the mysterious disappearance of the Classic Maya, the Spanish conquistadors arrived in the New World, quickly demolishing, with startling ease, the vast empires of the Incas and the Aztecs. As Jared Diamond writes in *Guns, Germs, and Steel,* the "decisive moment in the greatest collision of modern history" occurred on November 16, 1532, at the Peruvian highland town of Cajamarca, when Francisco Pizarro and his "ragtag group of 168 Spanish soldiers" captured the Incan emperor Atahuallpa and annihilated his army of eighty thousand men. Despite the vast disparity in their forces, "Pizarro captured Atahuallpa within a few minutes after the two leaders first set eyes on each other." In Diamond's account of this contest, the crucial ingredient was "the Spaniard's far superior armament." While the 168 Spaniards possessed some horses, a few erratic guns, steel swords, and steel armor, Atahuallpa's legions "could oppose only stone, bronze, or wooden clubs, maces, and hand axes, plus sling shots and quilted armor." According to Diamond, the military technology of the tiny Spanish force left the Inca defenseless.

From the account of one of the victorious Spaniards, however, it appears that military prowess had little to do with the Spanish victory. "All of the other Indian soldiers whom Atahuallpa had brought were a mile from Cajamarca ready for battle, but not one made a move, and during all this not one Indian raised a weapon against the Spaniards." Once Atahuallpa had been dragged from his raised dais and captured, the conquistadors, fighting for the "Holy Catholic Faith," proceeded to butcher their unresisting enemies for many hours, killing as many as half of the Inca soldiers and routing the rest. In the aftermath of this slaughter, Pizarro told Atahuallpa, "By reason of our good mission, God, the Creator of Heaven and Earth and of all things in them, permits this, in order that you may know Him and come out from the bestial and diabolical life that you lead. It is for this reason that we, being so few in numbers, subjugate that vast host." The Spaniard demanded a "king's ransom" in gold for the return of the emperor—"revered by the Incas as a sun god," exercising "absolute authority over his subjects," Diamond notes. When they received their loot, they executed Atahuallpa.

Despite Diamond's analysis, the technological advantage in armaments of the rapacious Spaniards does not seem sufficient to explain the results of this "decisive moment," which was repeated throughout the New World, and in far-flung regions of the globe, during the centuries of European imperialism—in Bali, for instance, the ruling nobility chose to march into the guns of their invaders, rather than offer resistance. Analyzing the parallel collapse of the Aztecs, the philosopher Jean Gebser proposes, instead, that the Europeans represented a new form of consciousness, for which the myth-based kingdoms of Mesoamerica had no defense. "The Spaniards' superiority, which compelled the Mexicans to surrender almost without a struggle, resulted primarily from their consciousness of individuality, not from their superior weaponry," Gebser wrote.

In his singularly forbidding tome, *The Ever-Present Origin,* the product of a lifetime's thought, Gebser presents a thesis of human development that fits the quantum-based model of consciousness—unfolding from a transcendent domain, evolving through sudden shifts or creative leaps into new contexts—presented by physicists such as Amit Goswami and F. David Peat. According to Gebser, we have passed through a number of "consciousness structures," each one a profoundly different realization of time and space. "Man's coming to awareness," he writes, "is inseparably bound to his consciousness of space and time." He defines four previous stages—the archaic, the magical, the mythical, and the mental-rational—and argues that we are currently on the verge of transitioning into a new stage, which he calls integral and aperspectival, characterized by the realization of time freedom and ego freedom. A new form of consciousness arises—as a sudden "mutation"—when the previous structure enters its ultimate crisis, having exhausted its possibilities. In a process perhaps similar to the appearance of language as a structure in the brain, the successive mutations he describes represent the spontaneous emergence of new capacities, which can only occur when the ground for them has been adequately prepared. "Every consciousness mutation is apparently a sudden and acute manifestation of latent possibilities present since origin."

If one could imagine a textual equivalent to the humming gray monolith that looms over the beginning and end of history in the film *2001, The Ever-Present Origin* would be it. Prototypically German and difficult, Gebser ranges over the entire history of civilization, from prehistoric cave painting to Cubism, from the chant-songs of aboriginals to the atonal music of Schoenberg and Bartók, from primitive spell-casting based on sympathetic magic and "the vegetative intertwining of all living things" to nuclear armaments. Originally published in the 1950s, his book encompasses modern developments in physics, psychology, biology, the social sciences, and other disciplines, tracking "the irruption of time" into awareness, which he considers the crucial element in our imminent shift into a new form, or "intensity," of consciousness. Gebser was an independent thinker, only loosely connected to contemporary schools of European philosophy. Like Rudolf Steiner, he employed the traditional tools of Western philosophy to elaborate a new model of consciousness, one aligned with mysticism, yet deeply evolutionary.

The archaic structure represents the primordial beginnings of humanity. The archaic mind is outside of time, lacking any concept of progress or evolution, undifferentiated from what Gebser calls "origin," a prespatial and pretemporal unity—pure presence or "suchness." In the archaic world, there is no awareness of individuality, only a tribal or group consciousness. Among Aboriginals in Australia—not identical to the archaic mind-set but one of the closest surviving links—the purpose of ritual is to maintain the creation in its original harmony, preserving the world's vibration or tone.

"For the Aborigines, there is no fall; paradise is the earth in its pristine beauty," writes Robert Lawlor in *Voices of the First Day: Awakening in the Aboriginal Dreamtime,* an extraordinary effort to recapture the Aboriginal worldview and way of life. "There is no part of this existence that needs to be transcended, repressed, or gone beyond." Aboriginal or archaic time is perpetual presence and celebration, in which every day is the "first day." The songs of Pygmies and Aboriginals are "manifestations without beginning and end, a chance intrusion of the voice and a chance ending: a sleep that has, as it were, become sound," writes Gebser.

The magical form of consciousness, for Gebser, is "pre-rational," characterized by instantaneity, egolessness, and "unitary merging," similar to sleep or trance. There is no distinct separation between psyche and world for "magic man," hence he feels a profound inherence and affinity between himself and the cosmos. "It is a world of pure but meaningful accident; a world in which all things and persons are interrelated, but the not-yet-centered Ego is dispersed over the world of phenomena," Gebser writes. The magic structure "lies before time, before our consciousness of time. How far back we may wish to place this magic time into prehistory is not only a question of one's predilection, but, on account of the time-less character of the magical, is essentially an illusion."

This is an important aspect of Gebser's thesis. The various structures he describes, "unfolding from origin," are expressions or possibilities of the transcendent consciousness, outside of space and time. Each of us contains all of the past forms within us, just as we preserve the older layers of cortical development within the physical brain. A regression to the deindividuated and prerational magic structure remains a present danger in our modern world—it is apparent in mass phenomena such as various forms of fanaticism and fundamentalism, in the trance-inducing powers of slogans and the media, as well as the behavior of crowds during sports spectacles and riots. "The experience of where the magic structure begins and how it works must be made by every person individually, and the peril lies "behind" the experience and threatens anyone unprepared with the loss of self," Gebser writes.

Through witchcraft and sorcery, totem and taboo, magic man sought independence from the primordial forces of nature, which pressed upon his awareness and threatened him with dissolution. "In these attempts to free himself from the grip and spell of nature, with which in the beginning he was still fused in unity, magic man begins the struggle for power, which has not ceased since; here man becomes the maker." According to Gebser, this ancient terror of the magic consciousness is still present in us, underlying the development of modern technology and contemporary power politics: "Nature, the surrounding world, other human beings must be ruled so that man is not ruled by them. This fear that man is compelled to rule the outside world—so as not to be ruled by it—is symptomatic of our times. Every individual who fails to realize that he must rule himself falls victim to that drive."

The oversaturation of the magic world by *mana*—vital energy, supernatural potential—eventually led to an "emergent awareness" of the soul, that daimonic and ambiguous inner realm, between body and spirit, described by Patrick Harpur. The myths expressed and articulated by the mythic consciousness "reveals the soul and, at the same time, an invisible and extended region of nature, the cosmos," Gebser writes. The mythic consciousness oriented itself around polarities—of dream and awakening, life and death, gods and men, Heaven and Earth. As represented in the Tai Chi symbol, one pole does not negate its antipode, but reflects and includes it. "When we speak of 'time' we are also speaking of 'soul,'" writes Gebser. The emergent awareness of the soul elicited an awareness of time, not in the modern sense of a linear progression, but as a cyclical process of change, becoming, and return, similar to the rhythms of dream and awakening. History, in our sense of the term, did not exist until the modern era; "the Egyptians of old knew only of annals or chronicles, not of historicity." Mythic man studied the stars swirling overhead, innately aware that their vast and intricate whirlings were precisely aligned to the patterns of human development on the Earth—an understanding summed up in the Hermetic dictum, "As above, so below."

Egyptian cosmology focused on the "Precession of the Equinoxes," a slow backward movement of the stars in relation to the Sun's rising on the equinox, which appears in a different constellation every twenty-two hundred years or so, requiring roughly twenty-six thousand years to complete an entire round. Precession is due to a backward wobble in the Earth's spin, caused by the gravitational pressure exerted on the Earth by the Sun and the Moon. As Robert Bauval discovered in *The Orion Mystery,* vertical tubes in the Egyptian pyramids were designed to align with particular stars at certain times, revealing the movement of precession. According to the Precession of the Equinoxes, we are currently at the end of the Age of Pisces, transitioning into the Age of Aquarius. Hindu or Vedic cosmology conceived of the Yugas, four periods, each shorter than the last, representing accelerating decline from the Satya Yuga, the Golden Age, when humanity was self-governed by knowledge of divine law. This is the Kali Yuga, the final epoch, which corresponds, in Greek myth, to the last of four ages—Golden, Silver, Bronze, and Iron. As discussed, Mesoamerican myth conceives of cycles within cycles, like wheels within wheels, continuing over vast periods. According to the Classic Maya, we are approaching the end of the "thirteenth baktun," completing a Long Count of 5,125 years.

"Anxiety is the great birth-giver," Gebser writes. A new form of consciousness emerges and realizes itself during a fatal crisis, when the prevailing structure has reached "the end of its expressive and effective possibilities, causing new powers to accumulate which, because they are thwarted, create a 'narrows' or constriction. At the culmination point of anxiety these powers liberate themselves, and this liberation is always synonymous with a new mutation." After mythic civilizations had elaborated their dream-like consciousness and inner awareness of the soul within temple cities and pantheistic cults, their cultures could only repeat themselves, becoming increasingly routinized and mechanical.

"The exhaustion of a consciousness structure has always manifested itself in an emptying of all values, with a consistent change of efficient qualitative to deficient quantitative values. It is as if life and spirit withdrew from those who are not coparticipants in the particular new mutation," Gebser writes. As mythic cultures declined, religious doctrine turned rigid and stagnant, locked into tradition rather than renewed through inner experience. The "deficient quantitative" mania that gripped the Aztecs during the last years of their empire led them to sacrifice increasing numbers, as many as 250,000 victims annually, according to some scholars, before the conquistadors crushed them.

The drastic shift—mutation or leap of quantum creativity—into the mental-rational structure was foretold by a myth: the birth of the goddess

Athena, who emerged from the painfully throbbing head of Zeus, split open by an ax. This blow was "accompanied by a terrible tumult throughout nature, as well as by the astonishment of the entire pantheon," writes Gebser, paraphrasing Pindar. Once sprung, Athena, goddess of knowledge and clear thought, bestowed her protective grace over Athens, cradle of the modern Western mind. In the movement from the mythic to the mental-rational mind-set, human thought was directed outward, discovering the external world, for the first time, as an object of inquiry in itself.

Like a lamp switched on in a dark room, the mental-rational mind illuminated the physical world, which revealed itself, first of all, as a spatial reality. "Magic occurs in the dark, indeed in darkness itself, while myth occurs in the night and dreams where a twilight is already present. But the mind or the mental presents itself in the brightness of daylight," Gebser writes. In dividing itself from its past, the mental mind abandoned the ambivalent and ambiguous polarities that characterized the mythic world, based on its projections of the soul and psyche, for duality, a "diminished and mentalized form" of thought, separating matter and mind, rationality and intuition, organic and inorganic, and so on.

Gebser argues that our current structure of consciousness began with the Greeks, reached its full flowering with the Renaissance and the discovery of perspective, and has since entered its "deficient" or decadent phase. During this period, mental-rational humanity became not only obsessed with space, but possessed by space—by the possibilities that developed from our increasing ability to transform matter and shape physical reality. We learned to see ourselves, for the first time, embedded in—and simultaneously alienated from—the three-dimensions surrounding us.

A stroll through the Metropolitan Museum of Art makes this clear. Works from tribal societies in African, Egyptian, Assyrian, Byzantine, and medieval periods display a hieratic flatness in which the figures are static and deindividualized. "The closed horizons of antiquity's celestial cave-like vault express a soul not yet awakened to spatial time-consciousness and temporal quantification," Gebser writes. The discovery of perspective in the Early Renaissance caused a profound shift in our awareness of the world. From the flat planes of medieval art, with their dematerialized saints fixated on the faraway, the Renaissance artists and scientists, from Giotto and Alberti to Leonardo and Rembrandt, learned to depict space in its full dimensions. This opening of space elicited, at the same time, the realization of the alienated modern self. By the time we reach the celebrated room of Rembrandt portraits, in our time-walk through the museum halls, the paintings have become mirrors reflecting our modern consciousness back at us—our mercantile cynicism and self-questioning irony, emerging out of darkness and shadow to confront our inner void. This static spatial conception began to break up in the late nineteenth century, when Impressionism and Post-Impressionism were followed by Cubism and Futurism, in which artists attempted to represent time as well as space on a flat surface. For Gebser, these modern movements symbolized the troubling "irruption of time" into consciousness, but not yet its integration in a new form of awareness.

Our changed realization of space and self was accompanied by a transformation in our relationship to time. Rejecting the vast cycles and "eternal return" of the mythological worldview, the modern mind saw "absolute time" as an unvarying linear extension, equivalent to space, as measurable as matter. Gebser writes: "Space and time do not exist for magic man. Even for mythical man, space is non-existence, despite his awareness of . . . self-contained, cycling seasonal time and its motion. . . . Whereas mythical man lived from this inner movement, mental man thought by virtue of spatial, external actualities: everything for him became space, including time." Modern thinking is "perspectival" in its essence, knowing "only walls and selective, divided objects" through conceptual schemes that must be "grasped," like objects in space. "Perspectival thinking spatializes, then employs what it has spatialized."

Possessed by space and matter, mental man spatialized and quantified everything, including time. Today this is clear in the metaphors we use when we refer to time. We talk about having enough time, running out of time, racing against time, wasting time, spending time, doing time, killing time, and so on. We speak of time as a quantity ("time is money") of which there can be enough or not enough. " 'Time' in our sense is an instrument we have created with which we are able to shape the three-dimensional perspectival world and permit it to become a reality."

The problem is that time, understood in its essence, is not comparable to spatial extensions, quantities, masses, or economic units. In fact, time is not akin to any kind of "amount" at all, and to conceive of it in this way is a deformation. Time, Gebser proposes, is not even a "fourth-dimension" to be added to the three-dimensional spatial world, as Einstein formulated it. "True time does not curve space; it is open and opens space through its capacity of rendering it transparent, and thereby supersedes nihilistic 'emptiness,' re-attaining openness in an intensified consciousness structure." The fourth-dimension, for Gebser, is not time, but "time freedom."

Our history since the Renaissance represents, for Gebser, the early foreshadowings and inchoate manifestations of an overwhelming crisis of time, which has now reached its acute and final stage. We can see this most clearly, perhaps, in the evolution of technology. Noting that James Watts's invention of the steam engine in 1782 preceded the French Revolution by only seven years, Gebser correlates these two events in the modern psyche. The French Revolution represented, in the social sphere, the same forces and underlying conditions that impelled mechanization. "What led to the invention of the machine? The breaking forth of time." Since time was viewed as a series of identical moments, akin to an unvarying spatial extension, the acceleration of production was considered good, in and of itself. With the French Revolution, the new bourgeois class of financiers and industrialists, driving the engine of material progress, became the ruling elite of society, as the aristocratic patronage culture gave way to a mass culture.

The French Revolution was the first revolt and mass assertion of "The Left," smashing the old aristocratic regime with its demands for *"liberte, egalite, fraternite."* The left-right division represents, for Gebser, one of the original dualisms underlying civilization. "Since ancient times, the left side has stood for the side of the unconscious or the unknown; the right side, by contrast, has represented the side of consciousness and wakefulness." Through the late twentieth century, the movements of the Left limited themselves to a materialist understanding of reality—exemplified by Marxism—demanding social justice and economic equity but not the restoration of intuition and the recognition of the hidden, qualitative dimensions of being suppressed by the mental-rational consciousness, narrowly focused on the quantifiable. The Left fought for the "rights" of man, while ignoring the "lefts" of man and woman.

In the late 1960s, the New Left in the United States and Europe attempted to redress this situation, seeking to integrate psychedelic, esoteric, and erotic aspects of liberation into their political program, but this undertaking could not be completed at the time. Like an unfinished process of mass-cultural initiation, the radical upsurge of the 1960s was short-circuited by the Dionysian rampages and violent outbursts marking that brief moment. Because the Left never mastered the suppressed dimensions of the psyche, the upsurge of the French Revolution—and all of the rebellions and revolutions that followed—collapsed into what Gebser called "uncontrolled intensities," reverting to systems of domination. In the same way, industrial mechanization led to "uncontrolled motoricity," an effort to overcome time through accelerated production and ever-faster communication, anxiously postponing a deeper reckoning with the essence of time, and the nature of being. Trapped by materialism, the modern consciousness fruitlessly attempted to dominate physical and temporal processes, incapable of mastering the underlying forces seeking realization, requiring a deeper relatedness and a more profound attunement. Since the "breaking forth of time" in industrial production was impelled by hidden impulses that were not consciously assimilated, "the motoricity of the machine arbitrarily began to dominate and compel man into its dependency," leading to the familiar horrors, and triumphs, of the Industrial Age, and its continuing consequences. In the modern age, we won our "rights"—hard-fought freedoms ever-threatened by the resurgence of right-wing authoritarianism—but we have not yet gained our "lefts."

Since then, we have attempted to overcome and subjugate time through mechanization, rather than confronting the essential nature of time and integrating it into our awareness. Current technology represents "unmastered time," Gebser notes: "Instead of intensifying time, man has quantified it by rational thinking into a cascading motion." Our obsession with speed and our desire to set ever-faster records for rockets and foot races "reveals the deep anxiety in the face of time; each new record is a further step toward the 'killing of time' (and thus of life). . . . The addiction to overcoming time negatively is everywhere evident." Anyone seeking to avoid, overcome, or outrun time—time misconceived as divider and destroyer—has failed at the outset. "Time is not an avenue," he points out. Therefore, the quest for exits or paths out of it is delusional, as is any attempt to overcome it.

In the modern era, humanity forfeited its laboriously acquired traditions of craftsmanship—he cites the medieval crafts guilds whose work culminated in the great Gothic cathedrals—which represented the acquisition of culture as a "second nature." In the process of modernity, we first became "denatured," cut off from our original relatedness to the natural world, and then "decultured," amputated from the skills and artifacts we had created in place of nature. "What counts now are the value-less facts, the material and the rational," Gebser notes. "All else is regarded with condescension as being of only sentimental value." With the rise of the bourgeoisie, we attempted to make the private sphere of intimate experience and family life into a new value to replace those we had forfeited. This has not proved satisfactory. As Walter Benjamin wrote: "A blind determination to save the prestige of personal existence, rather than, through an impartial disdain for its impotence and entanglement, at least to detach it from the background of universal delusion, is triumphing almost everywhere."

Time, as Gebser understands it, expresses itself in a spectrum of intensities and qualities, as "clock time, natural time, cosmic or sidereal time; as biological duration, rhythm, meter; as mutation, discontinuity, relativity; as vital dynamics, psychic energy . . . and as mental dividing. It manifests itself as the unity of past, present, and future; as the creative principle, the power of imagination, as work, and even as 'motoricity.'" These seemingly uncategorizable modalities represent aspects "of a basic phenomena devoid of any spatial character," and therefore defy systemization. "So long as we think that we can master such intensities as time by forcing them into a system, the intensities will simply burst such systems apart."

The new consciousness structure—our imminent mutation—requires a deeper realization of time, a conscious integration of its manifold forms and myriad expressions into an intensified awareness. We would no longer negate the previous forms of archaic, magic, mythic, and mental consciousness, but realize their concurrence. At this moment, now, we are embedded in the linear, spatialized time of the mental-rational mind-set; and we are spinning in the vast cyclic roller coasters of the mythic civilizations, approaching the end of the Kali Yuga and the Mayan Long Count, on the cusp between the Age of Pisces and the Age of Aquarius according to Precession; we are also in the instantaneous dream-time of the magic mind;

and we are secured in pretemporal and prespatial origin, celebrating the never-ending "first day" of the archaic consciousness. Liberated into a "satisfying all-at-onceness," we simultaneously realize and enfold conceptual time, cosmic time, biological time, historical time, creative time, the no-time of the photon, and the discontinuous time of the quantum event—within the here and now.

"Wherever man becomes conscious of the pre-given, pre-conscious, originary pretimelessness, he is in time-freedom, consciously recovering its presence. Where this is accomplished, origin and the present are integrated by the intensified consciousness," Gebser writes. From the integral vantage point, attuned to qualities rather than quantities, we would still find ourselves rooted in the physical world—mental-rational thought would still have its place, but it would forfeit its claim to exclusivity—but material reality would itself become transparent, diaphanous, as we telescoped time's full effectuality. When viewed through this "aperspectival" prism, all things as well as processes and concepts would be seen as following a predictable pattern, arising and then dissipating out of origin. Rudolf Steiner wrote: "It is owing to our limitations that a thing appears to us as single and separate when in truth it is not a separate thing at all." The Buddha noted, simply, "All is change." By consciously seeing "through" the things of the manifest world, we would recognize them, for the first time, in their true light.

In our current civilization, Benjamin noted, "Everyone is committed to the optical illusion of his isolated standpoint." The intensified consciousness of the integral structure leads, not to "ego-loss," a regression into the magic-mythic form, but "ego-freedom." Christ, for Gebser, provides a prototype of this ego-freedom. Christ demonstrated a way of being in which one is no longer threatened by "resubmersion" in the dark waters of the soul, or committed to the "optical illusion" of the ego, but capable of acting with the wholeness of humanity, and the unity of time, in mind. "In Buddhism the suspension of sorrow and suffering is realized by turning away from the world. For Christianity, the goal is to accept the ego, and the acceptance of sorrow and suffering is to be achieved by loving the world. Thus the perilous and difficult path along which the West must proceed is here prefigured, a course which it is following through untold hardship and misery," wrote Gebser, for whom the integral structure would also represent an "intensified Christianity."

The "superannuated spatial world" instituted by mental man is breaking apart, just as the old gods of the mythic world once met their end. Gebser believed that mental thought was being emptied of its content by the "robots of quantification," computers. He did not prognosticate a date for a conclusion, but suggested that the new form of consciousness had to be attained—at least by a small vanguard—before it was too late.

> Today there is still time to say this; but sooner perhaps than many suspect the end of time will come when it can no longer be said. We can only surmise in what form time-freedom will succeed time. In any event, the forms of this supersession already bear today the stamp of catastrophe and demise of the world. Just as the supersession of mythical temporicity by mental time was an end of the world—of the mythical world that had become deficient—it is today a question of the end of the mental world which has become deficient.

Considering this, one can only wonder how the seemingly subtle process of transforming the inner sphere of consciousness—no longer squandering "free time," but using it to attain "time freedom"—could provide any defense from the cataclysmic ravages and global meltdowns to be expected when our "superannuated spatial world" crashes and burns, "as everything that becomes lifeless and rigidified breaks apart," a collapse that appears to be approaching us at high speed. Gebser's answer to this is similar to Sri Aurobindo's, Jung's, Goswami's, and Steiner's: If "mind-stuff," rather than matter, is the fundamental ground of being, then a transformation of consciousness has, potentially, far-reaching effects—not just in the psychic world, but in the one we perceive to be physical, as well.

The distortions caused by our perspectival consciousness are evident when we look at modern phenomena such as nationalism, in which people are conditioned to accept the ideals of a particular society as somehow universal and immutable, worth fighting and dying for. An aperspectival view would see nation-states as "dynamic efflorescences" arising out of particular sociological and geographical conditions, never identifying with them. It is more difficult to conceive how the instruments and weaponry created by the mental mind are equally dependent on perspectival distortion, and could be overcome by a change in consciousness—but this is Gebser's viewpoint.

When the conquistadors arrived in the New World, the sorcery of the mythic consciousness had no power over them. Spell-casting, the most potent weapon of magic and mythic societies, "is effective only for members attuned to the group consciousness. It simply by-passes those who are not bound to, or sympathetic toward, the group." Mesoamerican sorcery operated on a level beneath waking consciousness, in the dream-like realm of "vegetative intertwining," and could not affect those who had attained ego-based individuality. He thought that a similar situation would take place when the deficient mental-rational mind-set was superseded by integral humanity. "Today this rational consciousness, with nuclear fission its strongest weapon, is confronted by a similar catastrophic situation of failure; consequently, it too can be vanquished by a new consciousness structure."

According to Gebser, the threat of nuclear weapons depends upon the mental-rational misconception of time. The attempt to incorporate "mental time"—"itself divisive and disjunctive"—as the fourth dimension, in Einstein's theories, led to a "consistent and obvious" dissolution or rupture in space-time. He theorized that the cause-and-effect, deterministic logic of modern technology would no longer threaten those who had "coalesced with origin," realizing the underlying acausal reality—attaining, through their own intensified (rather than expanded) awareness, the presence of time-freedom. "The new world reality, which is at the same time also a world unreality, is to a great extent free of causality." Someone who had attained this state would not exercise "a new kind of magic power" or "mental superiority," but their appearance would "effect new exfoliations and crystallizations which could be nowhere manifest without his or her presence."

IS IT CONCEIVABLE that Gebser's analysis is correct? Are we, currently, facing a perilous transition from one form of consciousness to another, with vast consequences for our imminent future? Have we suppressed the full dimensions of time, and by doing so, have we ensured the collapse of our current civilization and its accompanying mind-set?

Today—tomorrow or yesterday—I can visit Times Square and find a place to squat under the glowing LED displays of ever-changing stock prices, terrorism alerts, sports scores, and flashing news items, beneath huge screens streaming rapid-fire logos and advertisements. I can raise my eyes to the sleek surfaces of corporate skyscrapers, seeming to proclaim their desire to crush my puny individuality with their soaring might, as I inhale the exhaust fumes from the cars and rattling trucks and taxis zooming to their next destinations, using up the last drops of ancient sunlight extracted from fossil fuel deposits—old bones of dinosaurs, extinct ferns—deep under the Earth. In the midst of this familiar mayhem of blinking distractions, I can watch the dazed crowds—"preconditioned receptacles of long-standing," wrote Herbert Marcuse—loaded down with shopping bags from the superstores, their clothes fashioned by children's hands in Third World factories, as they rush to the next outlet or media event, flitting past me like phantoms.

O City, City, so rudely forced.

At the newsstand, I can peruse the latest issues of *Time* magazine and the *New York Times,* jarring me into new states of anxiety with stories of super bugs and smart bombs, flooded cities, and torture scandals. Or I can pick up the *Wall Street Journal*—idly recalling that Wall Street took its name from the original wall dividing the Dutch settlers from the indigenous Lenape, whose system of exchange was based on gift and ceremony rather than money—for competitive business tips.

Alas, Babylon, in one hour is thy judgment come.

Or I might put down my reading matter, take a good look around me, and wonder: Is it possible that our society, as Gebser suggests, suffers from "an ever-greater unconscious feeling of guilt about time, the neglected component of our manifest world"? Are "life and spirit" somehow withdrawing from this paradigm? Despite the concrete-and-steel solidity of all I see, is it possible that we are hovering on the brink of a new world reality? Life as an end is qualitatively different from life as a means. Time that needs to be filled—like an empty bucket, "devoid of any qualitative character"—

is different from time fulfilled. As a means of conceptualizing the passage of time, "history" itself has a history; it is a modern construct, and might be superseded by a different model of temporality. History, perhaps, is not just a deterministic accident, but a loom of resonance, a sorrowing dreamsong, a chant calling us into awakening. "The past carries with it a temporal index by which it is referred to redemption," wrote Benjamin, who believed that every generation was endowed with a "*weak* Messianic power, a power to which the past has a claim." The time has always been "not yet"—but that does not mean the time is never. It is not—is it?—inconceivable that a day might arrive when the time is now.

CHAPTER THREE

Supreme sincerity evokes resonance.

Taoist proverb

In February 2003, I visited José Argüelles at his home near the base of snow-covered Mount Hood in Oregon. The night before my flight, I dreamed I was wandering around a hilly city of brightly painted row houses, similar to San Francisco. The scene had the theatrical luminosity I had learned to associate with "big dreams," announcing their archetypal character. Lost in Chinatown, I walked down a dead-end street, underneath a vast suspension bridge resembling the Golden Gate. At the end of this cul-de-sac, I found three stately redbrick mansions that were the headquarters for a "College of Sacred Studies." Above one of the doorways, cast in weather-beaten bronze, was what appeared to be a gigantic Buddha head, smiling enigmatically. Across from this head, mounted above another entrance, was the patina'd sculpture of a dragon, serpentine and undulating. I peered into the display windows of the college, which featured rows of ornate samurai swords and antique figurines of monks and warriors. I was about to knock on the door when I awoke from the dream, looked at the clock, and hurried to catch my plane.

Argüelles and his girlfriend, Stephanie, picked me up from the Portland airport. They were relaxed, slightly hippieish, and buoyantly cheerful. A gray-haired man, sixty-three years old, Argüelles wore a dark reddish Nehru jacket and a Bob Marley T-shirt. He had an aristocratic hawk nose and sad eyes deep set in a distinctly Mexican face. We piled into their flatbed pickup truck. We spent a lot of the next few days scrunched together in the front seat of this vehicle, driving around mountains covered in pine forests and snowdrifts, discussing the Maya, Apocalypse, noospheric consciousness, the nature of time, and the movement that Argüelles had started—portentously titled the "World Thirteen Moon Calendar Change Peace Movement"—to replace the Gregorian calendar with a system he had devised based on the principles of the Mayan Day Count.

"So please tell me, why should we substitute your calendar for the current one? Doesn't the Gregorian work well enough?" I asked.

"Right now, we use an arbitrary system of twelve months of unequal days, cut off from natural cycles, and this conditions us to accept disorder and irrationality in all of our institutions," he answered. "The Gregorian is programmed for chaos and Apocalypse. This calendar would replace it with an instrument designed for perpetual order and harmony."

He explained that the Gregorian calendar—decreed in 1582 by Pope Gregory XIII and the Council of Trent, to correct errors in the Julian calendar used since Roman times—instituted a flawed model of time, based on ancient patriarchal traditions. Five thousand years ago, at the beginnings of civilization, the Sumerians—living in Uruk and other pot-sharded settlements of what is now the bombed-out, oil-rich, desert moonscape of Iraq—were the first to switch from a lunar to a solar calendar, based on an abstract principle: the division of a circle into twelve equal parts of thirty days each. Earlier calendars—used by archaic civilizations following the "Great Mother," found in relics and scratched onto the walls of prehistoric caves—followed the precise cycles of the moon (the word "month" descends from "moon"), which circles the Earth roughly thirteen times a year—as we shall see, the precise calibration is tricky. Along with dividing the year into twelve parts, the Sumerian high priests split up the day into twenty-four hours of sixty minutes each. "Mechanization was implicit in the first intellectual act of history," Argüelles said. Since there are 365 days in a solar year—a number that cannot be divided by twelve—the Sumerians added five extra days to the end of the year, which were considered unlucky. The Julian and Gregorian systems were adapted from this abstract model, based on a conception of the year as a flat circle, arbitrarily dividing the twelve months into mismatched measures of twenty-eight, thirty, and thirty-one days.

"I still don't get it," I said, when he was done. "I can see that the months of the Gregorian are somewhat arbitrary, but I don't see why changing the calendar would have any effect on the underlying patterns and structures of our society."

"The calendar is the macro-organizing principle of a culture, even if people don't realize it," he replied. "A new calendar would be like striking a new covenant. All of the institutions operating on the old timing frequency would be instantly delegitimized."

Considering it over time, I found this unfamiliar idea to be increasingly compelling. On a subliminal level, a calendar functions as a metaprogramming device for the human mind. Our calendar tells us when we are born and when we die, defines the rhythm of work and rest, creates a pattern of holidays and festival days, shaping our relationship to time in the most utilitarian and tactile way. Is it possible that, whether or not we are aware of it, the calendar functions as a "timing frequency," either attuning us to natural patterns, or instituting an arbitrary and artificial temporality that pulls us away from organic reality? Argüelles believes this is the case, proposing that calendars habituate us to a certain experience of time. "Time is mental in nature," he wrote in *Time & the Technosphere.* "Modern Western thought and science has been programmed and predisposed to limit its consciousness of time to such a degree that it cannot even perceive of time outside its inherently mechanized perception of it." He believes that our current calendar has trapped us in a feedback loop of accelerating desynchronization.

"How important is it, in your opinion, that people should switch to your thirteen-moon calendar?" I asked.

"Only harmony can unify," he said. "Condition the mind to an irregular standard and the mind will adjust to disorder and chaos as normal aspects of existence. Our civilization is based on false time, and artificial time has run out for humanity."

I asked him to elucidate the principles of his calendar that made it an instrument of "perpetual order and harmony." He explained that it was based on thirteen "moon-ths" of twenty-eight days each, following lunar cycles, with one "day out of time," for celebration and forgiveness of debts, at the end of each calendar year. "The moon circles the earth thirteen times a year, not twelve," he said. "For the Maya, thirteen was the lucky number of natural cycles and synchronization. The very act of replacing the order of twelve with the order of thirteen would be a profound one."

In *Time & the Technosphere,* he argued that the superstition over the number thirteen had deep roots in Western culture. "Could it be that the whole of civilized history is based on the fear of the number thirteen—epitomized in the superstition about Friday the 13th—and that, therefore, dealing with the true nature of time has been avoided altogether?" I grew up in an apartment building lacking a thirteenth floor, like most buildings in Manhattan, and when I was a child, I always found it eerie that our culture—while priding itself on its rational skepticism—clung to this antique phobia. Argüelles proposed that thirteen is actually the lucky number of natural and feminine cycles (according to Argüelles, the female menstrual cycle should be twenty-eight days), synchronicities, and harmony—for that reason, it was suppressed by the male patriarchal mind-set, increasingly committed to mechanized rationality.

Along with its thirteen regular "moon-ths," Argüelles's calendar also has a deeper level, following the cycle of 260 days, or "kin," of the Tzolkin, the 13-by-20 matrix with its "binary crossover pattern" that he considers the essential instrument of the Maya's sacred science. He elaborated a complex system, called the "Dreamspell," to help people enter into the Maya's fractal, fugue-like chronovision—to experience time as a tonal loom of synchronicity and resonance, on scales ranging from the personal to the cosmic. As he wrote in *The Mayan Factor:* "For the Maya what we call time is a function of the principle of harmonic resonance. Thus, days are actually tones, called kin, represented by corresponding numbers; sequences of days (kin) create harmonic cycles . . . and sequences of harmonic cycles taken as larger aggregates describe the harmonic frequencies or calibrations of a larger organic order, say, the harmonic pattern of planet Earth in relation to the Sun and the galaxy beyond." The cycles of thirteen days, twenty days, 260 days are fractals reflecting larger cycles and energetic patterns.

Each "tone" of the Dreamspell is made up of one of thirteen "pulsation rays" and one of twenty "solar seals," their positions recorded on the Tzolkin.

As an alternative to the Western horoscope with its astrological sign, your date of birth in the Dreamspell gives you a "Galactic Signature," mapping you onto the system. The thirteen-moon calendar deviates from the traditional Mayan scheme, offering a different interpretation for the system of symbols and numbers, taken from traditional glyphs, as well as changing the dates from the traditional Day Count. This alienates some indigenous Maya, who see it as an appropriation, and irritates other scholars who study the Tzolkin.

"I give my allegiance to the authentic Mayan tradition, not to re-designers," said John Major Jenkins, author of *Maya Cosmogenesis 2012*. "Following a new calendar cult seems more of a symptom of modern decadence than a solution to it."

Answering this criticism, Argüelles said, "I intended the names of the glyphs to be culturally odorless. My goal was to create a universal system." He makes a distinction between the "Galactic Maya" who built magnificent temple-cities such as Palenque and Tikal, and the indigenous Maya of today. "I wanted my system to have a galactic quality to it."

"In *Time & the Technosphere,* you say that you believe the technosphere is going to collapse before 2012. That sounds pretty cataclysmic. Does that mean there aren't going to be any planes or cars? How will people get around? How would I come to see you, for instance?"

Argüelles smiled knowingly. "When humanity is in the correct timing frequency, we will reactivate universal telepathy. We might just go into a telepathic seance and find that to be much more enjoyable."

"In some of your proclamations, you seem to suggest that there is some urgency in making your calendar the new global standard. Since this seems pretty unlikely, what is going to happen if people don't do it?"

"If the world does not adopt this calendar within the next few years, the human race is finished," he replied earnestly, without hesitation.

AT THE ZENITH of late-1960s psychedelic delirium, Jim Morrison sang, "Break on through to the other side." Argüelles, it seemed, had accomplished this. Two hundred pages deep in *Time & the Technosphere,* he announced, "Beginning in 1980 I became conscious that, from time to time, I had been receiving telepathic transmissions from the star Arcturus. . . .

The Arcturus Protectorate was established as a kind of protective time shield around the Earth." At the end of the book, he presented a long prophetic text he had received from the eighth-century Mayan king, Lord Pacal, whom Argüelles calls "Pacal Votan." Many people who consider his ideas come away with a seemingly reasonable impression that he has blown a fuse—or, in his terms, a "biopsychic circuit."

Argüelles has set himself up as a prophet for a postmodern world that has lost faith in prophets and prophecies. As he wrote in *Time & the Technosphere,* in his unusual science-fictional patois: "Prophecy is the release of information according to the psi bank timing program and in relation to degrees or levels of consciousness positioned at different points in time. The points in time are human channels." He believes he is one of those "points in time," a receiver attuned to the shamanic spectrum, beaming frequencies from the farthest edge of the cultural imagination. Meeting him, it was difficult to reconcile the stridency of some of his pronouncements with his gentle, soft-spoken, and reasonable demeanor.

Noting the difficulties in separating "signal" from "noise" in occult cosmologies, William Irwin Thompson thought that the new mythologies entering our postmodern world via the collective unconscious were necessarily distorted by the personality of the medium picking up the new pattern. "If a person is open to a new world view, it can often mean that he is not firmly rooted in the reality of the old world view; as a lunatic or alienated artist, his own neurotic traits can become magnified as they tremble with the new energy pouring in from the universal source." Thompson proposed that the relationship between information and noise was a "rich and complex one; it is at once symbiotic, like life and death, and relativistic, like time and space." In his ambition and willingness to operate prophetically, Argüelles had wrenched himself open to an extraordinary amount of information, as well as noise—and seeming nonsense. It was clear from meeting him that he was not a lunatic, but he operated at the extreme edge of the cultural imagination, where signal meets static. Even though I was attracted to many of his ideas, I found it extremely difficult to disentangle the two.

To complicate matters even further, I found myself mapped onto his prophetic grid—treated, upon arrival, as a long-lost and long-expected relation. When I contacted Argüelles's organization, the Foundation for the

Law of Time, to see if I could interview him, I was immediately asked for my date of birth. I e-mailed it to them: June 15, 1966. Stephanie e-mailed me to tell me my "Galactic Signature" was "Yellow Spectral Star." According to the Dreamspell, each kin has several counterparts, including an "antipode kin" and a "helping partner." Stephanie told me that José and I were perfect "helping partners"—his signature was "Blue Spectral Monkey." They took this as a good omen. I didn't subscribe to Argüelles's thirteen-moon calendar, and I didn't know what to make of this, but considering I was writing about him, it seemed an intriguing 1/260 accident.

I was told my birth date had another resonance, as well. The secret tomb of Pacal the Great, under the "Pyramid of the Inscriptions" in Palenque, had been discovered on June 15, 1952. This date is crucial for Argüelles, because the glyphs and sculptures found in this hidden mausoleum inspired his system. "The revelation of the prophecy of Pacal Votan on July 26, 1993, was the culmination of the event that occurred on June 15, 1952." On that date, the archaeologist Alberto Ruz Lhuillier pried open a hidden door beneath the Pyramid of the Inscriptions at Palenque, Chiapas, Mexico, "and beheld what had not been seen by human eyes for 1,260 years: the tomb of Pacal Votan," Argüelles wrote in *Time & the Technosphere.*

I asked him about this event.

"Alberto Ruz said that when he and his workers opened the tomb, they could actually feel the last thoughts of the people who had been in that chamber rushing out past them," he told me. "That chamber was hermetically sealed and the meditations of the last people in that room were highly charged and intentional thoughts. They anticipated the timing of the opening of the tomb. I believe those thoughts were meant to alter the planetary noosphere, and to accelerate the end-time."

At Argüelles and Stephanie's modest house on Mount Hood, surrounded by snow-dappled pine forest, Argüelles showed me the rainbow-hued mandala-patterned murals that he had painted on several doors, back in the late 1960s, when he was a graduate student in Paris, studying art history and dabbling in LSD. He played beautifully on his *shakuhachi,* a winsome wooden flute favored by Japanese monks. He brought me into his small meditation room, where the Quran was prominently displayed among Buddhist statues, crystals, modest relics, and the various elaborate gameboards his followers had designed to augment the Dreamspell. Most impressively, he handed me an intricately carved ceremonial staff, inlaid with crystals and stones, and carved with solemn obsidian faces of tribal deities, its hollow interior full of seeds that made a thunderous rustling when turned over. This weighty object, a product of careful craftsmanship, was presented to him, in 2002, by nine elders from different Mexican tribes, at the base of the Pyramid of the Sun in Teotihuacán, outside Mexico City. During a ceremony attended by seven hundred Indians in traditional dress, Argüelles was recognized as the "Closer of the Cycle," completing indigenous prophecies, weaving together Western knowledge and tribal wisdom.

"The indigenous elders made it clear to me that this was a very serious and sacred honor," he said. "In a way, the ceremony recognized the limits of indigenous culture in its own context. They understand that their wisdom needs the injection from a modern perspective that can integrate it and put it into a new form."

I asked David Freidel, coauthor of *Maya Cosmos* and one of the leading archaeologists of the Maya, what he thought of Argüelles's interpretation of Mayan culture. "I teach critical debunking," Freidel said. "There is history. Events that happened in the past are a reasonable guide to events that will happen in the future." The sacred culture of the Maya, for Freidel, was high-minded ornament, revealing their hopes and ideals, rather than direct meetings with metaphysical realities or galactic minds. "All civilized societies have a high culture which represents their most beautiful dreams. We have our paintings of George Washington carried up to heaven. We have our sacred space and our civil vision of the good." Argüelles's view of the Tzolkin-based calendar "hasn't anything to do with the Mayan understanding of reality."

"Decoding the glyphs is all based on the state of mind of the archaeologist," Argüelles countered. Mainstream academics such as Freidel and Linda Schele "interpret them as if they were State Department communiques." His perspective is that the dates on the glyphs are "synchronically coded information. An event in 600 AD will have very distant antecedents. They are giving the event in their present a cosmic pedigree, a cosmic resonance."

BORN IN ROCHESTER, Minnesota, in 1939, of a German-American mother and Mexican father, Argüelles split his childhood between Los Angeles and Mexico. He encountered his destiny when he was fourteen, during a family trip to Teotihuacán, the ancient pre-Mayan ruined city of still-present mysteries. Sitting atop the Pyramid of the Sun, he made a vow to himself, "Whatever it was that had happened here I would come to know it—not just as an outsider or an archaeologist, but as a true knower, a seer." Like McKenna, he was compelled by a recurring fantasy of a "cathartic and transfigurative experience" awaiting him in the jungle.

While on a 1964 visit to Xochicalco, a Mayan temple-city in the high mountains of Guerrero, "my intuition penetrated further into the mute stones." He felt a premonition, or a "recollection," that "gathered with disturbing intensity." The "historical" Quetzalcoatl was born in Xochicalco in AD 947, ruling over the Mexican kingdom of Tollan, ushering in a brief Golden Age. Defeated by the sorcerers of Tezcatlipoca, the dark god called "Smoking Mirror," Quetzalcoatl left on a raft of serpents, heading west, vowing his eventual return. The prophecy of Quetzalcoatl's return "on the day of 1 Reed, in the year 1 Reed, was vindicated by the arrival of Cortés on that very day, Good Friday on the Christian calendar, AD 1519." Argüelles was drawn to this mystery. "It was clear that Quetzalcoatl was not just a god, but a multiple god; not just a man, but many men, not just a religion but a mythic complex, a mental structure." He intuited Quetzalcoatl to be "an invisible and immanent force underlying and transcending the mythic fabric of mechanization."

As myth, legend, and historical account, Quetzalcoatl permeates ancient Mesoamerican thought, taking numerous forms. He participates in the creation of the cosmos, helping to separate earth and sky. The Plumed Serpent also "presides over another foundational act for the Mesoamerican peoples: the organization of time and space," writes Enrique Florescano, in *The Myth of Quetzalcoatl*. At the end of the Fourth Sun, the deity descended into Mitla, the Underworld, to gather up the bones of destroyed humanity, bringing them back to the surface after many travails, resurrecting them with his own blood. In the prehistorical age of heroes, Quetzalcoatl appeared as the bearded wisdom-teacher from across the seas, spreading the rudiments of knowledge to early civilizations (his almost identical counterpart in Incan legend is Viracoccha), establishing the Shambhala-like kingdom of Tullan. In these myths as well, the debauchment and disintegration of sacred order was accomplished by the magicians of Tezcatlipoca, who is an equally complex figure, Quetzalcoatl's twin-aspect and nemesis. The legend was repeated in historical time, when Ce Acatl Topiltzin, Quetzalcoatl's tenth-century incarnation, king of Tulla, banned human sacrifice and instituted a brief "Golden Age" before he was once again thwarted. As a last act, Ce Acatl oversaw the construction of Chichen Itza, the temple-city in which, according to Jenkins, Toltec and Mayan cosmologies were integrated into a single form and message, focused on the 2012 date, the birth of humanity's higher mind.

Like a wrinkle in space-time, Quetzalcoatl resonates with other cultural legends, such as King Arthur, the "Once and Future King" due to return to reestablish harmonic order. His task is identical to that of the Tzaddik, the "righteous one" in Qabalah, "the man who gives each thing its due, who puts each thing in its proper place," according to Gershom Scholem. The Tzaddik descends to the lowest level, becomes nothing, and by sacrificing himself, turns himself into a pure medium. "The Tzaddik is called a mirror, for everyone who looks at him sees himself as in a mirror," notes one Qaballistic text—and Quetzalcoatl was called "a mirror pierced from both sides." When God created the Earth—our "abysmal world of shards"—he shattered the vessel of higher mind, sprinkling fragments of lost wisdom across the globe. As the plumed serpent descends into the underworld to rescue and revive the bone splinters of humanity, the Tzaddik performs the "gathering up of the sparks," finding the hidden keys to spiritual knowledge to establish *shalom,* peace and harmony, on Earth. "The Tzaddik encounters evil by means of his descent, which he transforms by taking it and permeating it contemplatively," Scholem writes. With his rise, he raises up the fallen and redeems the lost. "For the essence of the perfect service of God consists in raising all the lower levels upwards." The task of the Tzaddik is "Tikkun Olam," the reparation of the world.

William Irwin Thompson also explored Quetzalcoatl as mythic complex and spiritual prototype. "The ancient prophetic calendar says that

Quetzalcoatl will return again, and that our era of the Fifth Sun is now drawing to its close and will end in earthquakes, volcanic fire, and famines. The age of chaos will be consummated in chaos, and then the spiral will turn, and a new age of gods with a new sun will be established," he wrote. He considered the "feathered serpent" to be an image of "the human race in the condition of enlightenment," having channeled the spiraling sex-energy of kundalini up into the higher circuits of the neocortex, utilizing yogic and Tantric discipline to initiate a "boddhisattvic race." But in our fallen age, Thompson pessimistically concluded, "there can be no question of a direct incarnation of the Feathered Serpent." Argüelles disagreed, realizing if "time-transcending mystical states of mind" were the basis of Quetzalcoatl's teachings, "then what was keeping me or anyone else . . . from entering those states of mind?"

Argüelles received his doctorate in art history from the University of Chicago in 1969, teaching at Princeton, UC Davis in California, and the University of San Francisco. He joined the founding faculty of the Naropa Institute in Boulder, Colorado, where he studied under Chogyam Trungpa Rinpoche, the celebrated Tibetan lama and author of *Shambhala: Sacred Path of the Warrior.* He pursued his interest in the Maya, visiting Palenque in 1976, struck by its "feeling of abandonment and human silence." Reaching the top of the Pyramid of the Inscriptions, he was greeted "by a double rainbow that seemed to emanate not far from us in the Temple of the Winds."

Over the next years, he studied the glyphs and binary patterns of the Tzolkin, uncovering what he felt to be a profound mytho-mathematical puzzle integrating the Maya's "harmonic module," the DNA code, and the *I Ching,* via the "magic square of 8," an 8-by-8 grid of numbers, in which each row, horizontally, vertically, and diagonally, adds up to 260, the number of kin in the Tzolkin. This magic square was found, "quite gratuitously," by Ben Franklin—leading Freemason, American revolutionary, and discoverer of electricity—around 1750. "For Franklin, the manifestation of these squares in his conscious mind was a matter of a highly curious but ultimately innocuous pastime, which seems to have led nowhere beyond mathematical parlor games," Argüelles wrote in *Earth Ascending* (1984), noting that the first magic square was attributed to the first emperor of China, Fu Hi, "the person who brought the binary system *I Ching* into the world."

The genetic code is formed by four amino acids, combined into triplets, making a sixty-four-part binary code. With an identical underlying structure, the *I Ching* consists of sixty-four hexagrams. "Based on binary permutations of eight primary triplet structures . . . *I Ching* may be considered to be the code of *biopsychic transformation,* much as the DNA codons are the code of more purely *biological transformations.*" He theorized that the Tzolkin was, similarly, an instrument revealing an underlying pattern, a principle of cosmic order. "The Tzolkin, like *I Ching,* is ultimately uninvented. Its provenance in the Mesoamerican heartland is as much a cosmic accident as that of *I Ching* in China." The 260 days of the 13-by-20 Tzolkin, the "Mayan factor," represent the average gestation time of the human fetus; the vast cycle of the Precession of the Equinoxes is close to twenty-six thousand years. According to modern astronomers, we are also twenty-six thousand light-years away from the center of our Milky Way galaxy.

"The Maya understood that whereas the 260-day sacred cycle is our period of individual gestation, the 26,000 year cycle is our collective gestation—our collective unfolding as a species," Jenkins wrote in *Maya Cosmogenesis 2012*. "Precession represents a 26,000-year cycle of biological unfolding—a type of spiritual gestation and birth—that Earth and its consciousness-endowed life forms undergo." The fractal model of time that Terence McKenna discovered—or downloaded—"quite unexpectedly," in the Colombian Amazon, linking the *I Ching* to the evolution of human consciousness in a quickening spiral, appears to be substantiated by a set of pristine relationships between the Mayan Tzolkin, the *I Ching,* astronomical observations, and the genetics code, elaborated by Argüelles, Jenkins, and others.

One month after the Harmonic Convergence of 1987—a global meditation signifying the arrival of "the race of galactic wizards on Earth," according to Argüelles—his only child, his eighteen-year-old son, was killed in a motorcycle crash. Although he mourned his loss, he also says, "Josh's death was my liberation. It freed me to end my involvement in the professional world. For me, there was a transcendent meaning to it."

Resigning his teaching post, Argüelles and his wife, Lloydine, embarked on an experiment, living in the "13-day, 20-day, 52-day, and 260-

day cycles" of the Mayan calendar. On December 10, 1989, they visited the Museum of Time, in Geneva, Switzerland, realizing, with the force of an epiphany, that the rise of modern civilization and its increasingly destructive "technosphere" was based on the institution of an "artificial timing frequency," and that the salvation of humanity required a return to natural time—the synchronic frequency of the Tzolkin, resulting in "the normal sensation and experience of the harmony of reality." This insight led to others, including a series of telepathic transmissions—titles include *The Heptagon of Mind* and *The 260 Postulates of the Laws of Time*—received by Argüelles over the next years. "During the early 1990s, I felt like a galactic fax machine," he recalls.

These events culminated in the summer of 1993. At a hotel in Cuernavaca, Argüelles had "a totally lucid clairvoyant waking dream," in which Pacal the Great appeared before him, and told him to go to a distant house and prepare to receive a prophecy. Argüelles found the house—a converted carport in a field near the village of Ocotitlan—and rented a room in it. Over the course of a nine-day period, beginning on July 26, 1993, he awoke before dawn every morning as a voice spoke to him, dictating the text of "Telektonon: Earth Spirit Speaking Tube," which he considers "the full realization of the Law of Time and the synchronic order."

"The experience of receiving the Telektonon prophecy altered my life more than any other," Argüelles recalls. "I realized—oh my God—I'm a messenger. I had a responsibility to communicate that message around the world. I knew that Quetzalcoatl and Pacal Votan wouldn't leave anything to chance. I finally had to accept that the messenger was me."

During the rest of the 1990s, Argüelles and his wife became "galactic gypsies," impoverished prophets without credit cards, traveling across the planet on rough roads, from Siberia to Japan and across South America, spreading the "good news" of the law of time. They found supporters—little nodes on their expanding network of apostates from the Gregorian faith—wherever they went. In the last decade, the movement has continued to expand. Offering a postmodern means of going "off the grid" of mechanized time, without retreating to a backwoods commune, the thirteen-moon calendar has caught on with a global subculture of ravers, Goa-trancers, New Agers, and neo-shamans. Dreamspell parties are held in countries across the world—Mexico, Israel, England, Brazil, and so on. Several hundred thousand of the calendars are printed each year and distributed around the globe—they are especially popular in Japan, parts of Russia, and South America. Followers of the new calendar get together to form nodes of a "Planetary Art Network," promoting the movement. Portland, Oregon, with seven hundred or so members, has the largest thirteen-moon community in the United States.

Surprisingly, Argüelles found an audience for his ideas among prominent members of the Nation of Islam. He has lectured with Louis Farrakhan, and is a frequent guest on Afro-American radio programs. He has traveled and held conferences with Tynetta Muhammad, the widow of Elijah Muhammad, founder of the Nation of Islam. "The ancients hold the keys but we can't turn them until we reach a higher frequency," Muhammad told me. "If we in the West can embrace other knowledge systems, we can make a huge leap. We can create a new avant-garde of knowledge." She lives in Mexico, where she studies indigenous traditions. Muhammad's enthusiasm offers support for Argüelles's belief that a new calendar could harmonize the world's religions, dissolving a major source of global conflict by making the cyclical and sacred nature of time self-evident to all.

But the moderate success of the thirteen-moon movement can only be disappointing for Argüelles, who believes this calendar alone holds the key to posthistorical harmony, or global death—activated noosphere or decimated necrosphere. "My whole strategy is to get wider and wider exposure—to push all the pressure points I can and leverage this into position," he told me, sitting at a massive old ski lodge and restaurant atop Mount Hood.

He has submitted formal petitions to the Vatican and the UN, receiving little response. Like a street-corner prophet of doom, he has issued a series of proclamations declaring that the world must adopt his "timing frequency" by a certain date or else—many of the dates have passed, and we are still here.

"Aren't you a bit concerned about the apparent impossibility of getting the world to change its calendar in the next few years?" I asked.

"I see myself as a messenger," he said. "I know what I have to do, whether I am perceived as a Don Quixote or a zealot or fanatic. From my position, I can't even think of the impossibility of it."

"So what do you foresee happening?"

"In a few years, we will start to dismantle the technosphere, taking the whole thing down. Once a huge mind-shift happens, everyone catches up. I think we will have the Internet until around 2009, and then pull the plug on it. We will realize that technology is just restricting our psychic abilities. After that, we will begin a series of global telepathic experiments. Very quickly, we will find telepathic solutions for our problems and restore the Earth to its pristine state. Humanity is going to be very happy back in the garden. I foresee a type of spiritual anarchy and people forming councils and really respecting each other's spiritual freedom as the highest way to peace. The nation-state concept is out of date."

"What about all of the signs that the world can't possibly make this deadline?" I asked. "If you are correct, isn't it more likely that we are headed for a complete breakdown?"

"I have to keep envisioning the most positive possible future," he answered. "That is my job as a visionary."

SINCE RECEIVING the Telektonon prophecy in 1993, Argüelles is convinced that he is Valum Votan, an incarnation, or emanation, of the "galactic agent" and time-wizard Pacal Votan, beamed from that ageless higher mind into our present tense, to recover and transmit the world-saving new paradigm of his thirteen-moon calendar and the law of time. But the Telektonon is a weird and unpleasant text to read or contemplate. Reminiscent of Crowley's *Book of the Law*, it sounds like something dictated by an audacious, arrogant pharaoh in a fit of pique.

> Telektonon, the perfection of time, is the only way for you to escape the fire that consumes the unrighteous. If you who know, who have followed the straight way, the good path, now falter in this challenge of unification in time, Telektonon, which has already been prepared for you, you will perish forever in the fire that consumes all unbelievers. God's command is in you. Do not listen to the evil one now.

> In this seal foursquare, all prophecies converge. The Beast of the G-7 stands revealed. The Babylonian Vatican is exposed. In the rainbow dream vision 144,000 of the elect are called again to meet, gathering together in circles to listen, to sing, and to dance to the song Telektonon.

One cannot read the Telektonon without suspecting that something has gone wrong—the capacious mind seeking to incorporate so many visionary vectors, in which "all prophecies converge," has been strained or singed by the effort. Argüelles's humorous and rather tender personality has been unnaturally displaced, pushed aside by some imperious and demanding force.

Similarly, the portentously titled "Law of Time"—the discovery of which, Argüelles informs us, was "only possible through a profound act of self-reflection," but once made conscious, "affects the entire medium of planetary consciousness, the noosphere, at first imperceptibly, but then building to a great point of climax coinciding with the climax of the biogeochemical combustion"—is revealed to be a simple equation of madcap compression: "T(E) = Art; Energy factored by Time equals Art." In a theorem that might not make Einstein spin in his grave with envy, the universe is revealed as a vast art project, a magic sleight of mind, precisely calibrated, in which everything arrives at the propitious moment, in just the nick of time.

"You will perish forever in the fire that consumes all unbelievers"—such a proposition appears a classic symptom of megalomaniac ego inflation, as McKenna recognized in his own story, although he had the humor to shrug it off. It lacks the sincere ring of Boddhisattvic compassion, suggesting some atavistic, imperious regression to older, moldier, arcane states of mind. It is a ruthless vision of maya as power, in which space-time is manipulated to create ambiguous and fascinating spectacles. In Valum Votan's repeated assertion that "God is a number," as channeled in the Telektonon, the human world has been reduced to a sorcerous equation, serving the magician's inhumanly detached purposes. Votan's dictates are reminiscent of Carl Jung's psychological portrait of the "jealous god," Yahweh, in the Old Testament: "The character thus revealed fits a personality who can only convince himself that he exists through his relation to an object. Such de-

pendence on the object is absolute when the subject is totally lacking in self-reflection and therefore has no insight into himself. It is as if he existed only by reason of the fact that he has an object which assures him that he is really there."

By thrusting himself deep into vision-space and bearing witness to the lunar currents of his unleashed imaginings, Argüelles has functioned as "antennae of the race" and culture-hero, bringing back extraordinary ideas and thought-complexes that could only be accomplished by someone stretching himself beyond the brink. According to Jung's disciple Edward Edinger, "The archetypes themselves cannot evolve into full consciousness without being routed through a mortal ego to bring that consciousness to realization." Edinger quoted Christ—who should have known—on the dangers of transducing the higher voltages of the archetypes. "Blessed is the lion which becomes man when consumed by man; and cursed is the man whom the lion consumes, and the lion becomes man," Christ said in the Gospel of Thomas. As with the McKennas, it was possible that archetypal material had, indeed, constellated within Argüelles's psyche—the demanding time-wizardry of Pacal the Great—but the lion had swallowed a piece of the man.

CHAPTER FOUR

By sheer genius, by sheer acuity, they got it done.

POPOL VUH

On the winter solstice of December 21, 2012, the Sun will rise within the dark rift at the center of our Milky Way galaxy, an event that occurs once every 25,800 years. As John Major Jenkins describes in *Maya Cosmogenesis 2012*, this alignment represents the "union of the Cosmic Mother (the Milky Way) with First Father (the December solstice sun)." Mayan hieroglyphs describe the center of this dark rift as the "Hole in the Sky," cosmic womb, or "black hole," through which their wizard-kings entered other dimensions, accessed sacred knowledge, or toured across vast reaches of the cosmos. In September 2002, astronomers verified the existence of a massive black hole at the center of the Milky Way, naming it "Sagittarius B." Jenkins writes:

> Something very profound and mysterious is going on here. Is it just a coincidence that lurking deep within the dark-rift "black hole" is the very real Black Hole at the center of our Galaxy? If not a coincidence, the dark-rift itself might indeed be the surface signifier of deeper cosmic mysteries, ones that the Maya were well aware of.

This black hole is "the cosmic womb from which new stars are born, and from which everything in our Galaxy, including humans, came." The dark rift through which the Sun will pass at the end of the Long Count is called, in the Quiche language, *xibalba be,* literally "underworld road."

For mainstream archaeologists studying the Maya, the encoding of this date is just an accidental by-product due to the Mayan method of counting cycles of time. "The Maya are not predicting any world change in 2012 AD," insists David Freidel. "That year marks a change in the cycle of 260 days and nothing more. It is like an odometer clicking over."

Jenkins reached a different conclusion. He found references to the solstice alignment of December 21, 2012, encoded in numerous temples, structures, and ballcourts. Friezes and statues show standing rulers, representing the Milky Way, holding "double-headed serpents" tilted at 14 degrees, symbolizing the "Plane of the Ecliptic," the path followed by the Sun, Moon, and planets. Jenkins thinks such icons represent the cosmic crossing of the 2012 Precession. For more than a thousand years, the intellectual genius of Mesoamerican civilization, from the Olmec through the Maya, was dedicated to determining this date, and, once it was established, inscribing it in their artifacts and monumental architecture in symbolic form.

Jenkins studied the great observational and ceremonial center of Izapa, in the southwestern corner of the Mexican state of Chiapas, founded in the first centuries BC, where the Long Count may have originated. "The Long Count calendar—a Galactic Cosmology—is the unique result of a shamanistic experiment seemingly conducted in secret, over perhaps three hundred years in the dimly understood Pre-Classic era." Astronomical observations, as well as psychedelic mushrooms and toad secretions, were utilized in this effort. "Izapa's monuments offered a mythic narrative, carved in stone, describing the astronomical alignment charted by the Long Count calendar." Jenkins considers Izapa to be the Mesoamerican Eleusis; the place where apprentice acolytes came for priestly initiation into the Mysteries, receiving instruction in astronomical knowledge encoded as myth, as well as direct revelation of the meaning of Precession and the 2012 birth date, induced by psychoactive substances such as DMT-containing snuffs, peyote, and psilocybin mushrooms, allowing direct perception of subtle energy fields and forces emanating from across the cosmos.

Jenkins theorizes that this astronomical event, in which the solstice meridian crosses over the Galactic equator, might induce a "field-effect reversal," just as magnetic forces operate in reverse form above and below the Earth's equator, causing tornadoes as well as toilets to swirl in the opposite direction. "As with a spinning magnetic top, the field effects on one side are different from those on the other, and Maya insights offer us the notion that a field-effect reversal occurs when the solstice meridian crosses over this line." At the moment of alignment, we cross the Galactic Equator, "the precise edge of our spiraling Galaxy," ushering in a new World Age, whose effects will be realized as a transformation of consciousness. The meeting with "Cosmic Mother" could be seen as a strange attractor, impelling the rapid development of technology and human populations and changes of consciousness that have accelerated in the last centuries as we approach it. Jenkins's conception is similar to Terence McKenna's vision of history as the shock waves sent back into time by the "Eschaton." If this thesis is correct, the Earth has been self-organizing in preparation for this geomantic juncture over the course of its history, just as the invisible lines of force of a magnet shape metal filings from a distance.

In *The Mayan Calendar and the Transformation of Consciousness,* the Swedish biologist Carl Johan Calleman, a cancer specialist and former adviser for the World Health Organization, raised the discourse on the ancient time-science of the Maya to a new plateau. According to Calleman's thesis, the nine levels of the most important Mayan pyramids—the Temple of the Inscriptions in Palenque, the Pyramid of the Jaguar in Tikal, and the Pyramid of Kukulcan (Quetzalcoatl) in Chichen Itza—represent a model of time, from the origin of the universe to the upcoming phase-shift, in which each step, or "Underworld," is twenty times more accelerated in linear time than the one preceding it.

"The nine-story Mayan pyramids are thus telling us that consciousness is created in a hierarchical way and that each Underworld stands on the foundation of another," writes Calleman. The initial level, starting thirteen hablatuns or 16.4 billion years ago, proceeds from the inception of matter in the "Big Bang," through the development of cellular life on Earth. During the second step, beginning thirteen alautuns, or 820 million years, ago, animal life evolved out of cells. The third underworld, starting thirteen kinchiltuns or 41 million years ago, saw the evolution of primates and the first, rudimentary use of tools by human ancestors. During the fourth underworld, beginning thirteen kalabtuns or 2 million years ago, tribal organization began among the ancestors of Homo sapiens. During the next underworld, starting thirteen piktuns or 102,000 years ago, Homo sapiens emerged, developing spoken language. The sixth underworld comprises the Great Cycle of thirteen baktuns, beginning 5,125 years before the approaching birth date, when we created patriarchal civilization, law, and written language—Calleman calls this the National Underworld. The seventh step, dubbed the Planetary Underworld, thirteen katuns or 256 years, beginning in AD 1755, introduced industrialization, electricity, technology, modern democracy, gene splicing, and the atom bomb. Our knowledge became Faustian power over the physical world. The eighth level—the Galactic Underworld—thirteen tuns or 12.8 years, began in 1999, with the development of the Internet into a global communications infrastructure. The final step, thirteen uinals or 260 days, will lead, Calleman believes, to the attainment of "nondual cosmic consciousness" across the Earth. By the end of this Universal Underworld, humanity will have crossed the threshold of the abyss, confronting the shadow projections of the Apocalypse, to become conscious cocreators of reality.

Each step on the pyramid "corresponds to a certain frequency of consciousness," in which evolution operates twenty times faster than the previous phase. According to Calleman's thesis, "In the Galactic Underworld, as much change must happen in a tun (360 days) as happened in a katun (19.7 years) during the Planetary Underworld, or in a baktun (394 years) of the National Underworld." Perhaps for this reason, the twenty-year "Generation Gap" of the previous era no longer seems applicable to our situation, which requires increasing flexibility and adaptability to navigate. Institutional structures formed over the last 250 years of the Planetary Underworld, from the nuclear family to the nation-state, seem to be losing their stability as we go deeper into the Galactic Underworld. At the same time, scientific breakthroughs proliferate, as we receive increasingly accurate data about the universe in which we are embedded. Calleman believes that understanding the spiral dynamics of evolution expressed through the Mayan calendar is, in itself, an aspect of the Divine Plan: "Taking responsibility as a cocreator with God presupposes a basic understanding of how creation works."

Confusingly, Calleman's interpretation of the Mayan calendar puts the crescendo of all human history back a year—even more breathtakingly close to our dangerously fragmented present tense. According to his thesis, based on his own interpretation of the sacred calendar's energetic shifts, global humanity will attain nondualistic enlightened consciousness by the end of October 2011. "It will simply not be possible not to be enlightened after October 28, 2011," he confidently asserts, "or at least from a certain time afterward when the new reality has definitely manifested." The final year of the calendar will be used, he conjectures, for celebrating and adjusting to our new circumstances of unlimited bliss and creative freedom.

In the thirty years since the publication of the McKenna brothers' book, *The Invisible Landscape,* a new outsider paradigm has crystallized, in which time reveals, in McKenna's words, "a built-in spiral structure." From this perspective, "2012" represents reconciliation as well as reversal. At that near-point in our future, science will reintegrate with aboriginal wisdom, rights will meet lefts, the carapace of modern technology will crumble as new support systems self-organize, causing a momentous polar shift in human thought and human values—from alienation to integration, from deformed and spatialized time to synchronic harmony, from either patriarchal or matriarchal dominance to true partnership, from ego-based delusions to global telepathy. As a dialectical synthesis of Eastern enlightenment and Western curiosity, the Maya—voyagers over obsidian oceans of time—depart the dusty dioramas of the past to await us up ahead, in the state of being called "time-freedom" by Jean Gebser; a new world reality that would be, as Gebser slyly suggested, also a new world unreality.

According to Calleman's study of the Mayan knowledge system, within each of the nine underworlds, there are cyclical pulsations of light and dark energies, which he calls "Days" and "Nights." Whether 16 billion years or 260 days long, each underworld contains a pattern of seven days and six nights, a partition into thirteen stages, each stage represented by a different Mayan deity. With its Seven Days of Creation, Calleman's reading of the Mayan calendar neatly enfolds the Judeo-Christian creation myth presented in Genesis. He argues that the energy shifts that manifest within each cycle become predictable, once the entire pattern is understood. According to Calleman, the crucial forward step in the evolution of consciousness takes place during the "Fifth Day" of each underworld, ruled by the energy of Quetzalcoatl. The previous form of consciousness asserts itself through final acts of destruction during the Fifth Night that follows, ruled by the energy of Tezcatlipoca. The 256-year cycle of thirteen katuns that began with the Industrial Revolution, which Calleman dubs the Planetary Underworld, reached its darkest point during the Fifth Night of 1932 to 1952, the period of Nazism, the Second World War, the Holocaust, and the dropping of atomic bombs on Japanese cities (in the 5,125-year National Cycle that preceded it, the Fifth Night, AD 434 to 829, corresponded to the collapse of the Roman Empire). The cycle of thirteen tuns that began in November 1999 will, likewise, reach its midnight hour, its involutionary crescendo, during the year 2008. Interestingly, various studies on the imminent peak oil crisis point to 2008 as the year when resources of energy, food, and water will become critically stressed.

Calleman proposes that this period could see a global financial and ecological collapse, accompanied by nightmarish misuses of power on the part of the ruling elite. "In Night Five, ruled by Tezcatlipoca, the lord of darkness . . . we will see the last desperate and at the same time most forceful, attempt to secure control by the forces seeking to maintain dominance." Such an interregnum could also provide the opportunity to circulate a new vision of what this world could be, and disseminate the tools and principles to implement it.

Right now, we are being forced to witness the shadow of the psyche projected into material form through systemic misuse of technology, biospheric destruction, and corrupt geopolitics based on entrenched egotism and greed. As predicted by McKenna, Argüelles, and Calleman, time itself seems to be changing form, accelerating and intensifying, as events follow each other at breakneck pace. A cynical or nihilistic perspective on the imminent fate of our species is, of course, plausible, but unproductive. In fact, if consciousness is intertwined with physical reality, then a nihilistic perspective actively helps to bring a nihilistic result into manifestation. The alternate hypothesis of this book sees the destruction of the biosphere and the development of technology as by-products, nondual and integral aspects, of our psycho-spiritual evolution. If the shadows appear to be growing darker, it is because the light that casts them is getting brighter.

Clearly, the shift to a higher form of consciousness could not happen passively. The transformation of consciousness requires not only personal work, but direct and surgically precise engagement with ecological, political, psychological, commercial, technological, and spiritual aspects of reality—as Nietzsche noted, the deed creates the doer, "as an afterthought." It is inconceivable that a movement to a state of higher consciousness could take place in a subconscious murk. Such a shift could take place in only one way: in the full light of consciousness. To accomplish this would require the complete engagement of our will and our higher cognitive faculties. We would have to decondition ourselves from negative programming, overcoming distractions and self-justifications and egocentric goals—to "abandon abandonment" and "escape escapism," as I was once told in a dream—in order to act for the greater good. At the moment, it is unclear how much sacrifice will be necessary. As futurist thinker Barbara Marx Hubbard writes in her book, *Conscious Evolution:*

> If the positive innovations connect exponentially before the massive breakdowns reinforce one another, the system can repattern itself to a higher order of consciousness and freedom without the predicted economic, environmental, or social collapse. . . . If the system could go either way, a slight intervention to assist the convergence of the positive can tip the scales of evolution in favor of the enhancement of life on Earth.

If our current civilization were to disintegrate—a possibility that should not seem outlandish, considering the New Orleans flood, peak oil theories, and numerous ecological factors—we might face an apocalyptic passage, forcing us to evolve at high speed. Breakdown and breakthrough may happen simultaneously.

According to Jenkins, in the ball games played in Mayan temple-cities, the ball entering the goal ring symbolized the Sun passing through *xibalba be,* the "underworld road," at the end of the cycle. Jenkins interprets the game players "as heroic semi-human deities whose job was to keep the sun rolling towards its meeting with the dark-rift." They were "cosmic midwives, or vision-helpers, who must facilitate the emergence of the next

World Age, the rebirth of the solar deity (and all life) into the galactic level." To make the alignment happen properly requires great effort on the part of the human players. That appears to be our contemporary role, which we can hopefully accomplish without losing our heads.

Like Argüelles, Jenkins suspects the Maya possessed actual methods of transdimensional travel. The superstrings, quantum jumps, and wormholes through space-time described by contemporary physicists might provide the basis for shamanic techniques of visiting other realms through the "serpent cords" depicted on their friezes. "We may propose a complex Maya science of shamanically invoking a 'wormhole' in local space-time, an opening to the transdimensional realm that ultimately gets its power from the Black Hole within the Galactic Center, and traveling through it to other worlds," Jenkins notes. "Is such a scenario just a fanciful fairy tale, or could it have involved the actual activities of Maya kings and shamans?" And if they practiced such techniques, can they be relearned by us?

Jenkins notes that, unlike our secular leaders, the "highest Maya political office required taking hallucinogens," asking us to imagine the U.S. president ingesting "psilocybin mushrooms ten hours before giving the State of the Union address." Such an idea seems far-fetched to the modern imagination, but it made sense to the Maya, and he sees logic to it. "The leaders of society *should* be able to journey into the deep psyche, to access the fount of all creativity and genius, to commune with the ancestors and beings from other realms and times, and to deliver into their country the organizing frequencies emanating from the cosmic source." Our incapacity to envision such a situation is part of an intrinsic incompatibility in worldviews. "Something very basic to the Western mind-set prevents us from understanding the full profundity of Mesoamerican cosmovision," he notes.

"THE HISTORY OF AN EPOCH," said Einstein, "is the history of its instruments." The modern mind instrumentalized time, fashioning it into a tool, as Jean Gebser noted, "to shape the three-dimensional perspectival world and permit it to become a reality." The Gregorian calendar is a medieval antique; while our science and the scope of our awareness have expanded exponentially since the sixteenth century, we are still using the temporal program shaped by that bygone epoch. To integrate a deeper and more accurate perspective on time, we require a new instrument—a new calendar or, more properly, a synchronometer—for navigating it. Argüelles may be correct when he writes: "For the new age, so longed for, is actually only realizable as a new time. A new time can only come about by the rejection of the instrument that holds in place the hallucination of the old time, replacing it instead with an instrument of such perfect harmony that it has no history, but is truly post-historical." Argüelles's "Dreamspell" may have validity as a channeled system—yet it is not logically satisfying or precise enough to become the basis of a new planetary civilization, requiring the scrupulous integration of rationality and intuition. However, the tonal, cyclical, synchronic infrastructure of the Mayan calendar—a new version of it satisfying both the aesthetic and technical requirements of contemporary humanity—might provide the basis for bootstrapping us into a new consciousness of temporality. Such an instrument would redefine our relationship to time. Instead of trapping us in a limited and linear history, a new calendar could integrate our earthly lives within a properly galactic frame of reference.

Like Teilhard de Chardin, Argüelles theorizes that the noosphere, "a function of the whole system of the Earth," already exists, in nascent form. Where the biosphere is the region "for transformation of cosmic energies," the noosphere would be the realm "for the reflection of cosmic consciousness and its mental programs." The functioning of the noosphere will be transformed and activated when humanity attunes with it through a global act of reflection. He suspects that the noosphere has a physical location, woven between the two Van Allen radiation belts in the Earth's upper atmosphere, girdling the Earth like a giant oroborus, the snake eating its own tail, a symbol in many sacred traditions. First discovered in 1958, the Van Allen belts absorb cosmic energies and dangerous rays from the Sun and other stellar bodies, shuttling them through the Earth's magnetic poles, shielding and transforming the biosphere. According to Argüelles's thesis, such interplanetary energies are also streams of information that the radiation belts transduce into forms that can be absorbed by the biosphere, altering chemical processes and accelerating transformations in the genetic code. Soon after the Van Allen belts were discovered, the U.S. military detonated a series of atomic bombs in them, creating artificial radiation belts that lasted for several years. Such drastic manipulations of little-understood forces are symptomatic of the current mind-set.

The institution of a new calendar—a new timing frequency—could be the self-reflective act, integrated, organically, with our collective raising of consciousness, initiating a new relation to time, space, and being, aligning us with the noosphere. Such a new covenant, however, could not be instituted by one visionary—it would require a collective realization and a global meeting of minds, synthesizing astronomy, physics, and the spiritual traditions of the world. Like Dean Radin, Argüelles suspects we are currently in the same situation regarding psychic phenomena that we were in the 1750s in relation to electricity. Before that time, people had been aware of lightning and static shocks, but they did not know that this power could be drawn down and made into a transformative force. Similarly, many of us are increasingly aware of synchronicity, telepathy, psychophysical interrelation, and so on, but we have not yet reached the intensified level of consciousness that would allow us to employ such psychic energies as a sustaining and transfiguring power, a new field of action, for our world. It might be that initiating the noosphere through this act of collaborative concordance would induce an exponential evolution of the psyche—thus explaining the telescoped last pyramid-step of 260 days, defined by the Mayan factor.

Argüelles, Jenkins, and Calleman are the three main proponents of the concept that the Mayan calendar is, as Calleman puts it, "fundamentally a time-schedule for the evolution of consciousness." While they agree on the big picture, they bicker over the finer points of how and when exactly this phase-shift will take place. Although Calleman believes we are moving toward nonduality, he insists that the Mayan calendar is based on "spiritual" cycles that have no relationship whatsoever to the physical movement of the stars and planets. "If today we are to embrace a worldview in which consciousness is more important than matter, we too need to base our timekeeping on the nonphysical, invisible reality rather than on the physical," he writes. Imposing an unneeded dualism between matter and spirit, Calleman dismisses John Major Jenkins's thesis relating the 2012 transition to an astronomical alignment, despite Jenkins's years of copious and careful research on the subject. Although Calleman absorbed many of Argüelles's ideas, he, like Jenkins, goes out of his way to dismiss Argüelles's work, especially the channeled Dreamspell calendar, which borrows elements of the sacred calendar of the Maya and rephrases it. Argüelles, meanwhile, remains steadfast that his idiosyncratic and imperfect calendar (which has, for instance, no coherent mechanism for dealing with the extra day of the leap year) is the only solution for humanity.

"For we know in part, and we prophecy in part . . ." St. Paul noted, in his Letter to the Corinthians. Each of these modern-day scholar prophets may have assembled an important piece of the puzzle, without reaching a complete or final answer. Perhaps there is no final answer. As Nietzsche noted, "Indeed, what forces us at all to suppose that there is an essential opposition of 'true' and 'false'? Is it not sufficient to assume degrees of apparentness and, as it were, lighter and darker shadows and shades of appearance—different 'values,' to use the language of painters?" In their study of the Mayan's sacred calendar and its applications, Jenkins, Calleman, and Argüelles have made valorous efforts to fathom an entire system of thought created in a vastly different, perhaps ultimately inaccessible, form of consciousness from our own. The truth may lie somewhere in between, or beyond, any exact model we can create. And yet if the Mayan calendar is indeed an artifact of galactic technology or the culmination of thousands of years of astronomical study and applied shamanic science, linking time to consciousness in ever-spiraling cycles, its rediscovery is a profound revelation—and our actions in the next few years may represent its application.

PART FIVE

THE DANCE OF KALI

It is the self within ourselves that we have to sacrifice.
It is our own heart that has to be torn out of the false being and offered to the light.

Pyramid of Fire: The Lost Aztec Codex

CHAPTER ONE

José Argüelles's visionary zeal proved infectious. My last night in Oregon, after he dropped me off at my hotel, as I lay in bed, my "third eye" geared up, projecting flickering images and hypnagogic communiqués into my cold-sober mind. I saw no Mayan princes in robes of state, proffering prophetic passwords to some new forked torque in time. I saw, instead, the Buddha on Mars. Like some gauzy blimp of peace, the calm "Awakened One," in cross-legged meditation, descended slowly into a quavering reddish realm of frenzied spirits, like angry-faced paramecium, rushing through him as if he were some kind of hologram. As I watched, their agitation slowly began to subside, absorbing the influence of the Buddha's calming vibration.

This was a pictogram torn from the pages of Rudolf Steiner's wild astral flights. In his cosmology, the material orbs of the planets and stars are only symbols of their inner reality—the sensible particle to their supersensible wave form. The spiritual sphere of Mars—responsible, in his cosmology, for iron in the blood, for ego-based ferocity, for the spread of materialist science, for the "mind-forg'd manacles" fettering us to public opinion—was undergoing a steady transmutation, Steiner declared in a lecture, through the work of the Buddha, who "incarnated" there after he left the Earth. "Whereas in earlier times Mars was said to be the planet of warlike traits, it is now the Buddha's task gradually to transform these warlike traits in such a way that they become the foundation of the sense for freedom and independence needed in the present age."

The Buddha, according to Steiner, had received an occult promotion, from boddhisattva to Christ-like savior, operating on another plane. "The moment a Being rises from one state or rank to another," Steiner wrote, "a new task is placed before him. And man, who has to fulfill his life's course on Earth, comes into touch during his time on Earth with Beings who, like the Christ, have from the beginning a cosmic task, and also with Beings who in their evolution upward leave the Earth and rise then to a cosmic task, as was the case with Buddha." Of course I had dismissed this fuzzy farrago of Western, Eastern, evolutionary, and arcane ideas when I read it—yet there it was, a film loop in my skull, as if Steiner's gauzy glimmers had infiltrated deeper filaments of my psyche.

This vision was followed by a telepathic chat with an alien intelligence—introducing itself as a representative of the enlightened hive-mind of the praying mantis. Several years in a row, at Burning Man—that occult nexus of avant-garde chaos—I had encountered singular specimens of this emerald green anthropod while out in the desert, engaging in short-term staring contests before they flitted away or expired. Since there are no bugs native to Black Rock's arid waste, these particular insects hitch rides from Burners to attend the festival, where they eventually die. During those strangely psychoactive moments, I had the presentiment that the mantis was appearing as some kind of diplomat or agent, witnessing and recording, as human consciousness gained traction.

"Yes," the mantis-mind informed me, "we are indeed emissaries from a galactic civilization that is peaceful and advanced." Given this rare opportunity, I popped a question that had always bothered me: Why does the female mantis consume the male after sex? "We find this sacrifice to be beautiful as well as pleasurable for the entire hive-mind, each time it takes place," it replied. I was told they looked forward to opening formal lines of communication with the human race once we had passed successfully through the dimensional portal. The interview over, and no more transmissions or delusions forthcoming, I fell asleep.

From Oregon, I flew to Oakland to meet a new friend, Steven, a television director who e-mailed me after reading my book. Addicted to heroin for over a decade, he wanted to know if the West African psychedelic iboga could cure him. An illegal psychedelic in the United States, ibogaine, as it is known, has become an underground legend for its anti-addictive properties. Following our e-mails, he had taken iboga at a clinic in Rosarito Beach, near Tijuana in Mexico, and it worked for him—he felt it had saved his life, diverting him from dissolution and incessant thoughts of suicide, and he had decided to dedicate himself to promoting the cause of iboga with a new convert's fervor. He wanted to make a documentary about iboga, for which I would write the script, but first, he wanted to take it again, with me, at the Rosarito clinic. I had agreed to this without too much thought. It had been five years since the brutal but transformative illumination of my Bwiti initiation in the damp jungles of secretive Gabon, and it seemed, abstractly, a good idea to check in with the ancestors and the hard-ass Iboga spirit once again. And the trip would be paid for.

We drove to Rosarito in Steven's white Lincoln Mark VIII—"a true white trash chariot," he proudly proclaimed—which smelled like a Skid Row cocktail lounge, littered with half-empty beer bottles, coffee cup ashtrays, sleeveless *Velvet Underground* CDs, and other detritus. Steven was one of the most fragmented talkers I had ever encountered, skittering wildly from tenderly describing the horses on his family's ranch in Yosemite to bragging of drug debauchery, from stories of fighting California brushfires to his fondness for *Barely Legal* magazine. He exuded an irrepressible good-natured mania of being, as well as droopy despair, familiar to me from other friends for whom drugs and booze were a means of evading reality while preserving a kernel of childhood innocence despite the most dedicated depravity. He regaled me with disjointed tales of sex and success from his Hollywood years. "You know, it's not all blondes and blow jobs," he said several times, his voice raspy from chain-smoking cigarettes.

I liked him. He reminded me, above all, of my friend Rob, the publisher of my literary magazine—like Steven, from an old, wealthy American family—who extinguished himself via overdose. "The pure products of America," the poet William Carlos Williams wrote, "go crazy." Like Rob, Steven seemed near the edge—victim of the schism between the old American pioneering spirit of his forefathers, with their Puritan rectitude and can-do attitude, and the post art-school quagmire of unmoored taboo and hipster nihilism, stale rock and roll, cable-channel-surfing, and B-grade cult flicks. Who could put a world together out of such chaos? The retreat into demon heroin's regressive death-drive, its surefire, one-size-fits-all cure for terror of the soul, was one solution—and a number of my friends had fallen for it.

Rosarito was a bland gated community pressed against the Pacific's pounding edge, on a rocky shelf above the beach, overlooking wide-angle sunsets. We stayed in a poured concrete house with Spanish trimming, tastefully decorated with Huichol yarn paintings and Buddhist statues. The director of the Ibogaine Association was Dr. Martin Polanco, a sweet-faced young man with a shy manner, who started the clinic after a relative was rescued from cocaine addiction through an underground treatment in Florida. Unlike the United States, there was no interdiction against the substance in Mexico. I was given a medical examination and a test dose of the drug, tapioca-colored gel-capped extracts of the root bark. As I took the rest of the pills, I felt nervous, shivery, light-headed—I realized how thoughtless, almost cavalier, I had been about this venture, suppressing the memory of the massive mind-shift induced by iboga, which had changed the trajectory of my life.

The nurse—Dr. Polanco's equally shy and sweet-faced wife—led me to my bed where she hooked me up to an EKG machine and headphones playing jaunty, ambient music. Like an airplane approaching takeoff, I felt a deep interior rumbling that amped up to a roar. My psychic receptors opened into what Carlos Castaneda called the "second attention," where intuitions and faint ideas shape-shift into images and firm directives. I closed my eyes to observe imprints emerging like patterns out of TV static—my iboga visions tended to be gray-scale and flickering, nothing like the prismatic rainbow holograms induced by LSD or psilocybin.

In vision-space, I saw a black man wearing a 1950s-looking dark suit and thin tie, resembling the actor Sidney Poitier. He held the hand of a five-year-old girl in flower-print dress, leading her up a staircase. I realized the girl in this vision was me. I had encountered myself-as-her before, in special dreams, including one that led from crowded cargo train to concentration camp gas chamber, touching a place of such dark and deep hurt that, as I awoke, I wondered if I had unscrolled a past life. It seemed to me that the black man represented the spirit of Iboga, and he was ushering me into his astral castle.

The internal roar had become a loud, ceaseless ringing in my ears that seemed like a dial tone, connecting me to the root's frequency, a level of vibration from which it could communicate while reorganizing underlying patterns of my psyche. I wondered what iboga really was—and an answer came back, telegraphed, shouted, into my mind—"PRIMORDIAL WISDOM TEACHER OF HUMANITY."

This was not my normal syntax—not my words at all. It seemed like a direct download from the logoidal consciousness of the plant. It substantiated something I had idly considered after my Bwiti initiation—that iboga could be the prototype for the original sacrament, not a fruit but a root, from the "Tree of Knowledge of Good and Evil," tasted by curious Eve and trepidatious Adam, as recorded in Genesis. Iboga grows in equatorial West Africa, considered the birthplace of humanity. It reveals a "good and evil" that is deeply personal, rooted in the individual character, meting out hard and adult lessons for those who can bear to evolve. If iboga was the original plant-teacher stirring primitive humanity from its bestial state to its uneasy ride toward world-transforming knowledge and power, if the Mayan calendar was right on schedule, it seemed a beautiful symmetry that the shrub had emerged into modern consciousness at this point in time, returning us to lost roots—so many tangled and fibrous meanings in that word—of our being.

My trip continued with journalistic interludes. When I formulated questions, I received emphatic answers that I would scrawl in trembling letters across the yellow-lined notepad next to my bed. I was shown, in excruciating detail, in herky-jerky scenes of myself filmed like old movie clips, my personal faults and lazy decadent habits. I asked what I should do about all of this sloppiness.

"GET IT STRAIGHT NOW," Iboga bluntly replied. This idea of straightness held particular meaning for me—since childhood, I have suffered from moderate scoliosis, a curvature of the spine. My posture is imperfect, and attaining uprightness a difficult achievement. I pondered on why I had such horrendous allergies to cats and dogs.

"FEAR OF ANIMALITY," the root bark barked. This response made intuitive sense to me, as it often seemed to be intellectual types such as myself, cut off from their bodily nature, who were afflicted with respiratory ailments, as if to punish them for their inattentive imbalance in relation to the physical world.

Piece by piece, I started to unpack the burden of guilt I had been carrying since my DPT trip, two years ago, leading to occult visions and poltergeist phenomena and the strangely continuing awareness that I had bonded with some fugitive daimon or Luciferic force in the trip's ominous aftermath. Even when I was straight, I suspected that the traumatic glamour of this episode had unleashed terrible events, with consequences that still afflicted my daily life—now, deep in the second attention, where the hidden hieroglyphs of my soul were illuminated, my suspicions seemed confirmed. Ever since that episode in the spring of 2001, while my partner was pregnant with our child, I felt as if reality had rippled into a new form. Deaths had multiplied around the event—my father before, her father after. Now I was forced to recall that in certain dreams in the months before I took DPT, Iboga had tried to protect me, sending me warnings—there was one, vibrant and eerie and ominously African in its tone, in which a swaddled infant spun around and around like a flashing police siren, while a voice in my dream intoned, *"Baby! Baby! Baby!"* In studies of shamanism, I had read that initiatory jolts, releasing currents of psychophysical energy, could be dangerous, even fatal, for the family and friends of the one undergoing them—I had read it, but I was still too much an "indifferent skeptic" to believe it, or to consider it a reason to interrupt my research. Only long after the fact did I fully accept that there were occult realms, and that mistakes made in them could have severe consequences in this reality.

Several months after DPT, the night before my partner went to get her second ultrasound, I had the most vivid and frightening dream of my life. I was a detective pursuing a case, stumbling upon a ceremony—ritual human sacrifice—in an amphitheater on the top floor of a glass office tower on the East Side. I leaped onstage, trying to prevent this murder, disrupting the audience in their crimson robes and hoods. The dream jump-cut to Times Square in sunstruck midday, where I was being pursued by two hit men, one with dyed-orange hair, the other a hulking, leering brute. I went into an old-fashioned barbershop, and the assassins followed me. Flashing a movie-villain grin, carrottop told me I had uncovered an occult conspiracy. They were going to get their revenge. "Tomorrow, we are going to kill you, your wife, and your baby," he said, sneering. I awoke as this dream noir ended—so lurid, so hyperreal, it sizzled in my mind. Beyond a dream, it seemed a direct message from astral-plane enemies. Shaken, I tried to conceal my anxiety from my partner.

When we met later that day, she sat down and burst into tears. The medical technicians had discovered a problem: A physical condition partially affecting our baby's mobility. This permanent condition could have been devastating, but in the case of our daughter, it was "mild," and she was not debilitated. In the following months, two other deaths took place around us—our cleaning woman's healthy five-month-old fetus suddenly died in the womb; the sister of my editor's assistant perished in a house fire. Of course I did not believe these were connected—yet at the time there seemed an odd symbolic parallelism. I felt as if I had inadvertently opened the gates of hell.

In our secular postmodern culture, who could even comprehend this psychic dark matter that had, it seemed, exploded around me? Anyone I knew would dismiss it as narcissistic delusion on my part—I had been forced to brood on these matters alone. At a little over one year old, our daughter already knew she was different. Why was this done to us? I asked Iboga, in rage and wondering sorrow.

"GOD IS JUST."

I could not resist this response; it reverberated through my core. In my altered state, I could sense, like invisible music, the karmic pattern behind the painful things that had happened to me in my own life, and to the people I knew and loved. I understood, and accepted, that the causes and conditions could be traced to previous incarnations—as well as preparations for whatever future we were unfolding toward. I had chosen to pursue the path of knowledge, and I had received a harsh teaching. At the same time, the reasons for it were inscrutably embedded in larger patterns that went far beyond us, like the eddying, interfering patterns of circles that raindrops diffused as they fell, one by one, into puddles.

Some of the faults that Iboga was forcing me to confront seemed ridiculously minor—and some were less so. In Berlin, my partner and I had learned a therapy for our daughter's condition, holding her in various positions until she cried and thrashed out, touching pressure points on her ankles and back, forcing her to activate underutilized chains of muscles, in the hopes of creating new neural imprints. We had done this several times a day, for more than a year, starting when she was only a few months old. We found it increasingly unbearable to upset her, day in and day out. Over time, I reduced the frequency of the therapy. Iboga criticized me for skipping these sessions—I argued with the plant-spirit, saying we didn't even know if it was helping her, and it was so difficult, and it didn't really matter—

"EVERYTHING MATTERS," Iboga insisted. Once again, I couldn't argue with this simple if unrelenting perspective. All I could do was promise to try harder—with everything—in the future.

Our normal human tendencies are distraction and dissipation. We begin one task, then get seduced by some other option, and lose our focus. We drift away from what is difficult and we know to be true, to what is comfortable and socially condoned. José Argüelles appeared in my visions as an example of the single-minded effort and lifelong dedication required in order to do what needs to be done. When nobody cared about what he had to say, when his ideas were rejected and scoffed at, he did not abandon his heartfelt mission. He and his wife walked across the world, in poverty, with nothing to their name, in order to convey their message of a new time and a new vision—I almost wept at this humble picture. For such a spirit, seemingly insurmountable obstacles are accepted, even welcomed, as part of the path. He had followed the call of his inner voice, not deviating from his goal—

"DO NOT DEVIATE," Iboga forcefully agreed. I decided that, in the future, in whatever I undertook, once I knew for myself the right path to follow, however difficult it was, I would follow it to the end.

Thinking of my love for my family, of the journeys and explorations I had been privileged to make, I felt happy, lucky—

"YOU ARE LUCKY," Iboga affirmed. I recalled the joy my partner and I shared in the first months after our daughter's birth. I felt overwhelming gratitude that I had been given the chance to live and love, to seek truth for myself, to try to understand so many things. I felt, also, how good it was to live at this momentous epoch in human history—at the end of one world age and the inception of another.

I considered the imminent collapse of world civilization—the robot wars, famines, plagues, freakish terrors that seemed the likely consequences of our societal and biospheric meltdown. I feared for the lives of myself and my family in New York City, vulnerable to the hate-filled whims of maniacs and Fundamentalists wielding weapons of mass destruction. I witnessed sheets of radioactive flame devouring cities, huge crowds of lost souls reduced to cinders. I asked Iboga if this was to be the tragic, traumatic fate of humanity.

"EVERYTHING IS SAFE IN GOD'S HANDS."

This telegraphed answer was startling—and reassuring. The message has stayed with me, sinking in, slowly, over time. It has liberated me from fear and anxiety. As I have let it penetrate into me, I have attuned with it, accepting its truth.

Iboga seemed to me to be a type of enlightenment mind, like a Buddha, that had chosen a different form, plant-spirit rather than incarnate teacher, to work with humanity. He imparted a cosmic message of "tough love." At one point in the journey, I seemed to fly through the solar system, passing through the heart of our Sun, where winged beings were spinning around the core at a tremendous rate. Up close, they looked like the elegant gold-tinged angels in frescoes of Giotto or Piero della Francesca. In this vision, the Sun was entirely made of consciousness, expressed by the harmonic orchestration of these implacably wise and faintly smiling figures, their choreographed light-speed maneuvers weaving our local star's vibrational field. Similarly, I conceived of humanity as an expression of the "Gaian mind" of the Earth, her sensory organs and self-reflective capacities, at her current state of development. If we were changing quickly right now, I realized, it was because the Earth was subtly shifting its frequencies, forcing humanity to evolve.

At moments, I felt the overwhelming DMT dimension hovering behind the frame of the ibogaine trip; it seemed that iboga was some kind of interface over the hyperspeed warp-and-weave of DMT—like the curtain in *The Wizard of Oz,* hiding the behind-the-scenes machinery. Thinking about DMT, I received the strong message or insight that iboga and ayahuasca were the best ways for human beings to access these realms at our current phase of development. The lightning strike of DMT is shocking overload—iboga and ayahuasca mediate between planes of being, transmuting higher voltages of cosmic wisdom so we can absorb slower messages at our current level of mind.

I was startled that these downloaded directives—like undergraduate lessons from some multiversity of supermind—spoke so directly of God, as if in the old monotheistic sense. I still didn't think they were indicating a god like the singular long-white-bearded individual stretching out his aged hand across curved space in Michelangelo's Sistine Chapel—the jealous patriarch we have, perhaps, spent many lifetimes trying to evade. It was more as if God was shorthand for the cosmological system that causes us to evolve. William Blake wrote: "God only Acts, & Is, in existing Beings and Men," but those beings, perhaps, included exalted hierarchies of higher intelligences—what Argüelles dubbed the "Galactic Ordering Directorate."

Late in the night, I retched and vomited out bitter root bark residue. I put on a CD of African drumming. Closing my eyes, I watched smiling Bwiti tribeswomen dancing around a jungle bonfire. The visions and voices in my head flittered away and dissipated—although I was already exhausted, it would be more than twenty-four hours before I could sleep.

Steven's trip was less visionary than mine. His faults were also paraded in front of him in repetitive loops that seemed endless—at one point in our dark night I heard him scream out, "No! No! No!" He was shown a potential future for himself if he didn't kick heroin, as a coffee shop dishwasher, sinking into dissolute old age with bad back and lumpy paunch. This grotesque scene flashed in his mind again and again, accompanied by mocking hurdy-gurdy-like carnival music. At another point, he asked Iboga what he could do to help the world.

"CLEAN UP YOUR ROOM," Iboga told him. Chastened afterward, Steven quipped, "Iboga is God's little way of saying, 'You're mine, bitch!' "

CHAPTER TWO

Love—in its methods, war, in its foundations, the mortal hatred of the sexes.

FRIEDRICH NIETZSCHE

Back in New York, I encountered new gravitational forces tugging at my psychic life, wayward currents that would impel me deeper along my path, whether I wanted it or not. Although I didn't know it yet, "the Mother" was summoning me. First, she sent her priestesses in my direction.

The priestesses had attended a talk I gave on shamanism and psychedelics, at Columbia University's usually buttoned-down journalism school in the spring of 2003, and they were not amused. "We are by no means specialists in this area, we are only two women that feel very deeply about the overwhelming darkness that shrouds our world right now and deeply long for and seek out means to contribute to life affirming actions," they e-mailed me after my lecture. "We could not help but feel disillusioned by the trivialization and glamorization of ayahuasca. Perhaps such experiences belong to the wordless realm." They noted that "there exists a higher intelligence that will give us quite a brutal teaching if these traditions and medicines are not held in the utmost sacred manner."

Surprised by the tone of these strangers' e-mail, as well as its use of the royal "we," I replied, noting that I was sorry my talk had upset them, but it was not meant to be the final word on the subject. I suggested the aggrieved women create their own event, with speakers more to their taste; if they did so, I promised I would try to attend. A friendlier back-and-forth followed this.

Several weeks later, I went to a party at the home of an art collector and private *galleriste,* in an apartment above the old-fashioned glamour of the National Arts Club in Gramercy Park. One of the aggrieved women introduced herself to me. In a chic velvet jacket, she was thin and strikingly beautiful, with long dark hair and a beautiful aquiline face. Sitting on a luxurious beige couch in front of a vast turquoise-lit tank of tropical fish embedded into the wall, surrounded by drawings of the Austrian Surrealist Hans Bellmer of twisted doll limbs and torsos, and large abstract paintings of trendy drips and blobs, the priestess told me she had been working with a woman healer who led a community of shamans on the West Coast. She spoke about her journeys in reverent, rapturous tones, her eyes flashing white light as she described forest ceremonies and plant-spirit visions and drumming circles raising pure vibrations for healing.

"In our ceremonies we are working with the Mother," she said, amidst the clinking of wineglasses and cocktail chatter. "It is so beautiful the way she comes to us. Sometimes the energies are so powerful—they overwhelm us. We have to learn how to hold her vibration." The priestess was an abstract artist, a painter of colored shimmers, but she said that her recent discovery of this sacred world had transformed her, shifting the focus of her life. I was simultaneously impressed by her intensity and put off by a vibe that sounded intimidatingly religious or devotional or perhaps cultish.

Once the first priestess had established contact, the second priestess made her move. A former television actress and current fundraiser for progressive causes, she invited me to lunch—macrobiotic, of course. Like the first priestess, the second one spoke about her relationship to the Mother in glowing but guarded and proprietary terms. She was organizing a healing retreat in a remote, rugged Hawaiian island that would involve yoga, sunbathing and ocean-splashing, delectable raw food, and the chance to work directly with her teacher, the healer who channeled this archaic maternal deity during shamanic seances. They wanted me to attend. Fees would be waived in consideration of my impecunious status. I was tempted, of course, but unconvinced. I didn't see how this voyage would fit into the book I was writing on prophecies. Second priestess set me straight. She explained that the teacher's work was directly related to indigenous foretellings, seeking to bring together the "Eagle of the North" and the "Condor of the South" by integrating traditions of sacred medicines—in ceremony, they worked with ayahuasca, mushrooms, peyote, marijuana, and iboga. The teacher's group was directly affiliated with the Native American Church, the Santo Daime religion of Brazil, and the Bwiti in Gabon.

Back home, I proposed to my partner that we visit Hawaii together, but she mistrusted my motives and rejected this idea. Our relationship had deteriorated since a period of domestic harmony accompanying the early months of our daughter's life. Whatever I did—or didn't do—seemed to cause her further affliction. She was enraged at me much of the time. Although I had my fair share of faults, I found her anger to be disproportionate. We had always teetered together on a precarious brink. On an early vacation together, lying beside a condominium pool in Miami's South Beach, she had suddenly turned to me with a smile. "You're smarter than me," she said. I protested this was not the case. "No," she replied. "You're smarter than me, and it is a good thing you are. Because if you weren't smarter than me, I would destroy you."

The force that impelled us together was a retractive force; beneath our opposite tendencies—her aesthetic perfection in material details, my obliviousness to them; her inherited formalism, my disregard of conventions—was a similar ferocity of personal will. If she had loved me for being a seeker, the extremist aspects of my nature also repelled her. Although I wrote about the validity of shamanic exploration, she often called me a "drug addict," inspecting my eyes for telltale signs of intoxication—she had never tried psychedelics herself. While I was obsessed with the iceberg of Apocalypse that our civilizational *Titanic* was unerringly navigating toward, she blamed me for not getting a normal, decent-paying job. In her darkest moods, she said I had never loved her, was just using her, and had "totally fucked her over."

Changing planes on my way to Hawaii, I spent a few hours in the hallucinatory Dallas airport. In that transitional void, where dead air circulated among the uncomfortably angular lounge benches and softly expiring potted ferns, all hell seemed on the verge of breaking loose. Rushing along the moving platforms were suburban families in Banana Republic khakis, their faces obscured with white masks to prevent contamination from the viral threat of SARS. A military squadron—short-haired young soldiers with the blank look of freshly sheared lambs—were standing guard in olive drab, with canteens on their belts and assault rifles over their shoulders. I stopped by one of the seemingly infinite television monitors bolted into the walls, braying nonstop infotainment into the void, 24/7. Just at that moment, tank battalions of the U.S. Army were penetrating defeated Baghdad, the final stage of "Operation Shock and Awe." Like the biblical "locusts with human faces" from the book of Revelation, devouring reality itself, the announcers for CNN and FoxNews could barely contain their chirping enthusiasm for this fantastic victory.

Ten hours later, I looked around, blinking, at Paradise. A wealthy couple had brilliantly purchased forty thousand acres of Hawaiian soil—a large chunk of a small island—turning it into a pristine eco-preserve. The estate included cloud-forest canyons, sculpted gardens, whispering palm trees, exotic plant nurseries, rocky cliffs, and silky-sanded beaches. With the land came a massive wood-framed lodge, built in the 1920s by a tycoon gambler from a grade of timber that barely exists anymore, repurposed as our ceremonial center. Outside, on the wide sloping lawn, birds scattered their trilling songs; botanical effulgence swayed in gentle breezes; enormous enameled spiders with yellow and black zebra stripes rested in magisterial stillness at the center of opalescent webs. My first night in my small room, I had a troubling dream in which the teacher appeared as a leader of a brainwashing cult, dispensing barbiturate patches instead of illuminations, numbing rather than enlightening us. The program began the next morning.

"While you are here, let the Mother take care of you," the teacher said, flanked by healers from her commune—all women, all wearing white—as we sat in a circle on the living room floor. "If you need any special attention, or special prayers said for you, you just ask for it. This is your time to heal yourselves and to get what you need to help you along on your path." One of her associates had set up a station for herself, of various dog-eared tarot cards and a profuse array of herbs and incense, which she precisely fine-tuned like a jet pilot's instrument panel, occasionally turning over a card and consulting it, or setting aflame a mound of myrrh or sheaf of sage with a soft-whispered prayer.

I didn't know what to make of this scene. Despite my interest in the esoteric and occult, I maintained the skepticism of my native New York—

a city where money talks and bullshit walks—and I had little patience for the sugary sentimentality of the wispy New Age. However, the teacher, an energetic lady in her fifties with sharp eyes peering out at us from beneath her white baseball cap, did not seem delusional, or anybody's fool. We were asked to go around the circle, to say what had brought us to this gathering. As I listened to the others speak—our group included lawyers, fashion designers, therapists, and so on—my feelings teetered between suspicion, cynicism, and some deeper interior surrender to whatever process was now taking place. When it was my turn, I spoke, to my own surprise, about my daughter's physical condition, how I feared it was my fault for unleashing occult energies through unprotected use of psychedelics. I admitted I felt I had come there for her, for whatever benefit might be had from prayer. The teacher asked me to put a photo of her on the altar they had set up, which consisted of photos and images of various gurus and spiritual leaders, large chunks of quartz and other crystals, Christian icons and Buddhist statues, and I did so.

The teacher said that in their ceremonial work, they asked the Mother to return, "to sit back down in her creation." She told us of messages she had received in trances—such as one where the Mother told her, "The mind will call this time chaos, when all possibilities seem open, when everything seems on the point of collapse. But I call it 'original ground.'" She said that she considered the shamanic plants to be "jewels in the Mother's basket. The Mother wants us to hold these jewels for now. Later, she will tell us what she wants us to do with them." She foresaw a future point when the shamanic medicines would disappear from the Earth—when we no longer had need of them.

That night we drank murky cups of mushroom broth and Syrian rue. Sitting cross-legged with eyes closed, I had a vision of a purple-caped phantasm with curlicued horns, poised on a cliff above a Gothic landscape, like a figure from one of Aleister Crowley's tarot cards. This seemed an emanation of the DPT daimon, now somehow tamed and integrated as an aspect of my psyche, working on my side as an ally. In the "second attention," I pondered deeper on the Maya. The correspondences that had embedded me within José Argüelles's Dreamspell system presented themselves to me as a teaching. I intuitively understood my connection with this galactic civilization, as Argüelles described them, beamed into this present tense to shepherd along the transition, to help the Earth reach harmonic alignment at the right juncture. My placement in his calendar seemed like a postdated postcard arriving from the future in my mental mailbox, a clue or "tell" left by some playful intelligence operating beyond our current conceptions of space and time.

Another thought clicked, like an internal gear turning. It was a secret concealed within the quaint riddle of my peculiar last name. As found in the dictionary, pinchbeck is "an alloy of zinc and copper used as imitation gold," invented by Christopher Pinchbeck, an eighteenth-century horologist—maker of elaborate mechanical timepieces for the English nobility—and an apprentice alchemist, or "puffer," who sought alchemical gold through chemical means. The word also came to mean "a cheap imitation," or, as adjective, "imitation, spurious." A dictionary definition of the word "maya" is "the illusory appearance of the sensible world." The world of maya could be considered a "pinchbeck" reality. The Maya, the search for alchemical gold, the illusory matrix of space and time—all of these appeared to be punned, by some dazzling and subtle mind, within my eccentric moniker. I thought again of McKenna's "cosmic giggle."

First priestess had catapulted into an extravagantly intense, theatrical journey. She took different postures, standing up and raising her arms, flapping them like an exotic bird-woman seeking to take flight. The teacher and her support staff went to help her. They brought her down to the ground, put hands on her back and abdomen, and sang harmonious chants—*"ah-nan-dee-ah, ah-nan-dee-ah"*—around her as she slowly unstiffened, settled back, and relaxed.

THE NEXT AFTERNOON, I met the teacher for a private conference underneath the gnarled branches of a flowering tree on the sparkling lawn, her female *consiglieries* by her side. She said she had read my book and considered me "one of the scribes" chronicling the transition to 2012. Since she had brought it up, I asked her what she thought this transition meant. "A new experience of time and space," she answered. She said her community "walked with the Lakota prophecies of Black Elk," foretelling the integration of shamanic medicine traditions—that was the secret, central purpose of her group. She had come to this work in a roundabout way, already plunged in a personal process of spiritual transformation when she had a potent dream in which a root was shown to her during a tribal ceremony. She was instructed to seek it out and drink it "to assist in your prophesying and healing work"—she later discovered the root was iboga.

The teacher was born in the South, teaching kindergarten in public school and then becoming superintendent of a shelter for abused and runaway girls. When she realized the impossibility of reforming the dysfunctional institutions that acted as a "dumping ground" for the courts and legal systems from within the system, she began to listen to an "inner voice" that told her, "You are throwing your stone into the wrong pond. You must now climb the ladder within your own labyrinth to reach the ocean of the collective." Following this intuitive prompting, she embarked on a long process of kundalini awakening that would reshape her life, as well as her body.

Kundalini, the "serpent power," is an occult energy that resides in the body, according to yogic texts. In most people, kundalini is "asleep" at the base of the spine, but when it begins to awaken, it travels upward, passing through the various chakra centers, burning out impurities and creating massive changes, often accompanied by illness and dangerous, even life-threatening, physical symptoms. According to *Sat Chakranirupana,* written more than five hundred years ago by a Bengali yogi, Purnanada:

> The sleeping kundalini is extremely fine, like the fiber of a lotus stalk. She is the world-bewilderer, gently covering the "door" to the central Great Axis. Like the spiral of a conch shell, her shiny snake-like form is coiled around three and a half times; her luster is like a strong flash of lightning; her sweet murmur is like an indistinct hum of swarms of love-mad bees. She maintains all beings of this world by means of inhalation and exhalation, and shines in the cavity of the sexual region.

The goal of yogic practice is to channel this energy, stage by stage, through the body and to the top of the head, symbolically uniting Shakti, the female principle of sexual energy, with Shiva, the masculine deity of universal consciousness. As the teacher wrote in a memoir: "In my experience Shakti found her way up my spinal column, opening each chakra, purging issues accumulated at each center, and then went over the top of my head, falling down into the third eye, the throat and filling the 'golden hara,' located in the belly region." She believes that humanity is currently undergoing a "collective kundalini awakening" that will "reshape and ultimately transform this dream we call reality." Citing an indigenous prophetic text, she calls our moment "the Time of Dreaming the World Awake."

Her archetypal and arduous journey of awakening led her from Texas to studies at the Jungian Institute in Switzerland; to visitations from the "Divine Mother" in the village of Medjugorje, Yugoslavia, where the "Blessed Virgin" had been appearing to a group of children at the local church; to an "etheric piercing" during a Native American Sundance in Hopiland; to the sheltering community of the guru Ananda Mai in India; and to Peru, where she and her husband relived a tragic past life among the ancient priest-caste of Machu Picchu. She described profound energetic shifts, in which a white light seemed to pour through her and she heard a loud buzzing in her ears, leading to visions of other times, places, and worlds, as well as voices conveying specific messages. After one such initiatory episode in the early 1980s that ended with a voice pronouncing, "Once your hands are on the plow, you can never look back"—many years before she started her study of shamanic substances—she and her husband found that "the bricks on the exterior wall of our garage had completely buckled out from the wall. . . . The repairman could not give us an explanation."

The rising kundalini energy "was electrifying [her] field, creating synchronicities that seemed to merge inner and outer realities." Newly arrived in Switzerland, she was instructed in a lucid dream to drive until she felt impelled to stop, then find a cross carved by "Indian hands." The next day, she did as instructed, and found herself entering an antique shop where she discovered a talisman that matched her dream description, and bought it. I had also experienced something similar after my DPT initiation: a specific vision of a red velvet suit which I sketched in my notebook, and then found, at cut-rate, at a basement sample sale next to my father's loft.

Over a number of years, as her psyche opened to other dimensions and orders of being, the teacher found herself acting from deep levels of intuition, drawing upon knowledge that appeared within her, spontaneously. During a trip to Rome, she found herself blacking out while standing before a sink in her hotel room. She entered a vision where "I was in a dark passage and a man was speaking to me in a language not from this planet. I seemed to know and understand him quite well. I spoke to him in an authoritative manner demanding that he leave—get out of me." Returning to normal awareness, she felt she had received an invitation to act as medium for this entity, "but some part of me had very emphatically declined. I was clear that if I was going to be asked to do such work, I wanted to be conscious simultaneously. But what did that mean?"

The culmination of an eight-year process of often agonizing bodily restructuring began with her first experiences of *samadhi,* a mystical immersion in pure radiance that temporarily obliterates normal consciousness. As a physical manifestation, she developed a permanently swollen "Buddha belly." While lying in bed one night, "something about the size of a small man's fist formed under my skin at my pelvic region and with a twisting movement, extended itself up to the middle of my abdomen, stopping at my heart region." As her husband and a close friend watched over her with much alarm, the process continued for hours. "When I awoke the next morning, my belly was extended to the point that I looked about five months pregnant." Given a reading by a psychic friend, she was told she was "merging realities. These merging realities have not coalesced a construct of any kind. There is no scaffolding. There is no mental hardware. We don't have any software in our computer that fits." Realizing she was being reshaped into a vessel of the Divine, experiencing physical anguish as well as boddhisattvic visions of deep love and bliss, she surrendered to the process. A voice whispered to her, in a vision, "What is birthing but breathing in. What is dying but breathing out. Isn't it time we put the two together?"

ENTERING THE TEACHER'S "archetypal field," I was experiencing some slight visionary uplinks of my own. Lying beside the pool next to the lodge, I entered a brief eyes-closed trance in which I found myself surrounded by squat spirit-figures with angular Hawaiian features chanting emphatically, *"Hai! Hai! Hai!"* At another point, I received a startlingly clear image of a long canoe or boat heading down a river, surrounded by deep jungle.

Sunbathing nude together on the private beach—white sand and glittering aquamarine ocean at the base of swooping cliffs—the first priestess told me of her mushroom visions. She had soared out of her body to meet beatific light-beings. Addressing her as "bride of Ashoka," the spirits explained she had been wedded to an Indian prince or Asian potentate with that name in a previous life. They brought her to meet Ashoka in the "spirit world"; he was profoundly grateful to her for introducing him to Buddhism, inducing him to commit to the enlightened path.

Time is compressed when you are doing shamanic work together in a small group, in a tropical setting. Over the next few days, first priestess and I became increasingly close. I had never met anyone like her before. She seemed to emit a pure crystalline tone. A natural channel who entered extraordinary vision-states in every session, she said that her out-of-body explorations were so beautiful she sometimes feared she wouldn't return—this reality seemed so paltry and drab in comparison.

We talked about our different ways of approaching visionary realms. I admitted I was still curious about the workings of ambiguous and even sinister forces. Raised Catholic, drawing inspiration from the mystical writings of Saint Teresa, first priestess was emphatic about her commitment to barring any intrusion from realms of darkness. "I think what we have to do is learn how to live in the Light," she said. She thought of her body as a vehicle for transmuting darker vibrations into healing energies.

She had grown up in modest circumstances, on a farm in a foreign land, but her elegant beauty and personal force destined her for greater things. After many life travails, including one failed marriage, she had created a successful career for herself as an artist. After Hawaii, she was returning home for the first time in many years, hoping to transfer the strength and blessings she had gained through her work with the Mother to healing abusive patterns in her fragmented and incest-ravaged family.

While I enjoyed my time with the priestess, it was not the focus of my week—my focus was a subtle interior adjustment, a quiet realignment of forces. It was as if an old layer of agitated and desirous and grasping thoughts was detaching, like an encrustation of barnacles, from the undersea hull of my psyche. At an evening ayahuasca session held on the beach, the teacher's songs almost obliterated by pounding surf and wind, I watched as every cynical and alienated idea that popped into my mind appeared as a literal cartoon demon or giggling gremlin in my visions. I realized that such negative thoughts were the results of failures of will, a kind of psychic slackness. I recalled what Rudolf Steiner wrote—that it was possible to reach a point in development where no idea entered your mind unless you allowed it to enter. Beyond the adolescent conception of "tripping," shamanic substances could be used as reprogramming devices for the soul, a means of training and strengthening the will—psychic yoga to attain a permanently heightened state of mind. At the beginning of the retreat, each of us had picked a divinatory card, without looking at it, to turn over at the end of the week. Mine said, in part, "All journeys are outward journeys, there is no inward journey. How could you journey inwards? You are already there." During the week, I felt I had contacted a forgotten core of my being that was purely still and untouched, wanting nothing, not even visions—this seemed a great gift. However, it was not yet a place I could remain.

My friendship with first priestess had developed to a level of physical—nonsexual—intimacy. On one of the last nights of the retreat, we lay together on pillows in the lodge, holding each other. Abruptly she turned to me and said, softly, "I love you." There are so many ways those words can be said—this declaration, as I considered it on the plane back to New York and in the weeks and months that followed, did not feel like a playful or simply friendly one; it seemed murmured from the depth of her soul. I didn't know what to do about it or what it meant—we were both in committed relationships. This simple statement seemed to impose a kind of duty on me; I felt the tug of some archaic, inescapable responsibility—not yet visualizing the train wreck this would cause.

During the closing circle, the healers from the teacher's community spoke about their work. They discussed how they handled their fear of persecution from the government for utilizing still-interdicted plant substances. One woman said she saw the traditional plant knowledge as "not secret, but sacred. Not hidden, but guarded." I was impressed with their steadfast acceptance of the risks they took to do what they felt needed to be done.

First priestess left the next morning. We hugged goodbye. "Last night, I dreamed of meeting your daughter as a young woman," she told me. "In the dream, she had healing hands."

CHAPTER THREE

A few months after my Hawaiian idyll—a brief interlude in paradise that would lead to a longer sojourn in purgatory—I returned to Glastonbury, seeking a clearer understanding of the crop circles. The exasperating conundrum of the patterns had remained indelibly etched in my mind since the previous summer's foray. For every article or book I read that supported their validity, I found an equally convincing text or hoaxer's Web site that undermined such a perspective. My partner rented a house for us on a winding tree-lined street leading to the misshapen mound of the Tor, but was forced to postpone her arrival with our daughter for several weeks, due to work.

My first excursion into the fields—with two coolly skeptical friends from the West Coast—was a dismal fizzle. The patterns we came upon seemed trodden down and lifeless, tattered from tourist traffic, and I feared I was devoting myself to a doomed cause. Taking the opportunity in my partner's absence, Michael Brownstein, my inveterately gruff older friend, the author of *World on Fire,* came to stay for several weeks, renting a car. Although Michael pursued more spiritual trends than anyone I had met—from raw food diets to Qigong workshops, Toltec shamanism to Tibetan Buddhism—even he had been initially suspicious, smirkingly dismissive, about the formations. However, he decided to keep an open mind, and take the time to look into them for himself.

Calling Michael Glickman—who described himself as the "air traffic controller" for crop circle research activity—one morning, we learned of a brand-new formation that had just materialized in a forbidden field owned by a farmer who hated the yearly appearance of the patterns on his land, according to Glickman, and refused to allow access to the formations. We drove the fifty miles to North Down on Wiltshire, then down the private road to the farmer's house anyway, encountering a mean-faced highly pregnant young woman screaming at a barking puppy. We offered to pay to see the new circle, but she told us to get out—then resumed yelling at her dog, threatening it with raised stick.

Possessed of encyclopedic knowledge of the region after a decade of crop circle forays, Glickman gave us precise directions for reaching the pattern surreptitiously. We parked the car along an embankment and stalked several miles into the gently rolling farmland, keeping our eyes peeled for combine patrols or pitchfork-wielding locals, along the still extant tracks of an old Roman road that crossed several fields and passed through a small wooded grove. In the distance, we saw several Neolithic mounds as Glickman had described, and walking out to them along tractor lines, found the formation.

Perhaps two hundred feet in diameter, the formation was extraordinary—I felt, as I had last summer in the "World Tree," a kind of raising of the psychic amplitude. It was a large circle with eleven concentric rings. The rings were not simple solid lines but each was patterned with an undulating "Egg and Dart" design; from above it looked like a Bronze Age shield. The complex crop circle was skillfully executed, and perfectly aligned with the Neolithic barrows a few hundred feet away. A scrupulous and coordinated effort had gone into this design—for what purpose? Considering the hostility of the local farmers, only a handful of people would ever see it in person. Without a word, the three of us—we were traveling with a Dutch woman whom Michael knew—went to different sections of the crop circle, lay down on the dried flattened wheat, and remained silent for quite a long time. Time seemed to distend. The atmosphere seemed to hold a tangible charge, a high-pitched pressure, forcing us down to the Earth and holding us there.

Shaking ourselves out of our trance, we returned to our car and drove to the five-thousand-year-old Avebury stone circles, jagged huge boulders that half-encircle—the other half had been pulled down by suspicious Christians during the Middle Ages—the ancient village. We walked the raised ridge along which the rocks were positioned to create a calendar marking the time for "festivals, celebrations, and other rituals" by "the eight-year cycle of soli-lunar synchronicity," according to Robin Heath's *Sun, Moon & Stonehenge.* "Aspects of the site, such as its latitude being one-seventh of the polar circumference of the Earth, the sheer scale of the ditch and bank, the proximity of Silbury Hill, and the scale and geometry of the outer ring, inform us that Avebury was once the center for a huge cultural development, with astronomy and geometry centrally placed on the agenda."

Looking across an expanse of green wheat fields, we saw the impress of another new crop formation, black dots of people moving inside, and headed toward it. The pattern revealed five pentagons conjoined in a neat geometrical arrangement, once again crafted with unerring precision, appearing as if a compressed force of some kind had made the design in one sudden sweep. Tourists were sitting beside the swirled "nest" at its center, listening to a Guinness-swilling local expert. "American Indians know about the five Star Nations," he said, pointing his beer in the general direction of the sky. "They are telling us they are watching us. We are being invited to join the galactic community—but our politicians are too stupid to realize this."

We visited Glickman, who told us how, in his long quest to understand the nature of the phenomenon, he had once visited a psychic in Los Angeles. The medium went into a long trance, then said he had contacted the beings responsible for the patterns, a group of higher intelligences linked telepathically, speaking in one voice in response to his queries. Glickman asked the circlemakers what they thought of humanity. "Why, we think you are hilarious!" the beings responded.

"What I got was that they love us, but they see us as ridiculously limited—sort of hapless and bumbling around, like characters in an old Laurel and Hardy movie," he recalled.

Glickman showed us the one extant video that appears to show a crop circle being made. Taken during the early morning hours of a night-long "crop-watch" at Oliver's Castle, on August 10, 1996, the clip is dramatically undramatic. Two of the infamous "balls of light" zip across the field, perhaps thirty feet above the ground. Underneath their path, the crop simply falls down as six circles are formed, then lines connecting them, in a snowflake pattern. The entire formation, several hundred feet across, appears in under ten seconds. Afterward, two more luminous orbs track across the field, as if checking on the first pair's handiwork—and that is it. Glickman considers the film a "gift" from the circlemaking intelligence; the man who shot it quickly vanished from the scene, never to be heard from again. Although there is nothing overtly fake about it, it could have been faked, and is insufficient as hard evidence.

The phenomenon of the crop circles appears to be a long-running expression of Terence McKenna's "cosmic giggle," in which the Mystery chooses to appear "in a form that casts doubt on itself," achieving "a more complete cognitive dissonance than if its seeming alienness were completely convincing." At the end of the 2002 season, after I had returned to New York, a spectacular, outrageously jarring "grand finale" had appeared. The imprint portrayed a gigantic face, immaculately made in dry crop, mapped out in thinning-and-thickening lines like the screen patterns of a video image—a technique never used in the fields before. This face, rendered with realistic shadows to give it a three-dimensional effect, was not human, but a familiar representation of a foreboding "Gray," with bug-eyes in wide sockets, big head, tiny nose, and grim slit mouth. Behind this visitor, three distant ellipses, apparent UFOs, were visible in the background. The Gray seemed to be holding a disk out toward the viewer. The disk was flat, protruding halfway out of the lower right corner of the rectangular frame. Resembling a computer CD, it was covered in spiraling dots and dashes that were quickly decoded as ASCII—a simple 8-bit substitution code, originally invented by computer scientists in the 1960s—spelling out a message:

> Beware the bearers of FALSE gifts & their BROKEN PROMISES. Much PAIN but still time. [Damaged Word] There is GOOD out there. We OPpose DECEPTION. Conduit CLOSING (BELL SOUND)

The damaged or obscured part was first interpreted as "EELIJ?E"; later it seemed to many interpreters that it intended to spell the word "BELIEVE."

The formation appeared on August 14, 2002, a year after the Chilbolton Face and Arecibo Reply, outside of Wiltshire, near Winchester, in Crabwood Farm.

For many dedicated researchers, the prospect that the virtuosic Crabwood Alien was not a hoax was as unsettling as the alternative hypothesis. If it was a fraud, it demonstrated an amazing level of craft on the part of human circlemakers, and called earlier aspects of the phenomenon into question. If it was not a hoax, then the circlemakers had unveiled a new level of shocking literalism. To make the situation more vexing, the Crabwood Alien appeared a few weeks before the debut of the movie *Signs,* a science-fiction film starring Mel Gibson that used the crop circles as a plot device, suggesting that the formation might be some kind of publicity stunt connected with the film. However, although the film featured aliens, they resembled reptilian musclemen, appearing nothing like the creepy Crabwood visage. Researchers nonetheless fretted before the film's release, wondering if it would have a negative impact on the crop circle phenomenon, compelling popular frenzy and government oversight. In the end, the film seemed to have no effect whatsoever—it was hyped, released, viewed by the distracted multitudes, and instantly forgotten. To add to the confusion, a month or two after the Crabwood Alien, the human group circlemakers Team Satan utilized a rudimentary version of the same video image–like style for a promotional campaign—crop glyphs made during daylight, with full crews, on a smaller scale. Either the Team Satanists were rushing to suggest they deserved the credit for the new style of the alien portrait, or they were indeed the perpetrators of the Crabwood Alien.

The Crabwood Alien was a masterpiece of subversion and shock—however, it fit perfectly with my interpretation of the abduction phenomenon, and the crop circles as a whole. The Crabwood Alien appeared to be a warning against the Grays, delivered in an intentionally cartoonish style. The Grays seem symbolically implicated in our accelerated evolution of Ahrimanic technologies; in the postmodern myth or pop-culture fable of the dealings between our government and these entities, they transfer technological knowledge for the right to continue their experiments from underground bases in New Mexico. Our increasingly invasive technologies—genetic modifications, surveillance systems, nanobot killing machines,

HAARP Projects, "tactical nukes," and so on—would be the "FALSE gifts" they proffer. The Mephistophelean bargain made in reduced states of consciousness would be the "BROKEN PROMISES." The relationship between the Grays and technology was indicated by the video screen format, as well as the digital CD they held out toward the viewer. Exactly a year earlier, the 2001 "Face" had parodied a halftone image like a newspaper photograph; this new image referenced videotape and computer disks. The circlemakers seemed to be calling attention to our rapid development of media and its uses for spreading information as well as propaganda.

If the thesis explored in this book is correct, if the crop circles are cueing us into a transformational process now under way, then "Much PAIN but still time" makes perfect sense. The transition we are facing in the next few years as resources become scarce may cause "Much PAIN" before we create new templates for a compassionate planetary culture. But, the message reassured us, there is "still time"—some kind of time—available to accomplish the change. "There is GOOD out there" might refer to other orders of galactic intelligence—perhaps what indigenous people call the "Star Nations"—awaiting our evolution. "We OPpose DECEPTION" cautions against the machinations of the Grays, who pretend to be benevolent watchers. "Conduit CLOSING" is more obscure—it might refer to the notion of a "dimensional gate" described by various mediums, like a portal or wormhole—or Mayan "serpent cord"—that gives the circlemakers an annual opportunity to leave such a message. The Crabwood Alien supports the *Star Wars*–like notion of a Manichean cosmic struggle reaching across planes of reality and illusion, with the illumination of psyche and soul as its purpose. The image did not neglect the comical, and grotesque, dimensions of the situation, which plays itself out in levels of dreamtime awareness, inaccessible to the rational ego locked in the one-dimensional mental structure of modern time. If the makers of the image "OPpose DECEPTION," they do so in a manner that is detached, watchful, even coolly amused.

Considering the otherwise flawless design of the formation, I conjectured that the damaged word—"EELIJ?E" or "BELIEVE"—was an intentional error. By creating a flaw in the word "BELIEVE," the circlemakers seemed to be pointing out the flaws in our linear logic and dualistic grasps at closure. As Carl Jung wrote: "I believe only what I *know.*"

After weeks of studying the crop circles, Michael overcame his own suspicions and skepticism, concluding that the phenomenon was not of human origin. He based his judgment on the awesome magnitude of the formations, the palpable energy shifts he felt inside new circles, and the flawless aesthetics they displayed. While I was away at a conference in California—coincidentally, on the day I landed in San Francisco, the *San Francisco Chronicle* was full of news of a large crop formation that had just landed outside of town, which they dismissed as a teenagers' prank, printing a guide on hoaxing one of your own—Michael visited a number of new circles in Wiltshire. Meditating inside one, he asked for some guidance to the meaning of the formations. He received a strong image of three dots hovering over two solid bars. "In my mind, I kept trying to change the image and move it around, but it wouldn't change," he said. He didn't recognize it—but I did. I told him it was the number thirteen, integer of transformation, written in Mayan characters.

DURING MY TIME in Glastonbury, I made my first visit to Stonehenge. That jagged array of ancient rocks, projecting like broken teeth from the Salisbury Plain, awkwardly trapped between two modern highways, is the source for whatever residue of magic and myth still stirs within unplumbed depths of the buttoned-down English psyche. Although I had never gone to the site, I had strong associations with it. The primordial forms in my father's late paintings resembled those rough-hewn boulders, sometimes arranged in simple post-and-lintel constructions like those used in the inner ring of the monument. My father came from England—growing up in the pretty, tacky seaside town of Brighton—yet I never thought to ask him if the stone circles inspired him. When I studied his work after he died, I felt he was channeling some ancestral heritage of megalith memories in the stacks of color-saturated paintings and shimmering drawings he left behind.

Pursuing his solitary artistic path since the 1960s, exiled from the art world in his cavernous SoHo loft, my father was fascinated with the concept of "Other Dimensions"—the title of a group show he curated in the 1980s. He was haunted by the intuition of planes or levels of being that he could not access, from which he felt exiled—as he put it in a notebook I found after his death: "A blade of grass, the suspended flight of a hummingbird. We are travelers in a land where signs elude us, and everything we think or do only magnifies our sense of loss."

While my father lived, I didn't take his yearning for other dimensions—like his interest in the space-time paradoxes of quantum physics—too seriously. It seemed part of his too-fervent denial, his too-gloomy rejection, of the world as it is, the world in which I was forced to make my way. As I pursued my inquiry into psychedelics, with its high-voltage, hyperspeed warpages of reality, as well as shamanism and crop circles and Mayan calendrics, I realized I was continuing the same quest by different means. The similarity of his burly shapes to the mute stones of Stonehenge and Avebury seemed like an iconic signifier, another clue to the arcane puzzle I had, apparently, stumbled upon—as if it was my task to enter the "land where signs elude us" and make them legible once again. His last large painting, a six-foot-tall image of a single emphatic shape, bore the title *Geomancer*—and geomancy, Earth magic, was certainly an essential element of the Neolithic culture.

Approached by car, the jutting tableaux of Stonehenge radiates a brooding strangeness across the otherwise subdued and sheep-dotted landscape. Emerging from the underpass leading to the monument, I was struck by the power of the archaic stones—ravaged time-travelers surveying the panorama of human history from their immobile locale, shadow-dappled by sun and cloud—as well as prickly déjà vu. Like secular pilgrims orbiting a Buddhist stupa, a steady stream of tourists circled the rock pile, separated from it by a large ring of green lawn and a knee-level rope fence.

According to Gerald Hawkins, author of *Stonehenge Decoded*—a 1963 bestseller that sparked the entire discipline of archaeoastronomy—Stonehenge was an astronomical "computer," printing out, *ad infinitum,* a coded message of solar-lunar integration. The temple is on a raised platform, encircled by fifty-six evenly spaced holes, known as the "Aubrey circle," that probably held wooden posts. By moving stones along these posts each day, Stonehenge functioned as an accurate predictor of lunar eclipses.

Its numerous alignments and sight lines followed, and synchronized, the movements of Sun and Moon over long cycles, lasting decades, requiring occasional reprogramming.

Many of the now-ragged bluestone blocks used to build Stonehenge came from the Preseli Mountains of West Wales, 135 miles away—the means used to transport them remain a matter of controversy. Within the outer circle, there stand remnants of five post-and-lintel constructs of a much greater size and heft—the upright stones weighing as much as fifty tons—in a horseshoe shape. These massive blocks were brought from twenty-four miles away at Flyfield Down, near Avebury. When the initial megalithic structures were built, between 3000 and 2600 BC, the climate of England was similar to today's northern Spain, and the material culture of its inhabitants was equivalent to that of Native Americans before colonialism. Historians and archaeologists have biased us to think of prehistoric or premodern cultures as societies of scarcity; this was certainly not the case in ancient Britain, as they had the time, as well as determination, to plan and build these enormous outdoor temples, constructing from five to ten thousand stone circles over a thousand-year period.

"The collection of stones, mounds, and ditches we collectively call Stonehenge evolved over a period of at least 1,500 years, an unmatched period of evolution for any cultural artifact," writes Robin Heath in *Sun, Moon & Stonehenge: Proof of High Culture in Ancient Britain*. The construction was continually modified as long as it remained in use. Although the monument is popularly identified with the Druids, its development, as well as decline, predated the druidic culture of Britain, which flourished during the period of the Roman Empire, by millennia. The practice of building stone circles continued for as long as two thousand years, before it ended "remarkably suddenly," notes Heath, a former civil engineer and head of the technology department at a Welsh college, and "the purpose and the meaning of the stones, and the culture which deemed them so important, apparently became lost to us."

Following Hawkins's work on Stonehenge by more than a generation, Heath's book is a remarkable achievement, recovering from the crude stones an extraordinary legacy—a prehistoric "high culture"—accomplishing for Stonehenge what "alternative archaeologists" such as Schwaller de Lubicz and Robert Bauval discovered through studies of the pyramids of Egypt. In Heath's analysis, Stonehenge is shown to encode an entire canon of sacred knowledge, including a precise astronomical understanding of the placement, shape, and movement of the Earth and its relationship to Sun and Moon, as well as pragmatic use of the "golden ratio," phi, which appears in many forms in organic life, encoded in the proportions of the human body itself. The Neolithic culture of prehistoric Britain was evidently obsessed with making accurate observations of the world around it, and systemizing its understanding for the benefit of future generations. "Perhaps the best one can do is to recognize that Stonehenge was built by people identical to ourselves, albeit very different in culture, yet perhaps superior in their ability to think about some fundamental issues concerning human life on Earth," he writes.

Heath's work builds upon the work of John Michell, who showed, in his *A New View Over Atlantis,* that Stonehenge included complex sevenfold geometry in its layout, and a near-perfect representation of the "squared circle." The problem of squaring the circle—in which the perimeter of a square equals the circumference of a circle—is an ancient one with important symbolic overtones. The square represents "Earth" and matter while the circle symbolizes "Heaven" and divinity; integrating these forms fascinated the ancient mind. This difficult problem cannot be solved—as a German mathematician proved in the late nineteenth century—through ordinary calculations or geometrical means involving a compass and straightedge. Michell found that the proportions of the squared circle implied by relationships within Stonehenge was identical to one produced, numinously enough, by bringing the Moon down to touch the Earth, then drawing a square that tangents the Earth, and a circle with its circumference defined by the Moon's centerpoint.

During the 1,500 years of its active life, Stonehenge went through many modifications, apparently switching from a lunar to a solar emphasis. "Stonehenge began as a lunar temple and became Apollo's temple, reflecting perfectly the changes in the social fabric of the societies which evolved the monument," Heath writes. In Heath's account, the placement of the temple reveals careful forethought. Along the fifty-six holes of the Aubrey circle, four boulders were placed to create a "station stone rectangle" that was necessary for charting the solstitial and equinoctial risings that were essential to the builders' purpose. "The right-angle arrangement between the Sun's solstitial and the Moon's extreme . . . risings is unique to the latitude of Stonehenge," Heath writes. In other words, if the monument was not placed at its precise spot—"located at a latitude which is one seventh of the circumference of the Earth referred to the equator"—the alignments would not have worked. The "station stone" rectangle would have been a crooked parallelogram. With his engineer's mind, Heath glides through abstruse areas of meters and metrics that elude simple parsing. What appears to have been deliberately indicated, in numerous ways, in the sighting and inner geometrics and alignments of Stonehenge, is the difficult problem of reconciling the 365.242 days of the solar year with the Moon's orbital cycles to make an accurate calendar.

José Argüelles argues that his "13 moon calendar" is perfectly harmonic because the Moon circles the Earth thirteen times a year—but this is not precisely the case. In actual fact, there are several different lunar cycles—the period from full Moon to full Moon, called a "lunation," is 29.6 days, occurring 12.368 times in each solar year, while the amount of time it takes the Moon to pass a fixed star—the "sidereal lunar month"—is 27.322 days. Although Argüelles makes a case for a twenty-eight-day "apsidal" cycle, "taken from the measure of when the Moon's axis is tilted farthest from the Earth," this cycle is not important for astronomers—and, apparently, still does not present the synchronized relationship envisioned by Argüelles. "The apsidal cycle is not 28 days but a 311 day overrun after 8 years," notes John Martineau, astronomer and author of *A Little Book of Coincidence.* Argüelles supports his calendar as our solution by noting that the female menstrual cycle naturally follows a twenty-eight-day rhythm. But this is not the case. According to biologists, this cycle, when synchronized, more closely approximates the 29.6-day interval that separates full moons. While Argüelles is correct that the twelve-month Gregorian calendar is a solar abstraction, which could be supplanted by a more accurate and harmonic calendar, the thirteen-moon system he proposes is a lunar abstraction, hence equally off-kilter. This may explain why the Dreamspell mainly attracts intuitive or "lunar" types to it. If we deemed it necessary to create a perfectly harmonized calendar, a more balanced solution to our timing problem would have to be found.

While it seems oddly remote and minor to our modern minds—certainly, growing up under the barely visible skies of Manhattan, I had never given it consideration—this problem recurs with startling frequency in our mythologies, fables, and fairy tales. As Heath notes, it is present in the fairy-tale classic *Sleeping Beauty,* originally known as *Briar Rose,* which begins when the king plans to invite the thirteen wise women in his land to a banquet to celebrate the birth of his daughter, but cannot invite the last one as he has only twelve golden plates. The uninvited thirteenth wise woman arrives in a fury, and curses the kingdom, sending the court and princess into a long sleep. The fairy tale offers "a prophetic warning concerning what *automatically* will happen when the feminine, and hence lunar, qualities of life are neglected."

The number twelve symbolizes masculine solar rationality; thirteen "is very much connected with the Moon and hence to matters matriarchal and the old Goddess religions," writes Heath. "The advent of Patriarchy, around 2000 BC, saw to it that all matters relating to lunar worship, and hence the Moon, were systematically and thoroughly eliminated from the new culture"—a mind-set that led to the medieval witch hunts and still permeates Western culture today.

In mythology, the interplay between solar twelve and lunar thirteen is indicated by the story of Jesus and the Twelve Disciples, King Arthur and his Twelve Knights, and Quetzalcoatl (Kukulcan), who was the leader of thirteen gods before his mysterious disappearance. "The two numbers twelve and thirteen, juxtaposed whenever our culture seeks to identify a superhero figure, such as Jesus, King Arthur or the Mayan Kukulcan, point to a cultural legacy which appears to be calendraic in origin, naturally fundamental to the immutable numbers of Sun and Moon," Heath writes. The archetypal pattern reveals an "enigmatic thirteenth savior figure, always with twelve Knights, disciples or kings, and always sacrificed to be then resurrected in order to save the culture."

In the Gospel of John, the risen Christ instructs his disciples to cast their net over the right side, into the water; they retrieve 153 fish. The square root of this peculiar number, Heath notes, is 12.369, the almost exact number of lunations in the year. Other references in the same scriptural passage relate to the ancient problem of solar-lunar integration. The number 12.369 is also the hypotenuse of a triangle made with the proportions 5:12:13, which Heath calls the "lunation triangle," used in the ancient World as "an astonishingly valuable tool for performing a range of astronomically precise tasks," including forecasts of solar positions and lunar phases. Heath believes that the site of Stonehenge was chosen because it created an enormous lunation triangle, stretching to the Preseli islands in Wales, from which the monument's boulders were removed, and to the sacred island of Lundy, directly east of the monument. This instrument of ancient measurement was thus inscribed, in massive scale, into the landscape, in perpetuity, to guard against its loss.

Again and again, these myths seem to suggest that the seemingly innocuous calendar is actually of fundamental importance, defining our relationship to reality—as Argüelles realized in his call for a new "timing frequency." A calendar that fails to balance solar and lunar elements, that does not correspond to natural cycles, will enforce, and self-reinforce, a deformed world. Heath writes: "Evolution is caused by the cycles of the luminaries and Man cannot detach from their importance if only because he is wholly formed (materially at least) by the effects of these two bodies and their ceaseless gyrations—their Sacred Marriage." What our science understands abstractly about the evolutionary engine of the "Sun, Moon, Earth system," Neolithic man may have realized intuitively and instinctively, codifying his understanding in monumental works of stone.

Why was Stonehenge built? Nobody knows for sure. As with the Mayan civilization, we seem to be separated from the operative mind-set of its creators by an unbridgeable cultural gap. It requires a tremendous leap for us to approach the "worlding" of their world. As Argüelles has noted, unlike our science-based civilization, myth-based cultures integrated all aspects of knowledge in a sacralized understanding of reality—the pattern of the past and the sensed or forecast future were also realized as aspects of the whole. As I circled the massive blocks, it seemed to me that Stonehenge was constructed so that the knowledge encoded in the site would survive to our present day. The rocks are scarred, weather-beaten; they have been smashed, pulled down, angrily chopped and chipped away by visitors over time—yet the plan of the monument endures, allowing modern astroarchaeologists to reconstruct its function, its sacred geometries and computer-like printouts of Sun and Moon synchronization. Perhaps the builders foresaw their synesthetic wisdom displaced by a different kind of thought, separating and conceptualizing, leading to our planetary crisis. Perhaps the monument was meant to outlast our amnesia about its meaning—to kiss awake the sleeping bride.

CHAPTER FOUR

If the thesis explored in this book is correct, if consciousness is the ground of being rather than an epiphenomenon of physical processes, we may find that a basic question asked by modern astronomy and space science—"Is there life out there?"—should be rephrased. Organic life, as well as intelligence, may already be a property enmeshed in the fabric of the cosmos, brought to fruition through the spiraling dynamics of the solar system and the galaxy, built into the structure of the universe itself. Such a perspective seems implicit in the harmonic, phi-based relationship between the orbits of the Earth and the cycles of Venus. Phi, the "golden ratio" of 1.618, found in many plants and in our human proportions, is mathematically derived through the Fibonacci sequence of numbers—0, 1, 1, 2, 3, 5, 8, 13, 21, etcetera, created by adding each pair of integers. When you divide the larger number by the preceding smaller one, you gradually approach the phi constant. Phi-based relationships are found not only in biological forms, but in the relations between planets orbiting around our Sun. As John Martineau writes in *A Little Book of Coincidence:* "Venus rotates extremely slowly on her own axis in the opposite direction to most rotations in the solar system. Her day is precisely two-thirds of an Earth year, a musical fifth. This exactly harmonizes . . . so that every time Venus and Earth kiss, Venus does so with the same face looking at the Earth." Eight Earth years equals, exactly, thirteen Venus years, the five kisses between them crafting a perfect pentagon, carved out of space. The numbers 5, 8, and 13 belong to the Fibonacci sequence, defining phi.

Every 243 years, Venus passes between the Earth and the Sun twice in eight years—a pair of "Venus transits," visible to the naked eye. In our own time, an initial transit occurred on June 8, 2004, to be matched by a second on June 6, 2012. The Venus transits also reflect the harmonic relation between the two orbits—in those 243 Earth years, exactly 365 Venus days will have passed. Venus was the planet of beauty for the ancients, because of these exquisite relationships; the transits symbolized the "hierosgamos," the sacred marriage, of masculine and feminine energies. Where the architects of Stonehenge and Avebury were obsessed with the patterns of Sun and Moon, Mayan stargazers were particularly fascinated by the cycles of Venus. The Swedish biologist and Mayan scholar Carl Johan Calleman proposes that the current pair of Venus transits directly relate to the prophecies of Quetzalcoatl's return, and the unfolding of consciousness predicted by the Mayan calendar. In any event, that the phi ratio—the blueprint for organic growth—extends from our terrestrial globe to larger patterns within the solar system suggests a larger-scaled evolutionary program, perhaps a galactic concordance.

Like the coiled arms of the galaxy, the development of consciousness appears to follow a spiral, sidereal motion, represented by the archetypal symbol of the mandala, which is universal in sacred art, from Navajo sand paintings to Tibetan Buddhist thangkas, Islamic geometries to Early Christian friezes. The anonymous sign-system of the crop circles, an enigmatic semiotic project, annually reiterates the mandala, icon of psychic wholeness. "The unconscious process moves spiral-wise around a center, gradually getting closer, while the characteristics of the center grow more and more distinct," wrote Carl Jung. Whether found in dreams or wheat fields, mandalas symbolize stages in a psychic process—the helical approach of the psyche toward integration of the ego and the self or higher self, through the difficult work of illuminating the dark matter within the unconscious. Like the stone circles, crop circles could be considered images of the "completely round" Hermetic vessel—Round Table or Holy Grail—the matrix or uterus that gives birth to the "miraculous stone" sought by the alchemists. According to Jung, the alchemist "experienced his projection as a property of matter; but what he was in reality experiencing was his own unconscious." The process of Apocalypse—uncovering or revealing—requires the realization, and reintegration, of our projections.

The meaning of the crop circles was still eluding my conceptual grasp. I spent feverish, sleepless nights, turning aspects of the phenomenon over and over in my mind. One friend of mine, Mark Pilkington, an English columnist for *The Fortean Times* and publisher of *Strange Attractor,* his own journal on matters abstruse and occult, assured me repeatedly that they were entirely human-made. He said he had hoaxed some of the patterns himself, and was friends with numerous circlemaking teams. "While I cannot, of course, account for every formation ever to have appeared in the fields, my research and discussions with others have identified the artists behind nearly all the formations to have made an impact on the wider culture, as well as many that will remain forgotten," he told me. When I begged for specifics, he replied, "Unfortunately I am double-bound not to reveal the names behind the formations—this is the prime directive of the unspoken circlemakers' code." I asked if he could connect me with one of these teams, so I could participate in the hoaxing of a formation while I was in the region, but he produced no leads on that score. He did, however, connect me with Rob Irving, a charter member of Team Satan, and one of the most vociferously defensive of the human circlemakers. We met for an unsettling tea in one of Glastonbury's vegetarian restaurants.

My family had joined me by that time, adding new layers of complexity to the situation. Obsessed with tracking the circles, I had barely spoken with my partner in New York, and her frostiness toward me had reached ice age proportions. Finding creative ways to express her glacial ire at my inattentive behavior, my partner had jumped to the skeptic's perspective, eagerly devouring Jim Schnabel's *Round in Circles* (1993*),* a witty and cynical take on the glyphs, giving voice to the hoaxer community. Schnabel claimed that "the crop circles, although made by humans, are not a hoax." They were a form of ephemeral artwork—an anonymous undermining of belief systems, perpetrated over several decades—as well as reinvented ritual. "The crop circles were being made by modern pagans either as some kind of complicated mystical rite, or, more likely, as a kind of cosmological terrorism, a surreptitious attempt to proselytize the larger population, to shake up its sclerotic, Piscean, Christian worldview, paving the way for an enlightened Aquarian one," Schnabel wrote. Himself an admitted deceiver—"It took me less than two months to move, on the cerealogical scale, from obvious hoaxer to superhuman intelligence"—Schnabel's decade-old book drove a deep wedge of discord into the community of crop circle researchers. His thesis was that the early circles had been created entirely by Doug Bower and Dave Chorley in the 1970s—the first one "an adolescent lark, a practical joke, and yet their delight seemed ineffable somehow, ancient, ageless: the pride of the creator, mingled with the pride of the spy"—and that the practice was then picked up and elaborated by younger teams of artist-pranksters—as well as, perhaps, Satanists and chaos magicians and even military groups—who were, collectively, responsible for the entire thing.

When we took a bus tour of recent formations, my partner curled up in the center of a pattern resembling a Native American design, and immediately fell asleep, as our daughter trotted around and chattered happily to middle-aged "croppies." I awoke her to point out a strange phenomenon in the skies: A giant white ring had formed around the Sun. She checked it out, then returned to her nap. She did not seem to find the circles to be convincing evidence of anything otherworldly, and seeing them through her eyes, I felt the same way. As if on cue, we encountered a string of the disillusioned and dismissive, German journalists and upper-crust conspiricists, once-committed researchers and "true believers" who had been converted to cynicism, sometimes after years of involvement, now mournfully convinced that the entire thing was a mean-spirited con job. These middle-aged men evinced the kind of psychic battle scars found in veterans of the 1960s, former idealists who had hoped and believed, only to have hope and belief robbed from them. But exiting a crop formation, we also spoke to a farmer, Tim Carson, who had as many as nine huge patterns show up on his land each season, and said, "They appear after rainy nights and in dry seasons. We never find a footprint, never hear a noise, or see a light. We never find any indication of trespassing."

A SELF-PROCLAIMED "famous artist," Rob Irving was hostile from the outset, almost bolting from the table when I mentioned I was open to the possibility that the circles were not entirely man-made. He was dogmatic in asserting that he knew for an absolute fact that all the formations were made by humans—and anyone who thought otherwise was a fool. When I described my synchronicity with the "World Tree," displaying a picture of it, he scoffed, exclaiming that he knew the artist who had made it. When I left the table to take our daughter outside, he immediately changed his tone. He softly asked my partner what year the formation had been made—he clearly knew nothing about it and hadn't seen it before, even though it was one of the best of the previous year's designs. He said that since the formation had "ridges," it was made by a local hoaxing team who specialized in ridges. When I returned, he announced he was seeking grant money to start an innovative program, teaching troubled kids to make crop circles as a way of redirecting their energies—considering that the clandestine patterns are illegal, this seemed an unusual plan. In the aftermath of this baffling encounter, my partner's skepticism was shaken; she noted that it almost seemed as if someone was telling Irving what to say. By the end of her stay, my partner was also possessed by the mystery of the circles, unable to sleep at night because of them. When I asked Irving whom he considered to be the mastermind behind the long-running phenomenon, he mentioned a Welsh astronomer, author, and book publisher, who had been one of the most diligent researchers in the early 1990s.

"If you really want to know about hoaxing, go ask John Martineau," he grumbled, as we parted.

Luckily, we had the opportunity to do just that, as we had been invited to visit Martineau at his farmhouse in the Welsh countryside. We went for dinner at an outdoor pub with a group of his friends, next to an old church and graveyard that hid the remains of an archaic stone circle beneath it—it was common practice in the Middle Ages to build Christian sanctuaries atop the Neolithic monuments, obscuring and disfiguring them, absorbing the numinous energies retained by the sites into its orthodox belief system. When I mentioned to Martineau that he was suspected of being the master hoaxer by Irving, he and his friends burst out in convincingly genuine laughter. "I used to insinuate that to wind Rob up," Martineau said. "I never hoaxed any formations."

In his early twenties, Martineau worked with a Wessex research group as a ground surveyor and geometrical analyst of the formations that appeared each summer, and contributed to a crop circle magazine, *Swirled News*. More than a decade later, he still seemed wistful for those early days. "The scene around the early circles was extremely fun and playful, like a movie," he recalled. "It was the apex of rave culture in Britain, and they seemed connected in some oblique way. I used to camp out in the formations, meeting dowsers, trippers, itinerant philosophers. We would awaken after dawn to watch military helicopters circling overhead. My first child was conceived in the 1990 Alton Barnes pictogram." Sleeping alone in an early pictogram, Martineau awoke one morning with a phrase in his head—"the passing of the great galactic interdimensional holospiral." The original crop circle research groups included retired members of the Ministry of Defense; Martineau asked one of them what they knew about the formations and often-reported UFOs. He was told, "All I can tell you is we don't know what it is. It plays with us. It has a sense of humor." Martineau is reluctant to speculate on the origin of the patterns. "I consider them great works of art. If they are made by a human being, you are dealing with a Mozart. I want to study under him."

With the influx of hoaxers, hecklers, and journalists, the crop circles became a "socially threatening environment," Martineau recalled. "I became concerned for my wife and children." From his interest in stone circles and crop circle geometry, he developed a fascination with what he calls "planetary harmonics." He had written a computer program to help him tabulate the geometrical relationships within the formations; for fun, he input the orbits of the planets instead, and was astonished at what he found—phi-based relationships and perfect geometrical figures of which the conjunctions of Venus and Earth are only the most dramatic. "I realized my work was the understanding of geometry—in music and math—beyond anything I could find in a book. I took a degree in astronomy, and grounded myself in Pythagorean and Platonic systems." He made a series of drawings of the planetary harmonics, self-publishing them in *A Little Book of Coincidence*, launching his own publishing company, Wooden Books. He realized that the crop formations had initiated him, naturally, into the sacred sciences of ancient Greece—perhaps ancient Britain as well—where it was said, "God is a geometer."

The researcher Allan Brown had been led on a similar trajectory, his obsession with the circles drawing him into the study of sacred geometry. At the 2003 Glastonbury Conference, Brown presented a talk that was the culmination of his years of research, proposing a solution to a central mystery: The hidden reason for the "quintuplet" patterns—four smaller circles arrayed around one large one—that had been among the earliest formations recorded, first showing up in 1978, and among the most frequent, appearing every year without fail. Over time, Brown realized that these quintuplets were related to the ancient conundrum of the squared circle. Implicit in their relationships, teased out through careful geometric analysis and the drawing of implicit tangent lines, was a series of previously unknown methods for performing this difficult intellectual feat that had obsessed the Classical mind.

"The circle is a symbol of spirit, of heaven, of the unmanifest, the immeasurable and the infinite, while the square is the symbol of the material, the Earth, the measurable and the finite," Brown said, noting that the symbolic essence of the problem was the reconciliation of seemingly opposing principles, and the resolution of dualities—"a sacred, cosmological act." Over twenty-five years, the differently proportioned quintuplets functioned as an instruction manual for the squared circle, suggesting dozens of new methods—some within 99.9999 percent accuracy—for solving this dilemma. "They were actually teaching a methodology, a complete system, through which a geometrical solution to the squaring of the circle could be attempted." Brown also proposed that this quarter-century project argued against human hoaxing. "Can so many formations, over so many years, adhere to such a sublime design principle, without anyone ever, in a quarter of a century, coming forward to hint at the geometric properties they were encoding into their designs?"

Brown stayed at the house we had rented during the week of the conference, and in our private conversations—held on the bramble-covered flat top of grandmotherly Silbury Hill and around the Tor's winding cow paths and other local spots—he admitted he remained deeply conflicted about the phenomenon. Despite his studious analysis and personal experiences of pristine new formations that seemed to belie the possibility of human construction—although he had staked years of intellectual effort on the endeavor—he often felt on the verge of giving up and concluding that the crop circles were, indeed, an extravagant sham. He had tried to stake out a "daimonic" position on the formations, in which the distinction between genuine and artificial was superseded. He cited Patrick Harpur, for whom the crop circles served as "a metaphor for deliteralizing, and a tool for deconstructing, our conscious viewpoints so that we come to understand that the world we imagine we are in is only one among many ways the world can be imagined." He stubbornly asserted that certain formations "will never be satisfactorily collapsed into a single objective truth"—yet remained aware that this vague claim was dissatisfying.

My family left for New York, and I stayed on in Glastonbury through the early weeks of August, hoping for another "grand finale" to the crop circle season, another jarring leap, like the Crabwood Alien of 2002, or the 2001 Windmill Hill spiral, Arecibo Reply, and halftone Face. Their departure lifted a weight from my shoulders—my partner's scorn had not abated during our time together. Although I tried to contribute to childcare, it was evident that from her perspective I wasn't doing nearly enough. I faced the uncomfortable realization that I had inherited bad habits from my father, who did not participate in my upbringing until I was much older—almost until I was old enough to discuss abstract painting with him, to be his "buddy" as much as his son. The disintegration of my relationship with my partner had deep roots in subliminal processes that I couldn't fully understand, or address—in my personal life, the integration of "Heaven" and "Earth," represented by the quintuplets and other crop patterns, seemed a mournfully distant and unreachable phantasm.

I traveled with my family to Wells, a larger nearby town, and put them on the bus to Heathrow, getting a final withering gaze from my partner as they departed. Wandering around the manicured gardens and lake beside the Wells cathedral, a stately Gothic masterpiece, and the Bishop's Palace, I felt liberated, disheartened, mildly desperate. Sitting on a bench overlooking green sheep pastures, I started a conversation with a young woman, a local shop attendant, giving cat-like licks to an ice cream cone. We took a walk together on a brambly path, through wooded hills above the town, then made out passionately while sitting on a tree stump. As I pressed against this pretty stranger, I felt I was breaching a dangerous taboo—what was I doing? I wondered, tingling with self-disgust and brute satisfaction.

Back in Glastonbury that night, the spinning matrix of my entire erotic history, its frenetic fizzles and drunken dissipations, arose in my mind like a fractured constellation. It was an area shrouded in personal darkness and deception, meteor-showered by unintegrated shards of longing. I admitted to myself something I already knew, but hadn't fully confronted—that I had made a compromise with myself to move in with my partner and suggest we have a child together. It was not that I didn't love her—I did love her—but to maintain monogamy and start a family, I had struggled to suppress the unfulfilled erotic curiosity that remained an integral part of my nature. I had put these socially unacceptable desires on the back burner, thinking that somehow I could face them at a later date—some imaginary point "up ahead" in spatialized time, that would be more "convenient." I was no longer sure this was possible—or desirable. It seemed to me that the tremendous heat of my partner's anger flared from secret jealous fires—she sixth-sensed these subliminal currents of my soul before I truly acknowledged them. I did not feel strong enough to face this reckoning, and had no desire to separate from my family—at the same time, we couldn't remain trapped together in such misery. I did not know what to do.

I pondered upon this as my days in Glastonbury dwindled, with no new masterpiece falling in the fields. There was, however, one spectacularly botched hoax—an attempt at a gigantic spiraling pattern of braided rope. During the construction, the hoaxers had gone off in the wrong direction in the dark, mangling their ambitious composition. Realizing their uncorrectable mistake, they had run off, creating a formation that resembled a systems crash. A few days later, however, another pattern appeared using the same braided elements as the flopped hoax, but orchestrating them elegantly. Brown and I visited it one morning soon after it was found. Inside it, we both felt the mysterious presence that seems to define "genuine" formations, with delicate, humorous touches in the way nests had been formed and in the "lay" of the crop, which had been carefully latticed in spots. We wondered if this formation was declaring open lines of communication between the circlemaking intelligence and the human hoaxers.

While in the pattern, we met a balding, middle-aged UFO researcher laden down with video cameras and tape recorders—an annual visitor to the cloud cuckoo land of the crop circles—who said he had watched a "ball of light" whizzing above the formation earlier that morning. As we spoke, a large black helicopter—exceptionally sleek and high-tech, presumably government-issue—shot over the hill and hovered over our heads, then gyrated back and forth across the fields, as if seeking out something. Allan and I exited this formation, laughing helplessly—our visit seemed to sum up the Chapel Perilous of the crop circles perfectly, which seemed to amplify cognitive dissonance as an operating principle.

As I tossed sleeplessly in my bed that night and into the dawn, I once again examined the phenomenon from every conceivable angle—and then, in one quick flash, I had my answer. The resolution was akin to a Zen satori, in which long meditation on an insoluble koan—"What is the sound of one hand clapping?"—leads to a sudden revelatory leap. I understood what the circles were, at least from my perspective. They were a teaching on the nature of reality—an instruction, and initiation, for the modern mind. This teaching was multilayered, many leveled, exquisitely subtle. One aspect, overtly indicated in many of the images as well as the geometrical analysis teased out by Allan Brown, pointed toward our difficult task of integrating "Heaven" and "Earth," spirit and matter, modern scientific knowledge and archaic esoteric wisdom, masculine and feminine, solar and lunar, in a truly nondualistic framework—the "Sacred Marriage" of alchemy, giving birth to the "divine hermaphrodite," melding Hermes and Aphrodite.

At the same time, the patterns were imparting specific cues on navigating the psychic nature of our new reality as we transitioned into it, as well as implicit guidance. Reviewing what I had gleaned, I noted how the formations were subtly and exactly responsive to the mind-set of each individual who investigated or became involved with them. Skeptics became further confirmed in their skepticism—they ended up in unconvincing crop circles, sought out and found the hacked or broken stems, uncovered hidden signs of hoaxing wherever they looked. Taking a mean-spirited pleasure in bamboozling others, hoaxers became more cynical and bullying in their attitude through their activities. "Believers" received clear, if subjective, evidence—paranormal occurrences, magical healings, synchronicities, sightings of balls of light, and so on—that the phenomenon was "real." New Age meditators and psychic channels found their light bodies and extra DNA strands activated, receiving "spiritual transmissions" from the patterns. Those who were open to the phenomenon but unsure about its origin were presented with incongruous aspects supporting both skeptical and nonskeptical perspectives, leading to deepening cognitive dissonance, and eventual illumination. The crop circles were demonstrating, in tangible form, how the individual creates and reinforces reality through his conscious intention and activity of interpretation.

They were also a teaching on discrimination, a lesson in the art of paying attention. It was only through meticulous study and dedication to details that the subtle nuances of the formations came into focus—such as the care put into the swirled "nests" at the center of the circles, which often appeared as if braided, strand by strand, by teams of itinerant fairies. As a teaching, the crop circles revealed a methodology, based on the essential importance of concrete details and specific qualities over preconceived ideas, suppositions, and abstract theories.

On many levels, they mirrored the perspective of the philosopher Jean Gebser, who prophesied a shift from mental and conceptual thought—the ingrained metaphysics of materialism—to a multidimensional realization, a translucent awareness, not denying this reality but taking into account the spaceless and timeless origin, Jungian Pleroma or aboriginal dreamtime, in which the fractal finite reflected the unfolding infinite. "Time-freedom is the conscious form of archaic, original pre-temporality," Gebser wrote. The deeper one explored their paradoxical nature, the more it appeared as if the crop circles toyed with our ideas of linear causality. They seemed to descend from another matrix, beyond our conventional understanding of space and time. The formations indicated that the emergent form of a new causality, a new temporality, would not be Newtonian cause-and-effect, but the synchronic domain of "transformations and unfoldings," described by the physicist F. David Peat—what Gebser termed "exfoliations and crystallizations."

According to Eastern thought—as verified by one interpretation of quantum physics—the universe is an emanation of mind. As human consciousness evolves in an accelerated spiral, we are being compelled to realize that our minds are manifesting reality to an ever-increasing extent—our collective shadow-projections of wasteful technologies, wars, and weaponry reflect the subtler interior regions of our psyche and the discordant deceptions in our intimate relationships. If this interpretation is valid, it forces upon us a concomitant responsibility, a grave burden. Gebser realized this with clarity:

> All work, the genuine work which we must achieve, is that which is most difficult and painful: the work on ourselves. If we do not freely take upon ourselves this pre-acceptance of the pain and torment, they will be visited upon us in an otherwise necessary individual and universal collapse. Anyone disassociated from his origin and his spiritually sensed task acts against origin. Anyone who acts against it has neither a today nor a tomorrow.

As an expression of the "daimonic realm" of the soul, the crop circles function as a mandalic gyre, mimicking the sidereal or spiraling movement, or return, of consciousness to the center, expressed by the ancient symbol of the ouroboros, the snake devouring its own tail. Through this same corkscrew motion, the patterns seem to be showing the requisite method for interacting with other levels of galactic intelligence—not through the direct transmission of deadening fact or technical statistic envisioned by SETI or NASA, but through allusion and intuition, humor and play. Wouldn't higher intelligence prefer to express itself through higher art—an art, perhaps, inseparable from life—rather than the dull technical details demanded by the reductive rational mind?

It also seemed to me that the formations offered instruction in how to work with paradox. Instead of negating the conflicting aspects of a paradox, you advance your understanding when you can hold both sides of a dichotomy in your mind at the same time, reconciling rather than negating. In my initial satori flash, I realized one could embrace paradox as an operative principle, rather than seeking to deny or dissolve or dilute it.

Prodded by this view of the crop circles, I saw a vast vista opening before us—a sacred investigation, merging science and art, dedicated to exploring the puzzle-prisms of psychic reality. Rather than seeking to resolve dualisms and institute some grand "Theory of Everything," the "science of the imagination" would embrace and explore paradox, going deeper into conundrums, relinquishing delusory attempts to achieve certainty. We may find, as Rudolf Steiner proposed, that thought and language are creative aspects of being, tools for transforming reality. Superstring physics describes a universe of nine, ten, or eleven dimensions. If the universe is actually a projection of subtler levels of the psyche—a loom of maya—then we may discover that the vibratory lattices of interpenetrating worlds elaborated by current physics are descriptions of the psyche itself, in its fully ensouled unfolding.

Perhaps we incorporate this manifold field of the psychocosmos into conscious awareness as the various intervals and intensities experienced in non-ordinary states—induced by psychoactive substances, meditations, lucid dreams, trances, and so on. As experiential instruments in this exploration, psychedelics—those disgraced and derided "toys of the hippie generation," according to *The New York Times*—open "lines of flight," ingressions across the extradimensional matrix that is the psyche itself. When we fully accept, and finally realize, the reality of the psyche, exploring these areas will be recognized as essential to expanding the parameters of human understanding. Instead of grasping at dry straws of closure, we will open new doorways and enter infinite new realms.

PART SIX

THE LORD OF THE DAWN

In the church of my heart, the choir is on fire.

MAYAKOVSKY

CHAPTER ONE

For a superior soul the stars are not only prognostics, but the soul is part of and evolves internally along with the All, in which it participates. . . . The adept persists in wanting that which he must do and, in this persistence, he will do only what he wills, and nothing else but that which derives from the idea in itself.

PLOTINUS

A week after returning home from Glastonbury, I went to Nevada for the dust-blown desert festival, the collective shamanic transformer, that is the one-week-a-year apparition of Black Rock City. I found myself there in a fraught, fugitive state. Like any street-corner prophet, I believed I had absorbed a complex of ideas that needed to be transmitted as quickly as possible, that were entirely crucial to the survival of the species, yet outside the synaptical firing range of what most people could even conceive. At the same time, I felt suspended above a personal vortex, despairing over my inherent incapacity for wholeness or happiness, plagued by excess desires that threatened the family life I had made.

Mars, planet of War and Will, was the surprise celebrity guest at Burning Man that summer. The red globe, laser-bright in the sky, outshining the waning Moon, was making its nearest approach to the Earth in sixty thousand years—reaching its ultimate point of proximity midweek. The alchemical vision considers the planets to be ontological states and aspects of the psyche that can be assimilated through esoteric techniques. Having wrenched myself open to the occult, I now seemed to be performing a series of precisely timed, almost ritualized actions through pure intuition, as if magnetized by the pulsations of the stellar bodies. In retrospect, everything that happened to me during that gathering, and for months afterward, seemed guided and ruled by that fierce deity, as if I were being forcefully initiated into his combative sphere.

I considered Burning Man to be a fulcrum for the evolution of consciousness on the planet, where cutting-edge scientists consorted with neo-pagan warlocks, and arcane bits of knowledge were exchanged at hyperspeed. But it was also a carnal carnival, an evanescent parade of erotic possibility. Nakedness, body painting, bizarre costumes, patterns of piercings, and every form of extreme self-modification became commonplace sights after a few hours at the gathering. As a laboratory of the contemporary freakster psyche, the gathering was not all sweetness and light. Much of the sexuality on display had a banal, burned-out character. Endless camps set up booths for making prints of women's breasts, organizing "beaver-eating" contests, anonymous feel-ups, and so on, while blaring loud techno or rock. This desperate edge—sleazy antipode to the easy avowals of New Age spirituality made by countless Hindu statues, Buddhist prayer flags, and Native American smudge-sticks—suggested one Kali Yuga end point, described by the occult writer Julius Evola, a stone guest in my mind, where "sex, being reduced to its content of mere sensation, will only be the misleading, obscure, and desperate alleviation of the existential disgust and anguish of him who has stumbled into a blind alley."

I stayed at a camp of West Coast journalists and programmers, dedicated day-trippers who spoke the abstruse lingo of obscure research chemicals and dosages, brain chemistry and the timewave, pranayama and bardo states. At a lecture series where I also spoke, the women in my camp gave a talk about an experiment they had made, administering psychoactive compounds through vaginal suppositories. When I tried to interest them in attending indigenous ceremonies, using plant-spirits as sacraments, they seemed to consider the idea oddly obsolete and uncool. Speaking with an organizer of cutting-edge conferences on altered states and technology, I described a Native American Church ceremony I had attended with my family, organized by the teacher—we had spent eighteen hours in a sheltering tepee, sitting around the fire, listening to the community's prayers and eagle-songs, feeling ourselves privileged to be woven into their sacred world. After the ceremony, my partner had experienced a deep healing, a release of the pain she felt over our daughter's condition. I told him how the "Road Man," the presiding shaman, addressed the peyote button with reverence and gratitude, saying, "Thank you, Peyote, for helping me, for healing me." I had wept, witnessing this—a grown man humbling himself before a gnarled stub of cactus, accepting its deeper wisdom.

"Yeah, but isn't that kind of like playing Indian?" the conference organizer sneered.

"Not really. I think it's more like playing human," I replied, surprised by his sneer.

He recalled his latest forays with "Special K," his ego-disperser-of-choice. He found ketamine's skewed cartooniverse fascinating and illuminating. I wasn't convinced. I felt my campmates were in danger of becoming know-it-alls without soul, reducing this essential area to a hedonic hobby, a means of mind-freaking, melting down boundaries while keeping the old-fashioned ego in charge, with no humility, no surrender to the sacred. The avowed spirituality of West Coast hipsters, which appeared so glamorous and enticing to me at first, increasingly seemed a shallow lifestyle choice—a new form of self-congratulatory consumerism, a better way to get laid. Tattooing the Buddha on your ass was easier than pursuing the eightfold path to enlightenment. Eating a diet of deliciously organic raw food hardly counted as an act of meaningful renunciation. Dressing in fake Day-Glo furs and dancing at all-night trance parties was not impressive as a spiritual discipline. I had the uneasy feeling that the fluid "temporary autonomous zone" created by Burning Man had congealed into a new form of social conscription—a detached hipster hedonism with grotesque and unearned spiritual pretensions. No longer challenging themselves, "Burners" had settled into a new comfort zone. Bicycling around the art-strewn desert, past half-submerged unicorn torsoes and giant urinals, for the first time I found the energy at the festival to be adolescent and stagnant—better than what was found in the mainstream, perhaps, but not good enough.

But what would be good enough? Why was everything so dissatisfying? Where was the secret road to the open heart—the resurrected human soul, the communion so longed for, ever postponed?

After my first visit to Burning Man, I had realized it represented something long lost to the Western psyche—an archaic revival and initiatory ritual, like the Eleusinian mysteries of ancient Greece, through which we were groping our way toward a different relationship to the cosmos and one another. Beyond its art-school pretensions and goofy gadgets, the festival seemed a dress rehearsal for some still-unknown and unnamable new way of being. Earlier attempts to restore an initiatory element in the Modern West had been made, and failed.

According to historian Francis Yates, the goal of the curtailed "Rosicrucian Enlightenment" of the seventeenth century was to reinstate a lost initiatory discipline within European Christianity. The movement was orchestrated by a coterie of alchemist-intellectuals—some of whom, such as Isaac Newton and Francis Bacon, became instrumental in founding the modern scientific method. "Rosicrucian thinkers were aware of the dangers of the new science, of its diabolical as well as its angelical possibilities, and they saw that its arrival should be accompanied by a general reformation of the whole wide world," wrote Yates. This brief Rosicrucian effort to return Christianity to its origins—as personal instrument for inner illumination—was quashed by the combined forces of Church and State. The alchemical quest was banished underground, barely surviving in the hermetic margins of the modern West. It rose aboveground again, in force, with the rediscovery of shamanic plants and psychedelic substances in the twentieth century, which forced the material-obsessed Western mind into a direct encounter with repressed psychic realities.

I had often meditated upon the discovery of LSD, which seemed to unveil a startling alchemical parable—one that was, I now suspected, synchronized with the activation of the noosphere in 2012. Lysergic acid diethylamide was first synthesized in 1938 in Basel, Switzerland—center of medieval alchemy, where Paracelsus practiced in the sixteenth century—by Albert Hofmann, a young chemist working for Sandoz. One of a series of compounds developed from ergot fungus, LSD-25 appeared to have no effect, and was shelved and forgotten. Five years later, Hofmann had "a peculiar presentiment," as well as dreams, about LSD-25, and went back and resynthesized it—the first time he had ever re-created a seemingly "uninteresting" alkaloid. Switzerland was an island of neutrality within Nazi-dominated Europe, and in 1943, the Final Solution was in full swing, with tens of thousands of Jews exterminated each month. On April 16, 1943, Hofmann's accidental ingestion or absorption of LSD-25 led to the first LSD trip, which began as he bicycled from his laboratory to his home. Three days later—the day the Nazis began the street-by-street evacuation of the Warsaw Ghetto—he took the compound again in an intentional self-experiment, definitively correlating the strange effects he unleashed with the ingestion of the alkaloid. At the moment of truth and tragedy for the modern European psyche, the seed was born that would briefly flower in the mystical renaissance of the 1960s, before dispersing to the winds yet again.

In this juxtaposition of gas ovens and gnosis, Eros and Thanatos, I seemed to see the sky-dance of Kali, her violent and voluptuous presence arching over our modern world. In the Hindu religion, Kali, the dark mother who liberates through decapitating, is considered the wrathful aspect of Shakti, deity of the sexual energy that powers all manifestation. Shakti is the consort of Shiva, representing immutable consciousness. As Julius Evola—himself a formidably shadowy figure, a Sicilian baron and former Dadaist who started an occult movement in Italy in the 1920s that worried Mussolini, was courted by the SS, and paralyzed by shrapnel while strolling in Vienna at the end of World War II—wrote: "Shiva is present in the unchangeable, conscious, spiritual, and stable aspect, while Shakti is present in the changeable, unconscious-vital, natural, and dynamic aspect of everything that exists." Shiva and Shakti maintain the balance of forces in the universe through their eternal act of copulation. Although I had contemplated the prophetic and destructive dimensions of the Kali Yuga—the end of the cycle of four long ages, according to Vedic lore—I had not fully connected this to the goddess Kali as active archetype and suprapersonal force. With its masks and fire swirls and suicide girls, its frenetic dispersions and kinetic dissipations, Burning Man seemed a playground for Kali—"all-powerful in the moment of the dissolution of the universe," according to Evola—an initiation into her wise and wicked ways.

On the night that Mars reached its closest approach, I took a fungal sacrament with an old friend of mine from New York, a writer and theorist on potential uses of the Internet to create "augmented social networks," linking progressive causes and affinity groups. After years of trying, I had finally prevailed upon Ken to leave his desk and visit the festival—he seemed stuck in psychic patterns that repeated like tape loops, alienated from women and other people by his obsessive relationship with his work, exhausting himself in front of his computer, day after day. Walking the perimeter of the stage-set city, in the "second attention" of our altered state, I tried to get him to see how he cut himself off. Before leaving New York, Ken dreamed repeatedly of an old bearded wise man who told him he was facing a crucial decision. He realized that I was presenting him with his choice—he could cling to old nostalgia and deep-set habit, or he could liberate himself, detach from the grip of his past. Our conversation seemed to trigger something—I brought him to a friend's camp, and he went off with a woman there, spending the rest of the week in her company.

Alone again under the Milky Way's cool parasol, I went back to my tent, then visited our neighboring village—a polyamorous camp of Silicon Valley software engineers and Microsoft project managers and their friends and lovers. Hours after the trip should have ended, I still found myself significantly altered; closing my eyes, I beheld a monstrous entity—a Lovecraftian caterpillar creature with multiple heads and mutable human faces—that recalled the Martian spirits I had witnessed months ago, at my hotel in Oregon, while visiting José Argüelles. A woman stretched out on a couch near me noted that her entire body seemed to be softly vibrating—and the same thing was happening to me, a subtle current, pulsing in my hands and all through me, that I had not known before. I went for another ride on my bicycle, drifting around various twirling sculptures, blaring rave camps, and swaying trance-dancers at distant intervals.

Circling the Man, a fifty-foot-tall statue at the center of the city, euphorically exploded on the gathering's last official night, I began to receive what seemed to be a cosmic transmission, related to the Kali Yuga and the Mars conjunction. Some Hindu sects believe the dark age stretches on for thousands or hundreds of thousands of years. I had heard that other traditions, especially among the Dravidians in the South, state that the entire cycle of four world-ages lasts sixty thousand years, and we are approaching the end of the Kali Yuga, which would unsheathe a new Golden Age. There had been a sixty-thousand-year interval since Mars last made its closest approach to the Earth—was the dramatic return of this luminary somehow linked to the completion of the cycle? It seemed to me that this could be the case; the pulses were a field-effect of this psychophysical event, changing the resonant frequency of the Earth, like the retuning of a harp. The return of Mars was a phase-shift, part of the process through which our planet was becoming, by subtly intensifying degrees, less materially dense and more psychically responsive. This was my transmission.

Dawn was spreading over the desert as I reached this conclusion—a Narnia sunrise of golden cloud fingers and taffeta swirls feather-spinning across the horizon. Still circling in my bicycle, I began to let out ecstatic calls—whoops and sibilant cries—that seemed to belong to some primordial language or nonterrestrial tongue unknown to me. I let the sounds pass my lips, observing them from a distanced level of mind, as if they were the products of an other. I wondered, idly, if I was losing my moorings. I felt triumphant, miserable, euphoric, alone.

Over the next days, I sought to intensify my far-out state. I stopped eating, gave up sleeping, and engaged in consecutive reenactments of Albert Hofmann's celebrated bicycle ride. I was coming to a realization of the pattern of my past, seen from a different angle than any I had explored before. I was tracking the wounds that Kali, the daimonic feminine current ruling our current aeon, had left like claw marks in my flesh, soul, and spirit. My partner's fury seemed only the latest example, one note in a long dissonant symphony. My inescapable longing for some form of ecstatic communion denied to me seemed shaped by my early history of melancholy deprivation.

As a baby, I was never breast-fed—my hardworking mother, pressed for time and lacking guidance, accepted the confident "wisdom" of modern medicine in the mid-1960s that formula might even be healthier, later significantly disproved. I broke bones frequently in my childhood and adolescence, developed allergies, and succumbed to a life-threatening bacterial infection of the spine that caused me permanent damage, cutting me off from the physical world. My early relationship with my grandmother also had lurid aspects that I now recognized as a form of sex abuse. During my frequent sleepovers at her house, she was compulsively obsessed with the regularity of my bowel movements, giving me enemas—over my strident objections, screams, and cries—when I did not evacuate according to her strict schedule. I sensed, lurking behind this, the suicide of her father, the deprivation of the Depression years, her frustrated intellect and unspoken yearning for what was denied her, her own loveless marriage and failure to control her rebellious daughter, culminating in a wraith's wrathful fury that demanded some form of sensual satiation even if it took the form of sadistic infliction upon her helpless grandson. Her dark craving for life—her fury—was impressed upon me, and never left.

In the feminist climate of the early 1970s, men were to blame for everything—for the patriarchy, for the dominator culture that warred and raped women and despoiled "Mother Earth," for their wandering eyes. Women were the helpless victims. Separating from my absent father, unable to find another partner, my mother was depressed—often desperate, sometimes despairing—throughout my childhood, and I internalized her grief, bore the full brunt of it. As a male child, I felt guilty for all the ancient failures of men. My father seemed a symbol of this, guilty for turning his back on us—it was only much later that I learned he had never wanted a child, warning my mother beforehand that he was not capable of changing or compromising his work or his lifestyle. In fact, the powerful adult figures in my life were all female. The men I knew—my father's painter buddies, my mother's writer friends—seemed centerless and disempowered, rootless compared to the women. I grew up without any meaningful vision, any convincing role model, of what it should mean to be a man.

Sexuality is a rhizome that draws energies from deep subterranean streams of the psyche, its hidden flows and vortices. My entry into sex and relationships was slow and awkward, tinged with shame, haunted and dispossessed. Feminism informed me that men and women should be equals—that gender was a social construction—and that male aggression was bad, yet in practice I had found it was the forceful and transgressive act, breaching and grasping for the feminine essence, often after obliterating my oppressive self-consciousness through alcohol, that had proved effective. The curiously deceptive and polarized elements of the feminine psyche seemed to compel such behavior from me—women expressed themselves one way in language, but continuously acted another way in reality. Shakti gave lip service and coos of sympathy to the man of good intentions and kind words, but surrendered herself without a thought to the one who brutally and contemptuously pushed his way forward. Far more often than not, the daimonic feminine deity draped herself over men who had amassed power and wealth in our unscrupulous world, but imparted condescending glances to those who sacrificed or risked themselves in the hopes of instituting a better one. When I peered across my life's panorama, I now saw how, through willfully unconscious misuse of Shakti, women were equally culpable for the disastrous world we had made.

At the end of a memoir, my mother wrote: "If time was like a passage of music, you could keep going back again until you got it right." I believed I still had time to get it right. When I reviewed my own past, my erotic life until that point seemed immature and one-dimensional. Inattentive, acting without consciousness, I had allowed the forces around me—the weight of social convention, the constricting models and limited language that defined current possibilities—to shape my romantic relationships. I had never defined my own truth about love. Sigmund Freud had asked, "What do women want?" I had never asked—what do I want? If my nature was deeply erotically curious, it was a metaphysical curiosity as well as a carnal one. I sensed a veiled and concealed essence lurking within Eros itself—something that needed to be brought to consciousness and confronted.

Sex is "the greatest magical force in nature," Evola wrote in *Eros and the Mysteries of Love: The Metaphysics of Sex,* a scathing, eye-opening tome about its potential uses and habitual misuses. He noted that sex was always not just a physical but a metaphysical act. Unlike animals, "Man can desire and make love at any time, and that is a natural extent of his love. It is in no way an artificial fact or 'corruption' derived from his 'separation from nature.'" Our freedom to love, or withhold love, reveals the spiritual, or suprapersonal, basis of the human sex drive, which can be utilized for reproduction, manipulated for malevolent ends, or driven inward and upward as the basis for illumination. He proposed that modern civilization was secretly "gynocratic" in character: "The modern world knows of women who wish to be emancipated materially and socially from man but not of men who feel they want to be emancipated inwardly and spiritually from women." Contemporary humanity believed, almost without question, in "this great ignominious dogma of the moral superiority of women."

When men behaved badly, they were like apostates from the faith, rebels but not revolutionaries—"in desecrating women, they still worship her, although negatively."

In tribal societies, boys must pass through initiatory ordeals in order to become men, confronting fear and accessing inner resources of courage and ferocity. In our culture, men do not receive initiation into non-ordinary states of consciousness and spiritual responsibilities. Tied down by the possessive entrapments of love, marriage, and children, the "family men" I knew forfeited any opportunity to explore the edge-realms of freedom and spirit that belonged to the solar masculine principle, fighting for truth and meeting danger with intellect, strength, and courage. They abandoned themselves, instead, to the cushioned realm of materialist comforts, numbing themselves with sports and media spectacles. This did not represent their "success" in the world, as they believed, but their capitulation to the negative pole of the feminine daimonic, Kali-as-destroyer. Noting that "materialism" and "mater"—"mother"—shared the same root, Jean Gebser had written: "It is almost as if the material-crazed man of today were the ultimate victim of the avenging mother—that *mater* whose chaotic immoderation is the driving force behind matter and materialistic supremacy."

I had always accepted, without question, the moral superiority of women. I believed that the yearnings for other women that predictably overwhelmed me, after a while, in any relationship, were, by their nature, wrong—a sign of my vile maleness. Like almost everyone I knew, I had been, by default, a "serial monogamist," staying with one person until my desires for others shattered my ethical scruples and overpowered my love for my partner—I had also been jealous and possessive toward my partners, seeking to control their desire. In some cases, I had cheated on them while doing my best to keep them from doing the same. I now realized there was nothing intrinsically wrong with these desires—or their fulfillment. What seemed wrong was a form of love based on possessiveness, jealousy, and mutual mistrust. Why couldn't love be based on generosity and openness, instead? Why couldn't we love someone for the way they were, instead of the way we wanted them to be? Why couldn't we even want the person we loved to have whatever experiences—erotic or otherwise—that they desired? Our model of love seemed hideously deformed and unsustainable—who was kidding who?

Of course what I was considering was outrageously utopian—everyone knew that "free love" was an experiment that failed, badly, in the 1960s, and in earlier utopian experiments in nineteenth-century America and elsewhere. However, as I mulled it over, what everyone forgot when they proclaimed this was the nature of experimentation—in love as in science, art, or politics, or anything else. The first experiments usually fail—the chemist blows up his laboratory, the artist ruins his canvas, the nation-state collapses—but over successive attempts, alterations are made, new combinations tried, until finally somebody gets it right. Suddenly something barely imagined before—plastic or Cubism or democracy—not only exists in the world but quickly becomes widely available. It seemed to me that at Burning Man there were communities making profound and positive steps in the right direction, deprogramming themselves from negative sexual imprints, such as the Seattle camp next to us.

In the grip of these raucous realizations, I found myself acting out in unpleasant ways. I sought out a woman, Megan, a Bay Area urban planner I had connected with at the last year's festival—a night that helped compel my slide down this slippery slope. We had traveled with a group of friends, riding with Albert Hofmann, and we had ended up tangled together in a "chill space," a pillow-strewn dome behind a rave camp. We did not go past a certain point—but even our limited intimacy, in that altered state, was incandescent, outrageous. I was entering different visionary states that seemed related to our positions and touch. When I closed my eyes, I entered a thrumming ocean of supersensible entities that seemed to be sharing in our communion, sensitive to it—even exulting in it. It was as though these spirits were pressing in on us, prismatically shape-shifting with our actions, from the other side of the psyche, the imaginal realm.

This inadvertent exploration had revealed the possibility of conscious use of sexuality to go beyond sexuality—into initiatory states, magical flights, synesthetic undercurrents. According to Evola, beyond ordinary Eros was Tantric practice, redirecting the most powerful force that shackled us to the physical plane—lust leading to procreation—for purposes of spiritual self-creation. As one Tantric text declared: "The world, being subject to passions, may achieve liberation only through them. As copper is transformed into gold through alchemical practices, likewise, those who have gained knowledge use passions as the key to liberation."

Outside of the East, Evola found traces of these practices in secret Qaballistic strains and in the ancient rituals of the Eleusinian Mysteries, represented, as well, by the state of the "divine hermaphrodite" sought in alchemy and Gnosticism. The goal would be the mastery of the sex impulse through esoteric techniques—"the purpose of such transformation is not to 'heal' a sexual neurotic struggling with his complexes, severe or mild. Instead, its purpose is the transcendence of the human state, real regeneration, and a change of ontological status. The transmuted force of sex should lead to this end." In contemporary "self help" and New Age manuals, Tantra had been debased into techniques for deeper orgasms and increased sensuality—this was the opposite of Tantra, in Evola's view. "Tantrism does not intend to empower human nature to the highest degree, but rather wishes to cauterize it," he wrote, "thereby consuming the individualistic I and its hubris in an attempt to overcome the human condition." This was the arena that I longed to explore further.

Burning Man seemed an inchoate pressing toward the left-hand path of Tantra, overcoming "the antithesis between enjoyment of the world and ascesis, or yoga, which is spiritual discipline aimed at liberation." While clearly not a path for everyone, the development of "a brotherhood of Tantrics" was prophesied in certain Hindu texts: "This brotherhood will awaken as the end of the Kali Yuga approaches. Recognizing the potent feminine principle of life, the Brotherhood of Tantra will transform this polluted world. Then, at the ecstatic moment when one Age transforms into the next, those faithful followers of the selfless path will reach their goal." Such a direction was also indicated by the "crazy masters" who arrived in America from the East in the 1970s, such as Bhagwan Rajneesh, more popularly known since his death as Osho, and Chogyam Trungpa, who founded the Shambhala Training. Their teachings pointed toward a discipline "that allows one to be free and invulnerable even while enjoying the world," as Evola put it, "or anything the world may offer." Osho argued that sexual liberation was a necessary prelude to spiritual progress—that enlightenment was inconceivable as long as seekers were hung up on sex.

Since last year's festival, I had passed through San Francisco several times, and attempted, each time, to see Megan. We ended up meeting in rushed gatherings with other friends, and it was clear she was not willing to spend time with me—I didn't blame her for this. She knew about my family in New York. Now, in the hypercharged container of Burning Man, amidst dancing multitudes and fiery spectacles, we set out again on an evening voyage, entering the "second attention," bicycling around Black Rock with her campmates. In my altered state, I could sense the volatile, flickering movement of Shakti within her, compelled toward me and then pulling away, desiring me and enraged at me at the same time. Out of frustration, I pushed too hard—grasping—and alienated her. The next afternoon, I went back to Megan's camp, asking her to take a walk with me. Strolling out to the Man at sunset, I apologized for my behavior and tried to explain what was going on with me—I felt I was spiraling through some perilous psychic passage that was out of my control—and I asked if she would help me, accompanying me for the rest of the festival. She said she felt it was "not safe." I couldn't disagree with her—yet I wondered if being "safe" was what Burning Man, or life, was all about.

During the last days of the gathering, I attained a uniquely far-out state of mind. I stopped eating as well as sleeping, pushing myself past the edge. Now and then, I would rest in a kind of hibernating state for an hour or two, as fizzing images traced across the flittering field of my closed eyes, until I was able to get up again. Since the night of the Mars conjunction, a new element of my psyche seemed to be taking shape. I was in a kind of agony, a torment of self-hate; at the same time, I never felt better. I had forced my consciousness into a previously unknown intensity. I felt more furious, and more free.

The Man was ceremonially exploded on Saturday night, shooting out showers of sparks, dust devils, and smoke clouds as it collapsed. I watched it, alone, from a far distance. In my first visits to the festival, I had been entranced by this spectacle—this time around I found it another distraction, another way that people got stuck, their desire for consuming radical experiences just one more desire to add to the pile. It seemed to me that the dangerous exchange of human truth would be far more explosive and incendiary.

As pink-feathered dawn brushed away the night sky, I found myself riding on an art car—a flatbed truck festooned with shiny steel flowers—with a bunch of friends from New York. A girl produced a fistful of pills and suggested I try them, pressing two into my hand. I asked her what it was and she said it was called Provigil, used by the U.S. military to induce wakefulness. She said it was very popular among college kids, as well as being given to our soldiers in Afghanistan and Iraq. I had been awake for days straight and knew it was a terrible idea to take this, but—what the hell—did so anyway.

Provigil sharpened me up but it had another effect—it shocked me with surges of high-voltage anger. I walked around with my friend Steven, with whom I had taken ibogaine in Mexico, in the burning midday heat. We ragged on the festival's pseudo-spiritual pretensions. "Fuck this place," I said. "Fuck these people." I repeated this several times until I heard myself—such language was out of my character. The pharmaceutical was working on me—the drug had a mechanistic, soulless vibration. It was evident that Provigil was perfectly calibrated to turn nineteen-year-old army conscripts into hostile killing machines. This crusade seemed to seal the martial/Martian ambience of my week.

I returned to my camp, centering my thoughts on my relationship with my partner; I realized I was going to have to change it. Clearly, she had reasons to be furious at me, and yet I still loved her. Nothing could work between us unless we could be truthful with each other. Part of my truth was that I was not monogamous, even if I wished that I was. Monogamy was not in my nature, and it never had been. Whether or not she could accept me as I was, I would have to find some other way to love.

The desert was displaying one of its finest illusions—a fierce windstorm, blowing an incredible blizzard of fine playa dust, the motes illuminated into blinding whiteness by a blazing Sun somewhere far above us. Ken and I biked to the Center Cafe, a beautiful, open-air structure at the center of the festival, serving coffee and chai. Unable to see more than four feet in front of us, we wobbled in the wind. We parked and entered the cafe, guided by sound more than sight—by the loud whipping of air through the tall banners circling the open interior. I told Ken I felt like I was dematerializing, turning into a phantom. An area near one shoulder was sore and painfully throbbing—when I touched it I found a strange lump. Painful shivers ran along my spine. As we approached the line for coffee, a tall blond woman, wrapped in a sky-blue sari, appeared in front of us, emerging from the white dust, like an apparition. She turned toward me and recognized me.

"We met exactly at this spot, exactly one year ago," she said. I recalled the event—Natalia was a Russian healer and energy worker, and I met her with her sister and her sister's boyfriend. I had told them about my book, and given them an excerpt. All of them were at the cafe once again, and we sat down together, as the wind was dying down. They said they had read the book, and it had inspired them to explore ayahuasca in Hawaii, where Natalia lived with a community of healers. She pulled out a large crystal orb that shimmered in her hands, twirling it in the air, and talked about the new planetary body, Sedna, that astronomers had just discovered on the distant edge of our solar system. I asked if she could look at my back. She traced the sore spot with her fingers.

"You have a blockage at your assemblage point," she said seriously. "You have to be very careful about this. If you don't take care of yourself, it could quickly explode and destroy your physical body. You need immediate rest and quiet and body work when you get home." She gave me a Reiki attunement, and her sister, another healer, also massaged me. I told Natalia about my family situation and my rejection of monogamy. She said that I had to do what was right for me—that I would discover resources I didn't suspect. I felt she was channeling, giving me permission from some higher source of intuitive feminine wisdom.

The assemblage point—a location on the back, near the upper spine, that allows the sorcerer to move between different levels of reality—was a concept from Carlos Castaneda's books, thought to be of suspicious provenance. Discussions of the factual or fictive basis for the assemblage point were now academic, however, as I could feel it, like a small whirling vortex near my spine—my whole orientation to reality seemed off-kilter, swirling around that sore point. I felt that Natalia was correct, and I was in a dangerously depleted condition. During the week, I gave talks on the crop circles, and on the implications of the end date of the Mayan calendar. Whenever I spoke with someone, I tried to move their "assemblage point," introducing them to the information I had amassed. I gave out far more energy to people than I allowed myself to receive in return. This, coupled with my rash overindulgence in certain substances, had caused damage.

Ken and I were planning to stay another night. I asked if we could leave right away. Seeing my overwrought state, he agreed. We returned to the campsite and packed up. Strange groanings started up deep in my gut, rising into chants and maniacal laughter. From deep within, I would pull up a small amount of phlegm and spit it out, repeating this activity again and again, as though I were seeking to dislodge some poison. My normal consciousness was not reduced during this, but seemed to hover above, allowing an older wisdom to work within me. I seemed to be channeling a technique of shamanic self-maintainance while under duress or magical attack—or tuning into something I had known, in some other form. I found myself praying out loud, in the style of the Road Man from the Native American Church, asking the "Creator Spirit" for support, apologizing for the way I had assaulted my body and mind.

"Thank you for this lesson, Creator," I said. "From now on, I will keep my sense of humor and my balance about things. I won't press so hard. If anyone wants to know what I think, I will tell them, Creator. But that is it. I won't force the issue anymore." Within a few hours, we had packed up and were on the road to "civilization"—the cardboard tackiness of Reno, where we spent a tranquil night at a Holiday Inn. Looking in the bathroom mirror, I laughed, beholding a wild-haired, weird-eyed, dust-covered scarecrow.

During my Bwiti initiation in Gabon, I had a vision of letters spinning around, rebus-like, to spell an obscure message: "Touchers Teach Too." The meaning of this message had now revealed itself. I was a thinker, and I had been teaching in my way—but touchers teach too. Now I had to integrate their lesson.

CHAPTER TWO

Sedna—the distant "planetoid" on the edge of our solar system, discovered by astronomers in 2003—was named for the Inuit goddess of the deep seas. In a literary story created by Clarissa Pinkola Estés, author of *Women Who Run with the Wolves,* Sedna becomes "Skeleton Woman," who is thrown from a cliff by her father, whom she has displeased. She sinks to the bottom of the sea, preserved as a skeleton, until a fisherman catches her and brings her up to the surface. In his attempt to flee from this frightening apparition, he only drags her along. Back at his home, he overcomes his fear of her. Feeling stirrings of compassion, he carefully untangles her bones and her long hair, and then falls asleep. While sleeping, he cries a single tear. Skeleton Woman crawls toward this tear and eagerly drinks it up. She reaches into the fisherman's chest, takes out his heart and beats upon it like a drum, calling out, "Flesh! Flesh! Flesh!" As she sings, the flesh returns to her bones; life returns to her. She sings the clothes off the fisherman, replaces his heart in his body, and lies down next to him, "skin to skin." They awaken the next morning, wrapped around each other tightly.

This startling and beautiful myth may be an essential story for our time—an instruction manual for the return of the goddess, denied and defiled by thousands of years of patriarchal civilization. It is interesting to note that the newly discovered planetoid, Sedna, is currently making its closest approach to the Earth in the last ten thousand years. During the last three centuries, women in the West and other modernizing countries finally escaped a status equivalent to slavery or property. We are quick to forget how far we have come in this short period of time.

We easily forget, as well, that across vast swathes of the Earth, women are still in bondage, treated like serfs, denied their basic rights and freedoms. They are systematically raped during genocidal wars such as the recent Bosnian conflict. In India, they are burned alive as brides over dowry disputes. In China, female babies are sometimes smothered by parents longing for a male heir. In the less developed world, "women's contribution is immeasurable, intense work that is never recognized as valid, never rewarded with money, and never even considered part of the economy," world hunger activist Lynne Twist writes in *The Soul of Money*. Even in the West, compensation for men and women remains inequitable, with traditionally female areas of employment such as nursing, childcare, and teaching given short shrift in salary terms, compared to work in the military and financial sectors.

The standard liberal, feminist, or left-wing critique of our society sees it as a patriarchal dominator culture, repressing the feminine and the natural. There is, of course, a great deal of truth to this. As discussed earlier, the lunar, feminine, and intuitive aspects of being were systematically suppressed as small-scale tribal societies gave way to agrarian and industrial civilization, prioritizing solar, masculine rationality. This shift was formalized by the change from a cyclical lunar calendar to a desynchronized solar one, based on the abstract division of a circle into twelve parts. The suppression of intuitive wisdom-traditions was often violent and barbaric, as in the witch-burnings of the Inquisition, or the colonial destruction of indigenous cultures oriented toward magic and nature-worship.

However, as Ken Wilber points out, the "patriarchy," emerging with agrarian civilization, "was a conscious co-creation of men and women in the face of largely brutal circumstances." The new situation was a complicit arrangement, equally bad for men, who worked the plows and manned the battlements during the wars that were the inevitable by-product of population growth and the creation of surplus value. The gender polarization created by civilization "was not a male plot, nor a female plot, for that matter, but was simply the best that these societies could do under the technological form of their organization at the time," Wilber writes. When industrial civilization arose, machines supplanted masculine brute force as the essential means of production. Almost immediately, as machines "removed the emphasis on male physical strength and replaced it with gender-neutral engines," the campaign for women's rights began—Mary Wollstonecraft wrote *A Vindication of the Rights of Women* in 1792, ten years after the discovery of the steam engine.

Two centuries later, we are still seeking to rebalance male and female energies, renegotiating gender roles and relationship patterns. This is one of the underlying tensions of modern culture, and it remains a difficult and demanding process. As high divorce rates and the gossip fixations of the media reveal, the constraints of traditional monogamy are chafing for many people, and alternative models of commitment seem necessary. At the same time, attempts to reach a more ideal or liberated erotic life (sometimes given the unsatisfying label "polyamory") are mired in endemic confusion, hurt feelings, and failure. We are being forced to confront programming and conditioning created by thousands of years of patriarchal domination and what the philosopher Herbert Marcuse called "repressive desublimation," as well as patterns that may be rooted in biology, or perhaps in deep-set patterns of evolutionary psychology.

According to Hinduism and Buddhism, we are currently in the Kali Yuga, ruled by Kali, the dark mother, Shakti's wrathful aspect, who liberates through decapitating. When we explore our current reality through this mythological, archetypal, and suprapersonal prism, we reach a level of insight beyond the stereotypical critique that sees modern civilization as patriarchal and women as victims. In the Kali Yuga, it is the feminine current of Shakti energy that has gone berserk—like an electric line that is cut, shooting out sparks—not the masculine Shiva principle of consciousness. As Nikolas and Zeena Schreck write in *Demons of the Flesh,* a study of sex magic: "During this Aeon, the lunar, sinister current of the Feminine Daimonic is at its zenith, a spiritual condition which allows for the breaking up of all boundaries and the free play of creative chaos, unrestricted by the male ordering principle." The Kali Yuga could be characterized as the goddess Shakti throwing a hissy fit. When such a powerful deity goes haywire, the entire Earth suffers and strains from her fury.

The 1960s provided a dress rehearsal for the return of the goddess. The opening of the Shakti current in movements for social justice and sexual liberation directly resulted from the nuclear arms race and the Cold War, culminating in the 1962 Cuban Missile Crisis, which threatened humanity with sudden annihilation. As the collective psyche recoiled from this prospect, the institution of a planetary culture based on passion and compassion, rather than fear and domination, appeared possible, even requisite, for a brief period, anthemized by the Beatles' "All You Need Is Love." In our own time, further disasters and cataclysms may, unfortunately, be required before Shakti can be fully awakened and freed from her "mind-forg'd manacles," releasing humanity from its negative programming to establish a civilization based on love, generosity, and trust.

While historical civilizations enshrined a monotheistic "God the Father," many prehistoric cultures worshipped the "Great Mother"—Indo-European examples include the pre-Hellenic goddess-worshipping Minoan culture on Crete, and Catal Huyuk, which flourished in Turkey, more than five thousand years before Christ—and the fertilizing forces represented by the Moon. As the polarity of Shakti/Kali suggests, there are two sides to the mother archetype: She can be nurturing, generative, and benevolent, or aggrieved, possessive, and devouring. In the modern world, we became obsessed with material goods, hypnotized by "false needs," possessed by our possessions. This obsession was caused by our subconscious enslavement by the bad mother archetype. Kali Yuga humanity, deprived of proper nurturing, became devious, desouled, insatiably greedy. Electronic culture created soulless replacements for connective rituals—television supplanted tribal legends told around the fire; "fast food" consumed in distraction took the place of a shared meal. We substituted matter for Mater, money for mother's milk, objects for emotional bonds.

As the occult philosopher Julius Evola noted, men tend to be "passively active," while women are "actively passive," impelling activity and erotic advances like magnets. Female impulsion tends to be subtler and more covert than frenetic male action—and the feminine shadow is harder to see, though no less of a force, than the destructive projections of the masculine psyche. If we entertain the hypothesis that civilization was built upon a complicit agreement between men and women, then our myths reveal a different character. According to the biblical fable of Genesis, for instance, it was the feminine current that drew humanity away from Paradise. Adam—archaic man—was perfectly satisfied in the Garden. It was woman, Eve, who bit the Apple of knowledge and desire—and she wants to keep biting.

In prehistory, tribal cultures recognized innate spiritual differences between men and women. This is reflected in the customs of the Australian Aboriginals, who retained vestiges of the archaic form of consciousness, in which every day is the "first day." The prelapsarian mind-set of these tribal people may seem impossibly distant from our own—however, according to Jean Gebser, our future state of integral consciousness, characterized by "time freedom," represents "the conscious form of archaic, original pre-temporality," regaining their stability and continuity. The Aboriginal way of life may have gone on, essentially unchanged until the arrival of the Europeans, for as long as forty thousand years.

The Aboriginals considered women to be naturally connected to the divine and creative forces in the universe through their fertility and menstruation cycles. Men, on the other hand, had to be consciously—culturally—conditioned in order to achieve full humanity. Men were considered child-like, essentially useless, until they completed their initiation. Undergoing a training that could last as long as ten years, men would develop the capacity to access the Dreamtime through a series of dangerous and difficult ordeals. Common initiatory practices included not only solitary fasts and walkabouts, but scarification and subcision—the infliction of wounds on the male member, that could be reopened at any time, as a kind of symbolic menstruation. Such practices artificially induced men into female status, giving them the sacralizing wound granted to women by nature.

In Aboriginal society, as in other enduring forms of nomadic cultures, women provided the cohering, stabilizing force for the tribe. These societies are known as "hunter-gatherer," but it would be more accurate to call them "gatherer-hunters." Traditionally, women were the gatherers, and their foraging satisfied the vast proportion of the tribe's nutritional needs. Aboriginal men often left on long hunting trips and shamanic walkabouts. While the men were absent, the women intensified their bonds with each other, which were necessary for the smooth functioning of the community. In the modern West, by comparison, there is little solidarity among women. Women tend to feel they are in competition for men and their "value" rests in their powers of physical attraction, making them possessive, jealous, and insecure. It is possible, as Lawlor suggests, that their innate propensity for gathering has reasserted itself as a cultural obsession with shopping. Similarly, for men, the obsessive focus required for hunting—as well as the energetic hunting of shamanism—has been deformed into workplace competition, fixation on sports statistics, and other trivia.

"The fully sensual life of the Aborigines, their deeply spiritual communication with the earth, and their unshakable belief in their Ancestral Laws created an Aboriginal psychology that was disinterested in acquiring and possessing material things," Lawlor writes in *Voices of the First Day*. Living without shelter or clothing on the southernmost continent, the Aboriginals stayed in continual contact with the planet's electromagnetic fields, its secret heartbeat. Maintaining a direct link to the Dreamtime, they never developed technologies to control or dominate nature. According to Lawlor, they embraced the unmediated challenges of their environment as a means of attaining shamanic discipline and supersensible perception.

Lawlor proposes that the Aboriginals also possessed a more liberated—though still male-dominated—attitude toward sexual relationships than we do. They allowed for monogamous couples and polygamous men, as well as extramarital liaisons. They considered sexually unsatisfied women to be dangerous for the harmony of the tribe, and would pair off young men undergoing initiation with older women who were widowed or alone. The wider range of relationship patterns among the Aboriginals, and many other human cultures, suggest that our culture's singular focus on monogamy is not natural, but socially conditioned.

Some of our more evolved and peaceful relatives in the animal world practice radical forms of sexual openness. Dolphins, with brains larger than ours, interact erotically freely, frequently, and playfully. And then there is the unofficial mascot of polyamory, the bonobos. Bonobos are close cousins to the chimpanzee, with DNA more than 96 percent identical to our own. In *Demonic Males: Apes and the Origins of Human Violence,* the evolutionary biologists Richard Wrangham and Dale Peterson traced the origin of human warfare and "alpha male" behavior to male chimps, who attack and kill foreign males in raids and territorial conflicts. Male chimpanzees routinely batter females into submission, proving their sexual dominance through violent displays and occasional rapes.

Across the Zaire River from the chimps live the bonobos, primate hippies who dedicate their lives to peace, love, and sex. "Bonobos use sex for much more than making babies," the authors note. "They have sex as a way of making friends. They have sex to calm someone who is tense. They have sex as a way to reconcile after aggression." Bonobos have frequent homosexual sex and condone sex between adults and children. When a bonobo group meets a group of unknown bonobos, they generally mate and socialize with them rather than try to kill them. Slight changes in food sources and feeding patterns may have allowed the bonobos to stay together in larger communities on their side of the river, unlike chimpanzees, who must break off into small parties to hunt for their favorite fruit and meat sources. In these larger and more stable groups, female bonobos were able to form permanent social bonds and resist the aggressive urges of the males. Female bonobos evolved to hide their ovulation patterns, which put them more in control of their biological destinies and made it less clear to males when mating would lead to offspring. The authors of *Demonic Males* conclude that, as with the bonobos, the potential for future human harmony lies in the increasing power of the female, which they see developing in the advanced Western democracies.

On the deepest level, men seem largely unchanged by history—they are the same soldiers, shamans, and duffers now as five, ten, or fifty thousand years ago. Women are the ones who are changing, struggling against millennia of male domination and negative programming. The transformation of the instinctual and intuitive feminine current is yet another process that is quickening in our time. According to the psychoanalyst Wilhelm Reich, "Sexually awakened women, affirmed and recognized as such, would mean the complete collapse of the patriarchy." In order to accomplish this, to bite deeper into the apple, "she," the archetypal feminine, embodying Shakti energy, requires recognition, permission, and affirmation from the masculine Shiva principle of ordering consciousness. She needs to know herself for what she is, and could be. She—the feminine daimonic—will continue to wreak havoc until she gets what she wants in the way that she wants it, which may have little to do with current societal values, moral codes, and sexual stereotypes. When this is achieved, Kali will, with the faintest trace of a Mona Lisa smile, retract her fangs, pull in her tongue, and liberate her victims. The goddess will return, and this time around, the apple will be eaten down to the core.

THE PERIOD of the Kali Yuga corresponds to the rise of civilization, writing, and recorded history. During the last five thousand years—the "Great Cycle" of thirteen baktuns, measured by the Mayan calendar—human consciousness realized its alienation from nature, and penetrated into matter through technology. European man's incessant probing of nature could be seen as a quest for knowledge of the lost and defiled goddess whose body is the world. As Francis Bacon put it at the beginning of the modern era of scientific progress: "We must torture Nature until she reveals her secrets." Modern science has never lost its sadistic tinge. Kali's wrath is a reaction to the denigration of nature and woman at the core of our culture, part of the historical movement to separate and seek transcendence from our bodily nature.

In the laboratory, the scientist sought to understand the wound inflicted on him by the aggrieved feminine. What Virginia Woolf called the "sterile brass beak of the male" pecked away at reality. The creations of the rational intellect, cut off from the fertilizing influence of feminine intuition, reflected our disconnection from nature; the masculine desire to supplant organic processes with technological ones culminated in the atomic bomb, known as "Oppenheimer's baby," that was "delivered" over Hiroshima.

Obsessed with the urge to overcome the limits of space-time through acceleration and expansion, modern man builds racing cars and rockets, skyscrapers and faster microprocessors. Acting out of subconscious rage, modern man splits the atom in an attempt to annihilate matter/mater. The oceanic feminine, palindromic Eve, waits for the wave to crash. Kali giggles. She whispers, "You do not know me yet, you man, you failed systemizer."

At the beginning of the twentieth century, science seemed the crowning achievement and triumph of the rational European intellect. Suddenly, the physicist discovered, to his horror, that matter was an illusion—

space-time was an illusion. There was only quantum foam, fluctuation, and flux. It was all feminine sinuous motion: Shakti. It was all relational—patterns of Dreamtime vibration. There was no hard fact, just seductive spectacle and uncertainty principle. At the deepest level, reality was inseparable from the act of conscious participation—there was no place for an outside observer removed from what he observes. Instead of a bedrock materiality, there was what the Hindus call "lila," divine play.

Similarly, when the scientific mind turned its attention to human sexuality, it discovered no normative standard, but endless difference. In the 1950s, the scandalous *Kinsey Report* revealed that human sexual variety was essentially infinite, and that each person was distinct in their erotic desires and patterns. Despite such evidence, our model of relationships continues to enforce a cookie-cutter approach, enshrining monogamy as the standard. According to Laura Kipnis, a professor of communications at Northwestern University, our failure to confront and articulate difficult issues around sexuality and romantic love reverberates across our social and political world. "Why not at least entertain the possibility that there could be forms of daily life based on something other than isolated households and sexually exclusive couples?" she asked in her 2004 polemic, *Against Love.* "Why not confront rather than ignore the reality of disappointment at the deadening routinization that pervades married households? Maybe confronting the flaws in married life would be a route to reforming a flawed society? Maybe reforming the fabric of individual relationships was the path towards political renewal?"

Kipnis may have hit upon something very profound here. The destructive friction between men and women—our failure to attain a mature communion—may underlie the various global political and socioeconomic crises afflicting our world. It may be that we cannot resolve the larger conflicts around us until we reweave the fabric of our intimate lives, finding the courage to admit our deceptions and release our delusions, forgiving ourselves and our partners. The institution of monogamy will have to be called into question. According to stereotype, women are naturally monogamous, while men are not. But if we were able to access and give expression to our essential nature, shorn of social conditioning, would this really be the case? Is monogamy ultimately any better, or more innate, for women—whose capacity for sexual pleasure is so much deeper—than it is for men? If monogamy were to give way to more flexible relationship patterns, based on deeper trust, this would have wide-ranging consequences in our institutions and even in the design of our cities and houses—our current systems and structures are oriented toward the nuclear family unit, rather than the extended family and communal life-patterns of tribal cultures.

If women want to do the work of integrating their shadow, they will have to follow the traces of Kali—addressing the subtle and unsubtle ways they have misused Shakti energy—within their own lives. Men, on the other hand, have to confront their catastrophic loss of soul, their forfeiting of intuitive and emotional depths. This condition is illustrated by another myth—the medieval legend of the Fisher King. The Fisher King suffers from a wound in the groin that will not heal, as his kingdom atrophies. According to the Jungian commentator Robert Johnson, the Fisher King legend describes the "wounded feeling function" that afflicts modern man, who suppressed intuition and emotion to attain technological rationality and one-sided dominance. Johnson notes that the English language reflects our emotional paucity. Ancient Persian and Sanskrit possessed more than eighty words for love, denoting different qualities and valences of communal and erotic feeling. Whether we want to proclaim our affection for Krispy Kreme doughnuts or our significant other, we are stuck with just the single word, obliterating differences and qualities. Locked in limited rationality, men tend to suffer from a profound incapacity to understand or explore their own emotional makeup.

Women are seeking to evolve to a higher state of consciousness, a deeper realization, of their creative and erotic nature, but this can take place only if it is accompanied by an equally profound shift in the masculine psyche, from brutality and control to patience and mastery, "in order that a world can come to be without maternal or paternal dominance, that is a non-masculinized world where man and woman together honor the human, and think not merely in terms of the human but of humankind in its entirety," Jean Gebser wrote. Masculine solar and feminine lunar currents can reintegrate to create a true partnership, neither patriarchal nor matriarchal. This new partnership culture could be symbolized by a new calendar—a new covenant—accurately attuned to sun and moon cycles, wiping away the distorted institutions and unjust legal codes produced by the dominator mind-set of the past.

From our present vantage point, it is difficult to envision what a truly sex-positive culture might be like. Sexuality is still shrouded in aggression and mistrust (the words "fuck" and "screw" are hurled as curse words), with young women's hypersexualized bodies endlessly used as props to sell products. Not only a deep revisioning of sexuality and relationship models, but a resacralizing of Eros, may be necessary, before we can hope to institute a planetary culture that is truly nonviolent and harmonious. With more time on their hands in a rationally organized post-work civilization, individuals could cultivate the art of love to levels hitherto undreamed, whatever their personal predilections.

IN SEDNA'S STORY, the tear that falls from the fisherman's eye is the realization of our divided nature, and the beginning of reconciliation. When he weeps, "he takes on his own medicine-making, he takes on the task of feeding the 'deleted other,'" Estés writes. It is the tear of compassion that draws Skeleton Woman to the fisherman. According to Estés, the tear is not just for her; it is for the heartbreaks he has suffered also. Once he sheds this tear, she knows he can be her mate.

Returning to New York, I confronted my personal manifestation of Kali in the form of my enraged partner—the Kali I had created through my deceits and incompetence and habitual guilt, also the Kali who had created herself through her recriminatory furies, psychic assaults, and habitual blame. Our battle began anew as soon as I walked through the door, and didn't abate until I gave in to her wish that I move out a month later. For a few weeks, I contemplated moving to San Francisco—I dreaded being a poor person in New York again, longed for the sun-drenched psychedelic playgrounds of the West Coast—but realized that this was a disastrously selfish idea. It would mean abandoning my child.

Years ago, I had written a feature for *The New York Times Magazine* on the adult children of the Beat Generation. I discovered them to be a traumatized group, carrying deep psychic wounds as a result of their parents' chaotic drives toward freedom. "Always do what you want," Kerouac told my mother repeatedly during their relationship. This idea led to his—and others'—ruin, rather than liberation. Kerouac's daughter, Jan, met her father only twice in her life, and followed his path into alcoholism and early death. William Burroughs's son became addicted to speed and similarly self-destructed. The testimonies I collected were a litany of misery. Coming to consciousness in the conformist 1940s, the Beats struggled to escape the constraints of the mainstream society—the "gynocratic" pressures to conform to family life—and fled conventional relationships and commitments, throwing off children and wives like beer bottles along the highway. Their outlaw rage against the prisonhouse of convention found real as well as symbolic expression—in Burroughs's "accidental" shooting of his wife through the head, in Kerouac's refusal to accept paternity for his daughter, and so on. The parents of these ruined children also ruined themselves in the end, becoming debauched caricatures, their quest for liberation ending in druggy dissolution or monotonous narcissism. They failed at the difficult task of reconciling freedom with responsibility—in pursuing their "lefts," they neglected the rights of their children and loved ones.

Knowing the consequences, I did not want to repeat such behavior. I wanted to be with my daughter, to see her grow up, to love her. I wanted to be with my partner also, if she would be willing to enter into a new form of relationship that was not based on the possessive urge to control each other's desires. I was still drawn to her. Obstinately, perhaps hopelessly, I felt that the misery we had caused each other was irrelevant, a passing cloud obscurring the connection between us, a deeper complicity, which remained unchanged.

During a temporary truce, I told my partner a story about the Buddha I had heard late one night at Burning Man. In the version told to me, one of the Buddha's disciples went to him and asked to be shown Heaven. The Buddha said, "If you want to see Heaven, you will have to see Hell first." The disciple agreed. The Buddha took them to Hell, where an enormous banquet table was set up, piled high with fabulously delicious food. Unfortunately, all of the diners had, instead of hands, enormously long forks on the end of their wrists, and they kept trying to get the food into their mouths, but could not reach them. They wailed and gnashed their teeth in misery. The Buddha then took his disciple to Heaven. Heaven was exactly the same situation—diners at a sumptuous banquet table, with long forks on their wrists instead of hands. The only difference was that, in Heaven, everybody was feeding each other.

THE DAY AFTER I returned to New York from Nevada, I went to visit first priestess at her painting studio in Chelsea. While I was in Glastonbury, she had sent me accounts of beautiful dreams she had about the two of us, performing Celtic rituals together in sylvan glades. I had thought of her often in England and at Burning Man—of her murmured declaration of love for me. While in her crisply white-walled studio—in my hypersensitized state, the room seemed to be swirling with currents of positive energies—I told her everything that had happened, and the realizations I had reached about monogamy and the misuse of sexual energy. We returned to our previous level of physical intimacy, holding each other, and exchanged one extremely promising kiss—I pulled away from this embrace. I did not want my separation from my partner to be dependent on some new relationship with someone else. I wanted to be clear with myself, as well as with my partner and the priestess, that it was both a deeply felt and a philosophical decision I was making.

Our meetings over the next months—difficult months for me, as I had left my family for an underheated loft space in South Williamsburg, Brooklyn, with several roommates and a cat that caused me sneezing fits—were brief and curtailed affairs. First priestess blithely glided off at the end of our dinners or teas to meet her boyfriend, ignoring any deeper undercurrents in our situation. Feeling hurt and discouraged, I sent her a long and ill-advised e-mail, suggesting that she make room in her life to be with me—without separating from her boyfriend, if she preferred—so that we could explore a Tantric practice together. She replied that I was completely out of line, that she was satisfied in her healthy relationship, and it was not up for discussion. She said she had been "confused" about her feelings previously, but had now achieved clarity. I felt betrayed and demoralized by this. She was the one who had declared love for me—but now, instead of exploring that connection, we were doomed to circle each other like characters in a Henry James novel. Despite our intimacy in Hawaii, our relationship back home was governed by convention, in which her commitment to monogamy was sacrosanct.

I found a strange satisfaction in my grim hermit's existence—the reduced circumstances somehow suited my inner daimon, who greeted each new humiliation with interior laughter. My new home was under the elevated JMZ subway line, its frequent passages rattling my window, with gloomy, mysterious Hasidim on one side—an old matzo factory with its wood-burning stoves just down the street—and working-class Hispanics on the other. Our apartment was robbed soon after I moved in, the thieves absconding with the few items I possessed that were of any obvious value.

After the protracted fighting with my partner, I had work to do to regain my daughter's trust—when I was bathing her one night, she turned to me and said, with a two-year-old's perfect candor, "I don't like you." She thought about this, then qualified it. "I love you, but I don't like you." As I spent more time with her over the next months, she slowly forgave me for the hurt I had caused by leaving home, telling me finally, and repeatedly, "I love you *and* I like you"—to my great relief.

One day, I received a call from the teacher. She was organizing a journey down to Brazil, to visit centers of the Santo Daime, a religion started in the Amazon in the 1920s, mixing indigenous and Christian elements, using ayahuasca as its sacrament. Having heard about my personal crisis from the priestesses, she wanted me to come with them on this trip. As part of her mission to fulfill the Lakota prophecies, they would be traveling with a group from the Native American Church, introducing the medicine traditions of the "Eagle" of the North to the "Condor" of the South. I was suspicious of the religious ambiance of the Daime—they called their ceremonies "Works," and from what I had heard, it did sound suspiciously like hard work. Santo Daime congregations sang Portuguese hymns under bright fluorescent lights, doing constricted little dance-shuffles back and forth, for as long as fourteen hours. But the prophetic dimensions of the trip fit the subject of my book—more than that, as soon as I heard the teacher's voice, I felt I needed to go with her. Since Burning Man, I had felt trapped between states—between the visionary ferocity I had accessed at the festival, and my squalid normal life—flailing to integrate new floods and flashes of inspiration. I hoped she could help me understand what was happening to me. I trusted—I hoped I could trust—in her sanity and wisdom.

CHAPTER THREE

Who, if I cried out, would hear me among the angels' hierarchies?

RAINER MARIA RILKE

Brasília, the capital of Brazil, is in its own way a visionary fabrication, constructed from scratch in the late 1950s and early 1960s as a utilitarian utopia, planned out and designed by the modernist architect Oscar Neimeyer. Imposed upon the sparsely populated Central Highland of the vast country—Brazil is as large as the continental United States, minus Alaska—Brasília's once-pristine blue and orange facades have faded and become weather-beaten over time. Built in the form of an airplane or bird with outstretched wings, the city fulfilled the prophecy of a nineteenth-century priest, who dreamed of a "grand civilization . . . a land of milk and honey" rising from the empty plain. The sun gleams off the tinted windows of government ministries, beats down mercilessly on the city's vast avenues, which have lost their pompous grandeur over time, giving way to a dusty dilapidation.

The early omens for our Brazilian voyage were not auspicious. At the last moment, the members of the Native American Church were denied their visas and could not accompany us. Nobody knew the reason for this, but it meant that the meeting of medicine traditions—the prophetic aspect of the voyage I had planned to report on—would not occur. We would be traveling with a much smaller group of eleven Europeans and Americans. Feeling cheated by this twist of fate, I had e-mailed the teacher's office a few days before we departed, seeking a refund, but received no answer.

Leaving New York, I resigned myself to an expensive trip I could barely afford, that I presumed would be of no use for my book.

Other portents were even more worrisome. A few weeks before leaving, I ran into first priestess at an art opening. Smiling oddly, she said she had a strong presentiment we would die during this journey—however, she did not seem particularly displeased at the prospect. Her dazed and dreamy attitude amped up my anticipatory dread. I faced brainlock when it came to preparing for this trip—I couldn't even bring myself to look at a map to see where we were going in the Amazon, only vaguely aware that it was quite near the center of that verdant mass. I did not visit the doctor to see if I needed to update my shots, and the only footwear I brought were a new pair of Campers, utterly inappropriate for the jungle, along with a pair of rubber K-Mart slippers that had survived the rigors of several Burning Mans.

Away from the imperial pomposity of the government center, Brasília was made up of mazes of small neighborhoods spread across gentle hills. We were staying at a Santo Daime community, Ceu do Planalto, off one of the wings. At the center of the village was a large hexagonal-shaped ceremonial space with a peaked thatched roof. Santo Daime is an officially sanctioned religion in Brazil—a country known for its proliferation of eclectic sects and cults and syncretic religious movements—and many of the members of this community had jobs in the government ministries, as press officers and Web site managers. At night, we gathered together in one of the members' houses and practiced the songs, the *hinarios* that are the core of the Daime practice, accompanied by guitar-strumming and hordes of circling gnats.

THE SANTO DAIME RELIGION originated in the visions of Raimundo Irineu Serra, a young Brazilian of African descent, who worked as a border guard in the Amazonian region in the 1920s. While drinking ayahuasca with an indigenous tribe, he was visited by the beautiful "Queen of the Forest," appearing to him again and again, teaching him the first of the hymns. To his utter surprise, he was told by this apparition—whom he believed to be a form of the Virgin Mary—that he had been given a sacred mission: to begin a church that would "replant the doctrine of Jesus Christ on Earth," wrote Jonathan Goldman, in his introduction to Alex Polari de Alverga's *Forest of Visions.* This doctrine was not "a set of rigid rules or an orthodox set of ideas," but "a living matrix of consciousness. Jesus Christ implanted a conscious seed in this world by his life and death," initiating "a vast change in human consciousness that is now beginning to come to fruition." The purpose of the Santo Daime is to call, "one by one, the many souls who are ready to rapidly awaken the seed that Jesus planted, the Christ Consciousness, in themselves."

Mestre Irineu conducted the first works in Rio Branco, his hometown in the Amazon, attracting followers from among the local rubbertappers. Over the next decades, the church slowly spread out, to other Brazilian centers. In the 1960s, Sebastiao Mota de Melo, a canoe maker, joined the Daime, becoming the leader of the church after Mestre Irineu's death in 1971. In photographs, Mestre Mota de Melo, Alfredo's father, resembles a Gandolfian wizard, with long wispy white beard and twinkling eyes, stirring huge cauldrons of bubbling Daime brew. According to de Alverga, Padrinho Sebastiao was a prophet, guided by spirit. "Prophets are those who, because they have arrived at their own self-knowledge and a larger knowledge of being, find themselves perfectly aligned with the Divine will to the extent that they wish nothing else but the fulfillment of Divine laws. They give themselves so fully to the prophetic truth revealed by faith that they draw a whole people to their destiny." Following visionary guidance, Mestre Sebastiao moved the center of the Santo Daime into the heart of the Brazilian rain forest, settling at Mapia and Jurua, two towns along the Jurua River. Today, the Daime describes itself as a "spiritual and ecological movement," seeking to create sustainable communities within the jungle. In the 1980s, the church grew in popularity across Brazil, taking root in many urban centers. Padrinho Sebastiao died in 1990. Under the leadership of Padrinhos Alfredo and Luis Fernando, Santo Daime has developed internationally, with tens of thousands of members, centers throughout South America, Europe, the United States, and even Japan.

The first evening ceremony we attended was held at Divina Luz, a Santo Daime center in another part of the city. This "work" seemed to confirm my negative preconceptions. The atmosphere was solemn and churchlike; the parishioners—former Catholics—were dressed in blue jackets and ties with stars pinned to their lapels like sheriff's badges, while those who were not part of the church dressed in all-white. Illustrations of American Indians posed in front of Gothic moonscapes were pasted on the rough wooden walls. Passing a dusty storeroom in the back, I was astonished to see an enormous supply of ayahuasca—called Daime, or "give me," in their tradition—in huge jugs, neatly stacked, each marked with its date and place of origin, like a munitions dump of revolutionary consciousness awaiting the igniting spark.

We lined up to get our sour cups of brownish Daime, men on one side and women on the other, and then returned to our separate sections as the *oracao,* the singing of the hymns, commenced, songs following one after another in rapid-fire Portuguese that I found impossible to follow, flipping the pages of my songbook haplessly. The bright fluorescent lights beaming into our skulls made private journeying impossible. Occasionally, the interminable series of hymns would be interrupted by equally interminable sermonizing. As overseas emissaries traveling with the teacher, we were obviously expected to behave in an appropriately respectful manner—but my entire being was rebelling against this rigid structure. I couldn't believe I had chosen to subject myself to three weeks of this. Finally unable to take any more, I reeled outside, into the soft star-filled night, and went to sit by the fire.

Casey—my fellow New Yorker, a somewhat manic investment analyst in his forties—was already sitting out there, another escapee or apostate. I was relieved that someone else was chafing at the bit. It was delicious to sit beside the quivering flames, around which I almost felt I could perceive dancing spirit-beings in my augmented state.

"Have you checked out the teacher? She is being put through the wringer tonight," said Casey, who had accompanied her on previous trips to the Amazon. I said I had noticed she looked a little woozy. "The teacher is a very spiritual woman, but in the end she is just a handful of this." He picked up a chunk of dirt and threw it into the fire.

Before Brazil, the teacher had visited Gabon, where one of her daughters-in-law was undergoing an intensive process of initiation to become a Bwiti priestess. Their community was planning to build a Bwiti temple, the first one outside of West Africa, on their land. Before this could happen, one of their members had to receive the full transmission of sacred knowledge and ancestral lore of the Bwiti, and her daughter-in-law had volunteered for this mission, which would take several years and numerous iboga sessions to complete. During one ritual, this young woman had been put into a possession trance, and a powerful "Thunder Spirit" called down to inhabit her body. The spirit demonstrated its presence by causing her to walk through a fire without injury, then marched her around the Bwiti temple, performing various ritual gestures and ablutions unknown to the Westerners. The Bwiti shamaness spoke to the Thunder Spirit directly, asking if it wanted to be the guardian of a temple in the United States, negotiating an almost contractual agreement with it. At some point during their stay, the teacher's daughter-in-law had contracted malaria. I wondered if it was really necessary to undergo this brutal entry into the heavy-handed ways of the African spirit-world to bring iboga to the West, but the teacher believed this was the only way to proceed.

"She's playing," I said to Casey, suddenly envisioning the teacher as a master shaman-strategist, gathering pieces on the Monopoly board of global transcendence. "She enjoys all the challenges. It's like a game to her."

"Wow," Casey said. "You are really cynical."

"Not at all. I think I am idealistic."

Guardians of the ceremony interrupted our talk, ushering us back inside. Sitting uncomfortably upright in the Divina Luz church, listening to the hymns, I felt the melancholy of the mestizo world, caught between the cruelly repressive mind-set of Catholicism and the lost indigenous past—the Daime was an attempt to knit back together these tattered strands. The teacher was introduced and asked to sing a few of her Native American Church songs, which she did in a manner that I found annoyingly theatrical and self-important, raising one arm above her head and shaking her hand. I was feeling a gnawing, growing suspicion about her. As in Hawaii, I wondered if she was really trustworthy or, in the end, just another New Age manipulator, overlaying her own sentimental and opportunistic trip on these native traditions. I questioned her insistence that "the Mother" demanded she always wear white outfits and keep her head covered. Why did she feel the need to single herself out with these costumes? I suspected she was thriving on the attention she got for her act.

I had been sending many of my close friends to the ceremonies she held in upstate New York—now I wondered if I was doing the right thing. After one ayahuasca session, the teacher told everyone present that they shouldn't look at the full Moon during an upcoming astronomical conjunction because she had heard from indigenous people that it might attract "bad spirits." My friends took this seriously, but I found it stupid. While I felt the need for an archaic revival—reconnecting us with sacred sources of aboriginal tradition—I didn't believe we could forfeit our modern pragmatism in the process, returning to the magical thought of the past. Without denying their validity, I considered "bad spirits" to be projected through our psyche, given energy by our thoughts and intentions. Regressing into a superstitious mind-set about them could only be disastrous for us.

Did the teacher's relationship with "the Mother" conceal a shadowy desire for power? After hearing my story about Burning Man and its aftermath, the teacher had publicly declared that I was in a crisis and needed a special healing during a Daime ritual. Pulling her aside, I told her I did not want one. I believed I had done my own alchemical work, transmuting the poison into medicine. She seemed disconcerted by this—people who attended her ceremonies and seminars were constantly begging her for special attention, and I wasn't interested. I disliked the New Age obsession with healing. Healers and psychoanalysts start with the notion that there is a sickness, there is a patient, there would be a long drawn-out process of recovery that might not ever quite end in a cure. It seemed to me that healers took control of the narrative of their patients in this way, stripping them of agency, feeding their egocentrism and fantasies of victimization. I preferred the Eastern perspective of Sri Nisargadatta, that "the world and the self are already perfect," that only our attitude is faulty and needs readjustment. According to Buddhism, each person is a Buddha who has forgotten their original nature. If we in the pampered West, having grown up with so many advantages, could not claim our own health and our agency, preferring to see ourselves as helpless victims, then who would do it? Who would take responsibility for the world?

TWO DAYS LATER, the night before the priestesses were due to arrive for our journey to the Amazon, the community where we were staying, Ceu do Planalto, held a Daime work. The members of this community were younger, more urbane and sophisticated than that of Divina Luz; they were thoughtful as well as soulful, with a soft sweetness to them. They described Santo Daime as a "collective shamanism" and explained that the ceremonies provided forums for working out interpersonal issues that arose in their tight-knit community, as well as connecting them to the Sacred. I appreciated their perspective, yet felt they had been led astray, locked into an imposed doctrine—in my mind I kept repeating, with grumpy approval, Mao Ze-dong's dictum, "Religion is poison."

After opening prayers led by Padrinho Luis Fernando—the second-in-command of the Santo Daime, he would be accompanying us into the Amazon—we lined up to get our bitter cupfuls of jungle murk. I was shown to my assigned seat in the men's section, facing the women. In the center of the round, voluminous space, the leaders of the service sat with the musicians around a hexagonal table. Candles, flowers, and photographs of the church's founders were displayed on the table. As I turned on to the medicine, the fluorescent lights seemed to grow brighter, beaming into me, unforgiving. I made an effort to follow along with the *hinarios,* but confused by the unfamiliar rhythm and pace of the tunes, incapable of following the words, I soon gave up.

Looking around at the faces of the fifty or so Daimeistas, singing for all they were worth, I felt sorry for them—for all of us, seeking some form of communion, some liberation from the oppressive modern wasteland, our abysmal world of shards. I sulked, withdrawing into myself. Seeking some diversion, I concentrated on the Maya and 2012—the improbable fable of conscious evolution, psychic escape from imminent meltdown. I recalled José Argüelles and his pipe dream that we could embrace a new "timing frequency," a new relationship to time and being, in the few years remaining—as hopeless as the teacher's mission of merging medicine traditions to raise the human spirit. Global civilization was headed for imminent breakdown and biospheric collapse—necrosphere, not noosphere. There could be no salvation.

And yet I considered the synchronicities leading me from one encounter to another, fitting my birth date into Argüelles's prophetic grid, drawing me into the crop circles, and the teacher's nexus. I recalled the magical events as well as supernatural torments inflicted upon me as I pursued my fascination with consciousness and prophecy, as if I were assembling pieces of a jigsaw puzzle, targeted toward some goal—the strange presentiment I perpetually had, that, as much as everything seemed to be collapsing, everything was also going seamlessly according to plan. I recalled the many masks of Quetzalcoatl, the *Deus Abscondus* who left the Mesoamerican civilizations in the lurch, abandoning them to the ravages of colonialism, but promising his eventual return. According to legend, the Plumed Serpent was meant to slither-flutter his way back to Earth, reestablishing "Sacred Order," reasserting harmonic concord amidst rampant discord, before the Great Cycle reached its end—and the time was growing short if he was going to beat the clock. As I wondered what this could mean, what form it might take, a voice piped up inside my head.

"Here I am," the voice said.

"Who are you?" I asked this intimate, alien presence.

"Quetzalcoatl," he replied.

As I pondered this, the voice began to dictate a message. Quetzalcoatl—or whatever spirit, daimonic entity, or disassociated shard of my overwrought psyche had taken up temporary residence inside my skull—explained that he was an immanent field of mind, a new level of consciousness, linking sacred traditions with modern thought-streams, melding modern empiricism with mercurial realms of the soul and higher pirouettes of spirit. The Plumed Serpent, winged avenger of Mayan myth, had chosen this moment to alight. I was a convenient reference point, a panoptic illusion, embedded at the necessary coordinates in our space-time matrix to transmit this noospheric news. Without knowing it, I had been living within a vortex of cryptic communications whose source was only now becoming apparent.

I left my chair and went outside to stroll in the night, walking the perimeter of the village, reeling from emotion, laughing at the absurdity of it, struggling against surges of anger and despair. A familiar contempt for human limitations rose within me. The voice dictated words and shards of prophecy, explicating its viewpoint. I wondered, idly, if this dispensation—if it was something other than delusion—came with new wizard powers. I tested the glossolalia I had accessed during DMT trips and the recent Mars conjunction, expelling insectile sound-streams from my larynx, but no helpful genie or will-of-the-wisp appeared to obey my commands.

"I am happy," I said out loud, to nobody—to the dark shapes of the trees.

I stretched out on a bench by the fire, enjoying the temperate breezes and star-strewn sky. One of the young mestizo men assigned to be "guardians" of the work asked me politely to come back inside. I refused. From this safe distance, the singing inside the church sounded almost beautiful—but there was no way I was retreating into that prison to mouth along with those treacly hymns for several more hours. I began to consider how I, as the vehicle of Quetzalcoatl's return, would establish colleges of sacred practices for exploring the proper use of ayahuasca, this shamanic technology for transforming reality, this archaic antidote for the sick modern soul.

One by one, and then in twos and threes, other guardians in blue jackets and ties came up to ask me to return to the ceremony. I told them I was happy where I was. They seemed bewildered that I wouldn't comply with their request, murmuring in the shadows like bumbling Keystone Kops. Apparently, my absence was creating a bit of a commotion.

"Padrinho Luis Fernando asks that you go back to your seat," I was told.

"If Luis Fernando wants me to return, why doesn't he come out here himself so we can discuss it?" I replied. "I like it out here."

One of the American travelers from our group left the ceremony to talk to me. Lou Gold, in his sixties, was an ecological activist from the Northwest who had joined the Santo Daime years ago, now living with them in Brazil. He struck me as something of a zealot, reverently following the dictates of the doctrine.

"You know, I always try to comply with the rules of any spiritual group I am visiting," he said.

"That is your choice. You prefer to give up your individuality to any sect that comes along," I said. "That is not my choice." I complained about the Daime and about the teacher. "She needs to take that scarf off her head and get real. I'll never recommend her trip to anyone again."

He told me about a Daime work he attended in Oregon that finally convinced him to join the church. "I was feeling the weight of what Christianity had done to indigenous people," he recalled. "I didn't think I could follow a Christian path. I decided to ask the medicine for guidance." He closed his eyes and, in his vision, he saw Christ and Mary appear, and then leap into his heart. "I was reeling after this, and I wanted to open my eyes to look around, but a voice spoke to me and said, 'Don't open your eyes. We put something in your heart for a reason. From now on, when you need an answer, just look into your heart.' That was the moment when I knew the Daime was the path for me."

"A nice story," I said, somewhat dismissively. He asked again if I would return to the work. I reiterated that I preferred to stay outside. He went back in to convey my answer to Luis Fernando.

Finally, one last guardian approached me. He was a large goonish fellow with messy black hair and pimples. Mustering his limited English, he adopted a slightly threatening tone. He told me if I didn't return to the ceremony, it would have "serious consequences."

"Oh really. What kind of consequences?"

"You won't be allowed to drink Daime here again," he replied.

This threat hit home. I immediately went back inside and took my seat. The ceremony was approaching its conclusion, and I fidgeted in my assigned place until the end. Outside afterward, I told the teacher about a famous DJ I had met, who described taking ayahuasca in Las Vegas, tripping out at strip clubs with his hired shaman, watching the dancers transform into shimmering Hindu deities. The teacher found this distasteful, even dangerous, as it might attract "bad spirits." She mentioned her years of social work, the negative effects of objectifying women, and that many strippers had been sexually abused in their youth.

"Why are you so afraid of bad spirits?" I asked her. "Maybe you're creating them, since you talk about them so much. I don't see how I can send my friends to you anymore. I don't want them influenced by your fears and superstitions. What are you so scared of?"

The teacher gave me a strained, shocked look.

That night, I could not sleep. The energy I contacted during the ceremony was keeping me wide awake, wired and wriggling from high-voltage jolts. Part of me was recoiling from the evening's revelation, struggling against an overpowering sense of fate. I yearned to retrieve my normal humanity from Quetzalcoatl's archetypal coils, as a cold sarcophagal solitude pressed against me.

At the same time, yet another alien voice—or disassociated aspect of my psyche—howled in my mind, with raw ferocity. The priestesses were due to arrive the next morning. The new voice insisted that I make love to first priestess while we were down in the Amazon. If she would not have me, this voice commanded, I should give up the charade of this present incarnation and walk alone into the jungle until I died. I tried to reason with this voice, but it was relentless in its demand. I feared it was stronger than me.

CHAPTER FOUR

God is pressure.

DION FORTUNE

According to Carl Jung, the Western "God-image" is a representation of the collective unconscious, an archetype of the psyche that undergoes a continual process of transformation through the Old and New Testaments—and beyond. The God-image evolves through its relationship to humanity. "Whoever knows God has an effect on him," Jung wrote. For the individual, knowing God, in Jungian terms, is the process of recognizing and assimilating the pressured and paradoxical contents of the self, which come to consciousness—seek incarnation—within the ego.

"Being chosen doesn't come out of a state of fullness, it comes out of a state of emptiness," Edward Edinger wrote. The Jews were not made the vehicle of the monotheistic God-image because of their magnificent power, but because of their weakness and puniness. They were poor peasant people, trapped between empires; due to circumstances, they had developed their inwardness, their self-consciousness, to a greater degree than their neighbors. Because of the intensified self-awareness created by suffering, Israel was "peculiarly suited to take upon itself the misery and dignity, the curse and blessing, of God's election." In Jung's interpretation, the relation between the Jews and Yahweh provides an archetypal model of the relation between Western ego and self.

Jung realized that Yahweh, the Old Testament God-image, was not simply a benign deity but given to "incalculable moods and devastating attacks of wrath." Yahweh had "a distinct personality, which differed from that of a more or less archaic king only in scope." The personal unconscious, with its power-drives and self-righteous vanities and imperious demands, still retains, to a great extent, this autocratic quality—contemporary equivalents, acting out of primordial and untrammeled instinct, include Mafia bosses, corporate plutocrats, and corrupt politicians.

The opposites contained in the unconscious run the gamut from love and compassion to vicious cruelty and sadism, the depths as well as the heights of our potential. "Yahweh is not split, but is an antinomy—a totality of inner opposites—and this is the indispensable condition for his tremendous dynamism," Jung wrote. The biblical narrative describes the humanization of the God-image, culminating in his incarnation as the half-human, half-divine Christ, a manifestation of the "good God." This incarnation required a dissociation from Satan, his darker half, who was witnessed, by Christ, falling from Heaven to Earth.

When archetypal material constellates in an individual psyche, according to the Jungian model, the experience is not one of overflowing potency, but of weakness and helpless rage. The level of anguish and humiliation involved in the descent of an archetype depends on the individual's level of awareness: "If there is little consciousness attached to the event, then the ego becomes the tragic victim of the archetype that it is constellating," Edinger wrote. "If there is more consciousness involved, then the ego does not have to be the tragic victim because it knows what is happening to it. It behaves in a much different way and can mediate the archetypal pattern much differently."

In the book of Job, written several centuries before the New Testament, Yahweh subjects his "faithful servant," Job, to a harrowing series of tests, after accepting a wager from Satan that Job's faith can be broken. "Job is no more than the outward occasion for an inward process of dialectic in God," wrote Jung. Like a scientist performing some cruel experiment on bacilli in a test tube, Yahweh kills Job's family, removes his land, riddles him with disease, and inflicts every imaginable form of ruin upon him. Job, however, remains steadfast. At the same time, he is determined to understand the reason for his plight. According to Jung, Job is the first man to comprehend the split inside Yahweh—that the God-image is an antinomy, comprising both the dark god of cruelty and the benevolent deity of love and justice; "in light of this realization his knowledge attains a divine numinosity." Confronted with archetypal injustice, Job insists on equalizing compassion, and eventually receives it, as his status in the world is restored.

Despite his overpowering might, the creator fears the judgment of his creature. "Yahweh projects on to Job a skeptic's face which is hateful to him because it is his own, and which gazes at him with an uncanny and critical eye," Jung noted. From the perspective of the God-image, Job had attained a higher state of knowledge than Yahweh through his travails, and this required a compensatory sacrifice, enacted, a few hundred years later, through the incarnation of Christ.

Jung realized that God intended to fully incarnate in the collective body of humanity, and that this time was quickly approaching. From his psychoanalytic and personal work and theoretical musings, he proposed that the Christian Trinity of Father, Son, and Holy Ghost was unfolding into a "quaternity," adding a fourth element that had been suppressed from the Western psyche. "The enigma of squaring the circle" was one representation of this quaternity, "an age-old and presumably pre-historic symbol, always associated with the idea of a world-creating deity." This aspect of divinity, now returning and requiring assimilation into consciousness, was the Devil, who had been dissociated from the Western psyche at the beginning of the Judeo-Christian aeon. Along with the Devil, the fourth element also represented natural wisdom, personified by the Gnostic deity Sophia, long exiled and excised from the canonical texts.

Since the creator is an antinomy, a totality of inner opposites, his creature reflects this schism. To descend into humanity, God must choose "the creaturely man filled with darkness—the natural man who is tainted with original sin," Jung wrote. "The guilty man is eminently suitable and is therefore chosen to become the vessel for the continuing incarnation, not the guiltless one who holds aloof from the world, and refuses to pay his tribute to life, for in him the dark God would find no room." The uniting of the opposites, the reconciliation of dark and light contained in the God-image, can only take place within the consciously realized "guilty man," not the sanctimonious, ascetic, or self-righteous one—anyone who denies their shadow will only project it in some new form.

"The archetypes themselves cannot evolve into full consciousness without being routed through a mortal ego to bring that consciousness into realization," Edinger wrote. If an archetypal god-form had chosen to route itself through my personal circuitry, I was not happy about it at all. I felt like a small squeaky mouse caught in a huge cosmic beak. I wished I could return to the dank comforts of my private hole, abandoning any quest for higher knowledge. It was clear to me that the universe had made an unfathomable clerical error—surely there were more exalted souls waiting around somewhere who could give this situation the attention it deserved? I did not want to be part of Quetzalcoatl's dialectic, a bit player in his cosmic saga. I was not cut out for the role.

AFTER BREAKFAST the next morning, the teacher and I had a talk outside. She said she was refreshed after spending the evening in "the Mother's washing machine." I said I felt there were other ways to use ayahuasca, that this devotional stuff is fine for some, "but I don't have the time for it."

We talked about working with archetypal material. Without mentioning my feathery serpent transmission, I asked her what would be the stage beyond individuation?

"Merging with the archetype," she said.

I didn't like that answer.

"Can't we create new archetypes instead?" I asked.

She said she thought we were saying the same thing with a different emphasis. I wasn't sure I agreed. We talked about her daughter-in-law's Bwiti initiations in Gabon. I told her I felt sorry for her daughter-in-law, forced into a possession-trance by a Thunder Spirit. Was this necessary? I quoted Christ from the Gospel of Thomas: "Blessed is the lion which becomes man when consumed by man; and cursed is the man whom the lion consumes, and the lion becomes man." By surrendering her consciousness to this atavistic ritual, wasn't this poor woman making a mistake—letting the lion devour her? How could the teacher let her make such a sacrifice?

"Why are you more critical of what we're doing than you are of that DJ taking ayahuasca to go to strip clubs?" the teacher countered.

"I don't think we should forfeit our modern selves to return to the spirit world."

"I believe it is necessary to connect with the original medicine lines," she replied.

"All right, maybe so. In that case I am just glad I didn't have to be the one to do it," I said. I told her I was worried that she was too attached to her persona, and that was why she wouldn't take off her hat. She said I had to respect what was going on with her body, and the guidance she received on how to protect herself. She complimented me on the fierceness of my spirit, and conjectured it would be an interesting experiment if we could "exchange hearts."

I was starting to feel better about her again. As we spoke, a large white ring appeared around the Sun—this was the second time in my life I had witnessed this phenomenon, after observing it once, the summer before, in a Wiltshire crop circle. A German woman who was part of our group, sitting nearby, said that a ring around the Sun was a symbol of the *hierosgamos,* the "sacred marriage," union of male and female energies.

A few hours later, the priestesses arrived—first priestess looked radiant, excited about the adventure ahead. We drove out to the Vale do Amanhecer, the Valley of the Dawn, a local center where spiritualists, mediums, and healers gathered to pray for a New Age of universal oneness, founded by a self-proclaimed clairvoyant truck driver in 1959. Along the route, the Daimeista who was driving us in her small car pointed out a place where a miracle had occurred some years ago—an apparition of the "Blessed Virgin" in the sky, showing herself to hundreds of astonished onlookers. Eager for miracles, our little team seemed deeply impressed by this.

"In the Islamic tradition, they warn you should be suspicious of miracles," I offered. "The Antichrist will also manifest his presence in the form of miracles."

As we approached our destination, the teacher said that in the future, those who survived the transition would return to living in the indigenous way, in small tribal groups, without technology. "We are going back to the original matrix," she said.

"I don't think you can dismiss the progress we have made in the last centuries," I replied. "Why would we go back when we can go forward?"

My attitude annoyed the teacher. As we parked, she said, "Even Terence McKenna predicted we were returning to the original matrix. He was receiving messages from the mushroom spirits, and they told him this."

"It all depends on what you mean by original matrix. I would think that original matrix is a state of mind, not a literal throwback to earlier conditions."

With cement pyramids, man-made lakes, and a huge wooden cutout of a blond billboard Jesus, the Vale do Amanhecer resembled an outtake from a Fellini flick. In the central plaza before a pseudo-Atlantean-looking temple, dozens of costumed mystics with gloomy faces, ranging in age from adolescent to ancient, were performing a laborious ritual. The style was Flash Gordon meets Mata Hari. They carried spears and swords, wore silver robes, bright-colored sashes, metal amulets, and veils for the ladies. I had worn shorts for the outing; because of this, I was told I would not be allowed into the sacred sanctum of the temple. As the others went inside, I wandered around, watching the ludicrous ceremony. A voice intoned orders through a loudspeaker, and the mystics rotated around a central circle, organized themselves into processional lines, the various ornaments on their costumes placing them in a hierarchy of sad-sack swamis. It seemed a scene from some low level of the astral plane, the mystics like uniformed wraiths doomed to wander there for eternity, hoisting their spears and jiggling their superstitious amulets. I had no doubt some of them experienced the occasional psychic flash or sudden healing that seemed to them like divine grace, but were just the rustlings of mischievous supersensible entities, mocking this sad world of ours.

But they were poor and undereducated people, doing the best they could with what they had—what was our excuse? Prayers and jangling amulets would not bring a new world into being, only consciously directed action could have that effect. The Vale do Amanhecer seemed to symbolize the dangerous path that the teacher was leading my friends along, as well as the priestesses, with their starry-eyed devotion to her magical mystery tour.

That night, once again, I couldn't sleep. The domineering voice returned, demanding I eject myself from this incarnation in the jungle if the first priestess would not be with me. I desperately tried to reason with it—

to no avail. If some barbaric segment of my psyche was demanding the priestess's subjection to my will, the more integrated aspects of my ego yearned for a woman I loved who could join me in this new realm I had entered—simultaneously in the dregs of suffering matter, facing the impersonal demands of divine or daimonic forces; and at the same time, exalted, blissful, seemingly in rapport with distant aspects of the cosmos; stuck, struck, despairing, broken open, stripped bare. Finally I drifted into unconsciousness, only to awaken in the predawn with hypnagogic visions of a jeweled serpent undulating through rainbow-hued underworlds. Fragments and phrases forced themselves into my mind, demanding to be transcribed—the "other" intended to make its presence known through a text. I was the vehicle of its self-expression.

WE FLEW TO CRUZEIRO DO SUL, our gateway to the Amazon, a six-hour flight, passing over sections of the rain forest resembling a huge sea of rippling kale. We were met at the airport by Padrinho Alfredo Gregorio de Melo, the head of the church, a tall, gaunt man in a T-shirt, with a thin beard and congenial smile. The local Daime community had a much different ambience than the ones in Brasília; it was poorer, calmer, and tropically slower-paced. The teacher distributed presents and hugs to the local kids, addressing some of them as her "godchildren." Our group strung up hammocks and mosquito nets in one large communal room. Outside our house, frail bushes of chacruna—*psychotria viridis,* their dark, shiny leaves containing DMT—were being cultivated in rows.

The community organized a work for us—a celebration, one of their ceremonies that involved all-night dancing. We were brought into their church, which was painted light blue, festively decorated with blue and white paper streamers twirled across the ceiling. I was shown to my spot—a little three-by-two-foot horizontal box outlined on the floor—and after Padrinho Alfredo's introduction, we drank the first cup of Daime. The dancing was a simple two-step shuffle, accompanied by the perpetual singing of the hymns. I kept at it for an hour or two, until my enthusiasm wilted. I left to sit by the fire. Casey, my fellow reprobate, was already there. In this peaceful village, the attitude was completely relaxed; there were no guardians prodding me back into the ceremony, just some children who kept approaching to ask me questions in Portuguese, then collapse into giggles at my answers. Padrinho Alfredo walked past where Casey and I were sitting. He smiled at us gently.

"Drink Daime?" he asked.

We nodded, and he gave us the thumbs-up sign, then returned to the church. I felt a wave of friendliness and trust toward him.

This batch of Daime seemed significantly stronger than what we drank in Brasília. I decided to sneak off to my hammock to see what was happening in my interior vision-space. I lay down in the empty room and closed my eyes. Almost immediately, I was surrounded by the "machine elves of hyperspace," described by Terence McKenna. These speedy critters kept pointing at objects they held in their hands that resembled multidimensional Rubik's Cubes, directing my attention into kaleidoscopic vortices and twisting tunnels. "See? This is how we make new realities—look down here, quick! It's another universe," they seemed to be telling me. Mutably Lego-like, they rotated their heads and limbs around, disappearing and reappearing. I began to get bored with their antics. They seemed to represent the churning activity of a certain level of mind that was analytical—intellectual thought able to break anything down into deconstructed splinters—but adolescent.

Padrinho Luis Fernando walked through the room. When he saw me, he sighed.

"Daniel, Daniel," he said softly, not stopping on his way.

What was his problem? I thought to myself. Doesn't he realize I am tired? It had been a long day of travel. How many hours did he expect me to go on doing that little two-step? It was absurd. I considered the teacher and the priestesses, who had been steadily shuffling away, on the women's side of the church, across from me. It occurred to me they were at least as fatigued as I was—the priestesses must be jet-lagged—but they weren't complaining, or lying down on the job. On the flight to the Amazon, I had talked with second priestess about the Daime works, and she said she approached the ceremonies as an exercise in yogic discipline. I closed my eyes again—the machine elves were still rummaging through their bag of tricks, pointing at things, skittering around in hyperspace.

Fuck these goddamn elves, I thought. I want to dance.

I went back to the work as they were lining up for another cup of the medicine. I drank it, shuddered, and returned to my place. For the first time, I gave myself up to the ceremony and allowed myself to "enter the current," as they say in the Daime. Overcoming my resistance, I forced myself to continue to shuffle, attempting to sing, while ideas and images flooded my mind. The room seemed to be pulsing with energy. When negative thoughts or neurotic impulses surged up, they would be sheared away by the repetitive simplicity of the practice, which seemed to center me in the healing stream of the medicine. The ceremony was like a mindfulness meditation, purifying and elevating.

I gave myself over to the Daime that night. I felt the doctrine enter my heart, overwhelming my intellectual defenses and resistances. I still do not know how this happened. I connected with the deep humility and unadorned truth of the tradition, abandoning all negative judgments. The Santo Daime was a necessary protection, I sensed, counterbalancing whatever arrogant self-inflation came with my Quetzalcoatl transmission. By the end of the ceremony, I knew that the Daime would remain an essential part of my personal path.

A central concept of the Daime is "firmness," and I was impressed by Padrinho Alfredo's gentle yet firm manner as he conducted the work. There were several interruptions. One woman seemed to become possessed, shouting incoherently. She was led outside, where a group of musicians gathered around her, serenading her in an effort to calm her down. Another man demanded to sing several hymns out of sequence. Alfredo stopped the ceremony, patiently waiting until the disturbances subsided, and then began again.

The faces of the parishioners ranged a vast spectrum from jet-black to mestizo to Indian to our European contingent. I watched tiny wizened indigenous women—perhaps born into rain forest tribes, in a different world—singing the hymns and dancing the two-step. There were fishermen, workers, and farmers, their countenances hard and dignified, ravaged and soulful—leading inner lives unimaginable to me. There were children—adolescents and younger—who held their place from beginning to end, with no smirking or fidgeting. Lou Gold told me later, "Most of them are just here looking for a better life." Through Daime, they had found it.

Until that night, I had not understood the essence of devotion as part of a spiritual practice. Returning to the empty church the next morning, under the flittering streamers, still deeply affected by the mood of the ceremony, I realized that devotional prayer produced a pure tone, a heart-centered vibration of humility. At the deepest level, prayer was not some sentimental frill or old-woman consolation. It was an energy necessary for maintaining the order and purpose of creation. Without this binding tone underlying reality, it seemed to me, the world would simply crumble away.

CHAPTER FIVE

Our boat trip—three days and nights down the muddy brown Jurua River, surrounded by impenetrable jungle, to the Santo Daime village of Estorrões, also known as Céu do Jurua—was one of the great experiences of my life. The old, weathered two-story boat chugged loudly as it put-putted along. Casey and I strung our hammocks up on the top, outside, protected by a protruding roof from the worst of the rain. I had started smoking cigarettes in Brasília, and Casey gave me my own pack—the brand was Hollywood, a local product of Brazil. Reclining in the hammock and smoking a cigarette, heading deeper into the secretive greenery that opened on both sides of the river, was a sublime treat. I considered how nobody among us—not myself or the teacher or Alfredo—really knew anything at all, no matter what we pretended. Who could say how we had arrived at this place, or where we were ultimately heading? We were fellow travelers floating across the surface of the void, carried along by the current.

There was little required of us except to get acclimated to the dense humidity, and eat tasty meals of river fish and beans prepared by Luis Fernando's wife. Each night on deck, we met to rehearse the *hinarios*—the more the songs were drummed into my head, the more I was starting to appreciate their simple but unusual melodies, heartfelt poetry, and skittering rhythms. The teacher barely moved from her hammock, resting in the downstairs cabin with the others.

I awoke past three A.M. one night to find the river fog-shrouded, trees on the bank invisible in mist, and noted a startling silence. In the afternoon, the motor had developed a cough, sputtering, cutting out, then coaxed back to life. Now the motor had died and our boat was no longer going forward, just circling aimlessly. I wondered how many hours we had been in this state. I heard nervous voices down below as the crew struggled with the problem. I walked to the front of the boat where the spotlight illuminated ghost branches reaching from darkness. I returned to my hammock and tried to rest.

On the verge of sleep, I entered a visionary state. With eyes closed, I was approached by a group of forest spirits—elementals, some squat and some thin and long, with geometrically patterned faces and sad owl eyes, resembling the carved and painted images on sculptures, masks, and shields of Amazonian tribes. They conveyed they were worn out from sustaining the planet for us for so long, on the verge of giving up. I promised, when we got our act together, we would come back to this river and perform Daime rituals along the banks to thank them and celebrate them. They seemed to appreciate this. As the hallucination ended, the motor sputtered into life again.

The next afternoon, the priestesses lay on Casey's hammock, swaying next to me. Second priestess mocked my new ideals of erotic freedom, which I had explained to her over lunch in New York. We had been told that Padrinho Alfredo had several wives in different Daime communities.

"I'll bet you would like that," she said.

"Only if his wives can have other lovers," I said.

I explained the anguished thoughts and introspective surveys of my own personal life that led to my new understanding. As we spoke, I realized that the second priestess understood more than I thought—her mocking of me was, perhaps, a defense mechanism. After a while, I became impatient—with the first priestess present, I could no longer keep the discussion on a theoretical plane.

"For instance, I love the first priestess," I said. "I felt she gave strong indications she wanted to be with me. When I changed my life and it became possible for us to be together, she acted as if I was completely mistaken. She took no responsibility for having conveyed any other message or intention. When it became inconvenient for her, she acted as if I was utterly deluded."

"I never treated you with anything except generosity and love and courtesy," first priestess said. "I thought our friendship was beautiful."

"I don't find it beautiful to have a nonsexual relationship with a woman I desire, who has said she loves me. I find it constraining and ugly."

"She is in a monogamous relationship with someone else," second priestess said. "You have to accept that."

"I do accept it. At the same time, I don't think the first priestess knows, or wants to know, her own nature."

"I can't listen to this," first priestess said. She left the hammock to consult with the teacher.

LATER THE FIRST PRIESTESS returned and asked if we could sit together. She said my attitude had brought up buried issues from her childhood and her past relationships with men. We meditated together in silence. Looking for safe ground, we talked about the ceremonies. She said she loved the healing that took place during the Works.

"It is wonderful how we are learning to transmute suffering into light," she said.

"Why don't you transmute my suffering into light?" I asked.

I then made the abysmal, embarrassing mistake of trying to use rational argument to resolve matters of the heart; at the same time I was unable to conceal my anger. I talked about the negative manifestation of the Shakti current, finding her inconstant behavior a perfect example of it. I explained that women, through use of Shakti, were choosing what type of male behavior was socially condoned and supported. I railed against prisons of monogamy and convention.

"I am sick of the cowardice and deceit of women. You want to help save the world? Why don't you make your men into warriors, into fighters for human freedom?" I had met her boyfriend, a nightclub promoter, and sarcastically paraphrased his underlying attitude during a conversation we had: "Alcohol is a great drug. Cocaine is cool. It's cool that most people are programmed robots."

"I can't listen to this," she said again, and left.

Again that night, I couldn't sleep for many hours as the warlike voice howled in my skull, demanding some unlikely consummation with the first priestess. When I finally lost consciousness, I entered a dream that took me far beyond the reaches of language. In the dream, I seemed to perceive our space-time realm, with its finite dimensions, and the atemporal transcendent realm of the origin and the archetype as two vibrating matrices, shooting sparks of lightning between each other. A voice in my dream spoke one emphatic, enigmatic line:

"WHAT MANIFESTS OUTWARD FROM THE GROUND OF BEING IS FREEDOM IN TIME, AND FREEDOM FROM TIME."

I awoke in pitch blackness, these words ringing in my ears, and staggered downstairs to the bathroom where there was light, scrawling it out on a discarded tube of toilet paper. It was part of the Quetzalcoatl transmission, which kept coming into my mind in fragments and phrases, like puzzle pieces to be arranged.

Our boat finally reached its landing, and we disembarked. Immediately, we were confronted with a change in plan. From Cruzeiro do Sul, we had brought with us a large cargo of chacruna leaves, perhaps twenty sacks. On the ship, we had helped to prevent the leaves from fermenting in the heat by pouring them out on the deck and rotating them daily, discarding those that had turned brown. Suddenly, the Padrinhos decided that we would clean the leaves before hiking to the town. They marched us to a large one-room barn or storehouse above the muddy riverbank. The chacruna leaves were emptied in the middle of the floor, making an impressive pile. The leaves, we were told, had to be cleaned by hand, one by one. The teacher cackled at this surprise. She said it was typical of the Daime. It seemed a kind of test.

As part of our week-long initiation process in the jungle, we would help to make the medicine. Ayahuasca is brewed from two plants, combining the "force" of the Banisteriopsis caapi vine, containing beta-carbolines, with the "light" of the chacruna leaves, full of DMT. Usually, caring for the two components of the Daime are gender-specific assignments, with women cleaning the leaves, while men gather, scrape, and pound the curlicued vine. For some unknown reason, Alfredo decreed that the men would also work with the leaves that day. When women are "mooning"—menstruating—they are not allowed to do this work. Except for the teacher, almost all of the women in our group were in their cycle. They sat off to the side and watched.

We were instructed to separate out a clump, pulling a pile in front of us. As our small group began this seemingly Sisyphean labor, we were given cups of Daime to drink. As I turned on to the medicine, the task quickly transmuted from an arduous chore to the world's most delightful occupation. Each leaf, as I picked it up, presented itself to me as a precious feminine deity, a little princess, preening, demanding to be attended to in just the right way. The medicine seemed to instruct me in how to take care of it, putting me into a trance of receptivity. After an hour or so, my personal pile had barely shrunk. I looked around and estimated; it was clear, at our present rate, we would be working on these leaves throughout the night or far longer. I mentioned this to the teacher, and she laughed. She said they might keep us here for days until all of the leaves were done.

Most of the men in the group, including Casey and Lou, faded after a while, taking naps or wandering around aimlessly. Since the night of the dancing, my attitude had changed as my suspicions had lifted. Realizing there was a deeper wisdom at work, I decided to dedicate myself to whatever task I was assigned. In the melting heat, tempers were beginning to rise. Lou was traveling with a fellow Oregonian, Michael, an older man with a salt-and-pepper beard who belonged to the same Daime chapter. Michael sang hymns while we cleaned, then broke off to launch into long cliché-riddled sermons, as if mimicking a dissolute backroads preacher from the Old West. The teacher asked if he would be willing to sing without talking, and he snapped that he didn't like taking orders from a woman. I nudged Casey, who was loafing on a bench, and asked if he would consider returning to help us with this chore. He snarled at me angrily. Padrino Alfredo, who was cleaning the leaves with us, started to sing a soft and beautiful Daime hymn, dispelling the tension. Some of the local women also arrived and joined us, working at a much faster pace.

After several hours, we broke off for lunch. They fed us at the nearby house, a rudimentary structure, its interior wooden walls covered by a collage of colorful pages from Brazilian fashion magazines, a few scrawny chickens pecking their way through the rooms and wandering outside. The villagers sat with us quietly. They were models of impassivity. Whenever they were not required for some specific task, they conserved energy by remaining still, demonstrating the best way to survive in the sweltering jungle. When our meal was over, we were told that it was time to hike to the town.

Shlupp, shlupp, shlupp—the sound of my K-Mart slippers sinking into jungle muck. We were passing through a botanical wonderland—the greatest profusion of species of flora to be found on the planet—and it was all I could do to keep my focus on my footing, reaching down every few steps to pull one or the other of my slippers back on, preventing the mud from claiming it. We crossed swampy patches and streams where moss-covered logs had been hacked down to make slippery bridges of sorts, and other areas where there were no bridges, and we waded knee-deep through brackish bogs. When I did have the opportunity to look around, the forest seemed overpowering, yet surprisingly still—peaceful rather than ominous, though Lou Gold told me that if you stepped a few feet off the path, you could easily get lost forever, and if you got lost, you would "go from forest floor to canopy in under forty-eight hours."

We arrived at Céu do Jurua just before dusk. The town was primitive, lacking comforts, with houses strung along winding paths like beads on a string. There was no plumbing, no phone, and no electricity, except for one generator for the obligatory bright light during ceremonies. At the center of the village was a small cluster of buildings including a kitchen and eating space, and two ritual centers; a beautiful hexagonal church, and an outdoor ceremonial area with a large white cross at its center. The Santo Daime symbol modified the Christian icon with a second, smaller horizontal line below the traditional crossbeam. The teacher explained to me that this second line symbolized the "Second Coming" of the Christ, not as individual being but as collective realization. "The Daime believe that the Second Coming occurs within our hearts," she said.

Our group was bunked in an undivided house that was essentially an oversized shack, swarming with mosquitoes. Various other annoying insects were also plentiful, including biting flies and hornets and invisible "peons," minuscule menaces that left tiny blood spots that itched like crazy when scratched. These irritants hardly bothered me, but attacked other members of our group. After a few days, the teacher's ankles had swollen to elephantine proportions from their assault. In semidarkness, we strung up our hammocks and mosquito nets, then investigated the interiors with our flashlights, hunting down mosquitoes trapped inside our protective cocoons.

AFTER BREAKFAST, the men of our group, myself among them, were taken to the *feitio,* where the Daime is made in a semi-industrialized manufacturing method, probably instituted by rubbertappers in the 1920s, and unchanged since. Sitting on the ground in a circle, we carved out the moldy and rotted parts of thick chunks of vine, given Daime to drink first. The medicine again seemed to guide the process. I concentrated on whittling away the reddish-brown bark while observing my thoughts spreading out in different directions. I appreciated the perfectly imperfect nature of this organic matter; each piece of liana required an extended series of minute calculations, slices, and chops before the work could be considered satisfactorily complete. Entranced by this labor, I lost all concept of time. I meditated upon my usually abstract and distant relationship to physical reality—my intellectual's curse. I longed to heal this deep separation within me, and this task seemed to indicate how any skill could be mastered with attentiveness, patience, and self-discipline. I hoped I could maintain this attitude when I returned home.

Learning how to work with my knife, I nicked myself a few times, drawing blood. One of the Germans in our group was also bleeding. Recalling we had been warned that open cuts became infected very quickly in the fetid jungle, I told Luis Fernando I was going to get the first aid kit. I realized I wasn't sure which path to take back to the village—the direction was indicated vaguely. All of the villagers watched this exchange with smiles—once again, it seemed a kind of test. I found the right road, walking over a swaying wood bridge suspended over a local stream, and retrieved the kit from our house. Returning, I realized how high I was. Every leaf, every branch, seemed to be pulsing, electrified by its vital current. On my way back, I took a wrong turn without noticing. After a few hundred yards, I stopped and looked around, exhilarated by fear, noticing several possible branching paths behind me. The foreboding immensity of the jungle pressed down upon me. I recalled what I had been told, that if you went even a few feet off the path, you could be lost forever. I stopped my swirling thoughts, focused my attention, and retraced my way back to the *feitio,* dressing the cuts suffered by myself and the German.

The teacher said that the first priestess and I were caught in some complex process in which our "shadow material was tangled up together." She visited the village herbalist and retrieved a second medicine for the two of us to take during the week. The tincture was called Saracura, sharp but sweet-tasting, made from the blossoms, branches, and roots of a rain forest tree. She said it helped "clear up things caught in the energy fields." She and her husband had used it, during a difficult juncture in their relationship, while in Peru. She moderated a meeting between us in the empty church. We were still at a communicative impasse. First priestess said she did feel love for me "but it is more of a heart thing."

"I just feel like . . . I miss you," I said, surprising myself.

First priestess started to weep. She took my hand. I asked her to explain her emotion.

"I'm not thinking," she said. "I am just feeling."

FOR OUR FIRST JURUA CEREMONY, the Padrinhos trotted out their heavy artillery—the teacher had warned us the medicine would get stronger as we continued on our journey, and this was experientially confirmed that night, when the cosmos seemed to break open in a phosphorescent downpour, razzle-dazzling us with its loving warmth. I was now an enthusiastic participant in the singing, taking pleasure in mastering the unusual Portuguese locutions. I felt an expansive embrace toward the parishioners gathered around me—the beautiful, serious children so patient in their rows, the mestizo men and women with their hard and tender faces, my fellow fragile travelers. The medicine of the songs seemed to twirl the gyres of astral machinery—"pollen of the flowering godhead,/joints of pure light, corridors, stairways, thrones,/space formed from essence, shields made of ecstasy, storms/of emotion whirled into rapture . . ." the poet Rainer Maria Rilke wrote, as if he had been there—that became perceptible in the air around us.

We paused for a long silent meditation. With eyes closed, celestial deities, dakinis, winged Apollonian messengers whirled and rushed toward me, bearing offerings of flowers and fruit. I noticed the presence of a darker figure by my left side reaching toward me—I was scared to look in his direction. Was this another sorcerous emissary from some twilight demon zone? Finally, I peeked. There, in my vision, was Padrinho Alfredo himself, smiling at me softly, waiting to shake my hand. I mentally extended my hand to him and he took it, then disappeared. A storm had started, pattering across the roof, inducing synesthetic effects. As we resumed our singing, I saw and felt the presence of a vast being—feminine, angelic—twirling above us, her incorporeal body formed from patterns of rain and particles of astral light. I understood this was the Mother of the Forest, blessing and nurturing us.

Later in the night, I began to lose my focus; the visions became less paradisiacal and more ambiguous as my energy shifted downward. Padrinho Alfredo interrupted the singing and spoke, with Padrinho Luis Fernando translating, about this, exactly. Again and again, he used the word "firmness"—it was necessary to remain firm, to concentrate the mind's attention upward, toward the light. I felt his words resonating in my inner depths, instinctively raising my hands to my forehead in a gesture of prayer.

LONG AFTER THE CEREMONY had ended and our group had returned to the room to sleep, I sat in front of the hexagonal table at the middle of the ceremonial space, watching moths and mystery bugs circling and sometimes incinerating themselves in the sputtering candles that illuminated framed photographs of Mestre Irinieu, Mestre Mota de Melo, and other leaders of the church, as the rain fell outside. I felt I was reaching an understanding of different forms of human evolution taking place concurrently. In my urbane and cynical culture, people had developed their cognitive faculties and discriminatory skills to a refined degree, able to deconstruct a literary text or feel the thread count on a fitted sheet. In the mestizo and indigenous world, the emphasis of evolution was on the deepening of soul-qualities, relating to nature and the slow cycles of natural time, persevering despite suffering and oppression. If I had, perhaps, developed my intellectual skills more than Padrinho Alfredo, he had attained a deeper level of soul development, which was a different wisdom.

I received another startling insight, rising from the depths of my psyche. In one flash of understanding, I seemed to apprehend the origin of my painfully tangled situation with the first priestess. I recalled our trip to Hawaii, our first mushroom ceremony, where she had been told that in a past life, she was the bride of Ashoka, a Buddhist potentate in India. In her visions she had spoken with him. I had not thought much of this at the time—it seemed part of the magical realism of our idyllic week together. I had never believed in specific past lives, particularly if they were associated with celebrated personages. That kind of thing seemed to belong to the naive spiritual trappings of the overeager New Age.

All of a sudden, I beheld faint visions and scenes—the priestess as veiled dancer, myself as Asiatic potentate watching her—and felt the emotional imprint of a long-lost bond. I seemed to access an older form of my consciousness, as psychic frequency or vibrational tone. One level of my awareness was sheared down to its essence, isolating something like a background hum, which I felt but could never identify before. What remained was a warrior mentality, almost too chilled for my modern personality to accept. Ashoka's spirit—if that was what it was—was like a knife that cut to the core, coolly and accurately. The martial ferocity of this energy had returned to me during the Mars conjunction at Burning Man, introducing a new perspective on the "war of the sexes," focusing my attention on the Kali energy dominating the planet, and leading me to review my life through the prism of Kali's wrath.

According to Amit Goswami's thesis on reincarnation, certain lives are correlated via "quantum nonlocality," having privileged access to each other through "nonlocal information transfer." As I recognized what seemed to be a past incarnation, I received an intuitive download about what had transpired in that long-ago epoch. What I recovered was that first priestess had been captured during a war and forcibly taken as one of my brides. She had introduced me to Buddhism, as well as Tantra, and helped to change my path away from military conquest to the Buddhist pursuit of enlightenment. Confronting the horrors of war, Ashoka had realized the truth of the Buddha's dictum, "If one man kills a hundred men, and another man masters himself, that second man is the much greater warrior." He had dedicated the rest of his life to this goal.

As bizarre as this seemed, it explained the savagery of the voice that screamed in my head, night after night, demanding that I be with the priestess. The voice seemed a residue of that past incarnation, expressing the erotic entitlement of an Indian emperor from the third century BC, demanding the return of his consort. Perhaps it also explained the needlessly brutal e-mail I had sent the priestess in New York, proposing we explore a Tantric practice—at that point, I had no idea I actually meant we should resume one. Whether we had tripped into a literal past life, or accessed some archetypal pattern, I found it extraordinary that the first priestess was fulfilling a role in the present situation that was directly analogous to the one she would have played in our previous life, when she had brought me to Buddhism. This time around, she had brought me to the teacher and, through the teacher, to the Santo Daime.

THE NEXT MORNING, I pulled the teacher aside and asked if we could confer in private. Sitting by the cool bathing stream near our shack, I told her about my ongoing Quetzalcoatl experience, and the prophetic transmission I was receiving. She asked to hear it, and I read it to her. When I was finished, she smiled. She told me that, many years ago, she had had a series of dreams where she was a girl and Quetzalcoatl, in the form of a gigantic feathered serpent, came to visit her, swirling around her like a Chinese dragon. The mood of these dreams was friendly, even gentle. "I was surprised in the dreams that such a powerful being could be so playful," she recalled.

"I am surprised you aren't more surprised by my story," I said.

"I knew you were in a major process," she said. "That's why I wanted you on this trip." She said she had watched me writing while we were on the boat, organizing the text of the transmission, and sensed a powerful force enveloping me. She worked with other people who were receiving and mediating archetypal material—above all, she had her transformative experiences with the Mother. She also recalled witnessing the ring around the Sun that morning in Brasília and interpreted it—as I did—as a sign.

"When I go back to New York and try to tell people about all of this, who is going to be able to understand?" I asked.

"Some will understand," she said, "but you will see others look up for a moment while you are speaking, and then fold back down again as soon as you are done. When you see them folding over—that is their karma at work."

I asked if we could meet with the first priestess again, and the teacher went to get her. I shared the Quetzalcoatl transmission, and then told her of my insight about our possible past life together. The energy immediately cleared between us. The priestess said she felt "greatly relieved" to have reached this understanding. I was as well—although it did not liberate me from my tormenting desire for her, at least I could attribute it to a theoretical origin.

THERE WAS ANOTHER DAY of working at the *feitio.* They taught us the *batisan,* the rhythmic and heavily physical labor of pounding the vine into strands with heavy mallets, before it was boiled with the leaves in huge metal vats set in concrete. Each component in the making of the medicine was made sacred with singing and prayer. I focused my frustration into the ritualistic smashing of the vines. Even though I had a deeper appreciation of the situation, I still felt cheated out of my private self by the arrival of this alien archetype within my psyche, and its drastic directives about our immediate future. And yet this revelation was not entirely surprising—I could see how I had been carefully prepared, sneakily impelled toward it.

Our final work before leaving the jungle was a healing ceremony. My intention was to work on mending my mediocre relationship to the physical world. It seemed to me I had never completely incarnated, come down to Earth, since a protracted illness pulled me away from my bodily nature in early adolescence, and I had never fully reentered it. Just before leaving New York, after more than twenty years, I had been contacted by one of the children I had known during my long residence in a hospital, when I was eleven. Although permanently confined to a wheelchair by crippling scoliosis, my long-lost friend told me he had become a lawyer, working for the rights of the handicapped. It was almost more than I could imagine, the suffering he must have endured. When I told the teacher about this, she noted that the timing was not coincidental; part of my healing work during our journey was "soul retrieval," revolving around this critical point in my past.

Once again, the visionary energies released during this ceremony, using the medicine we had made over the course of the week, seemed monumentally vast. The Mother of the Forest was called in by the songs:

Que procurer esta casa
Que aqui nela chegar
Encontra com a Virgem Maria
Sua saude ela da

twirling over our heads with her arms outstretched, sheltering us beneath her vast umbrella. A fountain of light appeared to rise from the center of the room, a column of energy, pulsing with the harmonic mesh of the singing. I had never felt so close to the presence of the Sacred—not as an idea or a theory, but as something that filled my whole being.

I perceived my past impersonally, as if it belonged to somebody else, realizing how each small strand of it—the deepest wounds, the most dissolute and despairing aspects—was a necessary part of the weaving, required to bring me to this place, to integrate what appeared to be prophetic knowledge, with the teacher as spiritual midwife. Without her mediating influence, I would have been lost. Late in the night, I seemed to commune with cosmic entities of pure thought, beaming a message at me. Their words hovered in my mind, then scattered away. I retrieved my notebook and scrawled them down before they vanished from my memory: "YOU GO DEEPER INTO THE PHYSICAL TO GET TO THE INFINITE." I promised myself, in the future, I would find the will to follow that dictate.

We left Céu do Jurua the next day, accompanied by further portents. Instead of hiking, we took a motorized longboat through tributaries to meet up with the sputtering craft that had brought us downriver and would carry us back toward civilization. For more than half an hour, as our motorboat put-putted through estuaries and across dream-like lakes, an eagle flew before us, gliding from branch to branch, as if protecting us.

During our return voyage, I received, and wrote out, the rest of the transmission announcing Quetzalcoatl's return:

> I am an avatar and messenger sent at the end of a kalpa, a world age, to bring a new dispensation for humanity—a new covenant, and a new consciousness.
>
> I am the same spirit who appeared here, in the Mayan period, as Quetzalcoatl and incarnated at various other points in human history. Like Avalokiteshvara, the Tibetan Buddha of Compassion, Quetzalcoatl is an archetypal "god form" that occasionally takes human rebirth to accomplish a specific mission. As foretold, I am also the Tzaddik—"the righteous one" and the "gatherer of the sparks" of the Qabalah—as well as the "Once and Future King" promised by Arthurian legend.
>
> I do not let anything interrupt me in my quest for truth—neither fear nor indifference, poverty nor cynicism. In the realm of thought, I practice warrior discipline. As gravity draws matter to it, I have pulled myself back into manifestation in this realm, from the depths of cosmic space, piece by piece and bit by bit, reassembling the component parts, the sparks of thought, that make up my being—which is, primarily, a form or vibrational level of consciousness.
>
> Soon there will be a great change to your world.
>
> The material reality that surrounds you is beginning to crack apart, and with it all of your illusions. The global capitalist system that is currently devouring your planetary resources will soon self-destruct, leaving many of you bereft.
>
> But understand the nature of paradox: For those who follow my words and open their hearts and their minds—for those who

have "ears to hear"—there is no problem whatsoever. What is false must die so what is true can be born.

You are, right now, living at the time of revelation, Apocalypse, and the fulfillment of prophecy. Let there be no doubt. You stand at the edge of the Abyss. What are those shadows that crowd around you? They are the unintegrated aspects of your own psyche, projected into material form. The word "Apocalypse" means "uncovering"—and in these last clock ticks of this world age, all must be revealed, uncovered, so that all can be known.

You have just a few years yet remaining to prepare the vehicle for your higher self. Use them preciously. For those who have gained knowledge of the nature of time, a few years—even a few days, a minute—can be an eternity. For those sleepwalking through reality, time exists only to be wasted—as they too will be wasted, in their turn.

"Reality," as you currently experience it, is something like a waking dream. It is a projection, or let us say an interface, disguising deeper and more intensified levels of being and knowing. For those who are ready and willing, the doors to those other levels now stand open.

Those who have expended their lives in the pursuit of egocentric and material gains—without courage or originality, without fighting for human freedom or the preservation of the planetary environment—will also receive the rewards that they deserve.

The materiality of your universe is a solid-state illusion. What is this universe? It is a poem that writes itself. It is a song that sings itself into being. This universe has no origin and no end.

What you are currently experiencing as the accelerated evolution of technology can now be recognized for what it is: a transition between two forms of consciousness, and two planetary states. Consciousness is technology—the only technology that exists. Everything in this universe is conscious at its own level, and in the process of transformation to higher or lower states.

The first principle of my being is unconditional love. As a rational intelligence, I accept the logic and necessity of the Christ consciousness, that we should love one another as we are loved. Love and devotion are vibrational frequencies that maintain reality. Love can only be given in freedom; therefore, to be human is to be free.

I resonate, at the same time, with the essence of Islam. Islam means submission, surrender, to the will of God—a more polite way of saying this is "Go with the flow." But either formulation is correct. Whatever you do, in fact, resist as you think you might, you are always submitting to God's will. So why not give the process your joyful assent?

I am in complete harmony, as well, with the Tibetan tradition of Dzogchen. Ultimately, there are no entities—there is neither being, nor nonbeing. From the perspective of nondual awareness, samsara is nirvana. The Apocalypse, the Kali Yuga, the Golden Age—these are all states of mind. Hell is a state of mind. When you eliminate fear and attachment, when you self-liberate, you attain the Golden Age.

This universe spontaneously self-organizes into higher levels of consciousness and wisdom. Underlying all are great cosmic entities or vibrational fields, alternately at play or at rest. Not satisfied with mere enlightenment, the god-form Quetzalcoatl still seeks to puzzle out the workings of these deeper forces—hence the reason for his return to your realm. He and his kind have been granted this world for their continued exploration—made with loving reverence—of the many layers of galactic intelligence, cosmic illusion, daimonic beauty, and telluric transformation. All are invited to participate with them.

The current transition is, simultaneously, a return to origin. The original matrix of this new world reality is the ecstatic limitlessness of your own being. This world—any world—is the ground for a certain level of being. What manifests outward from the ground of being is freedom in time, and freedom from time.

My "doctrine" is not transcendent, but immanent. It is not "somewhere out there." It is here and now. The task of human existence is to transform the Earth, to reconcile spirit and matter in this realm. We go deeper into the physical to reach the infinite. As there are no conceivable limits to this task, God, in his greatness, has granted us a project that is without limit and without end.

Thinking is a part of reality. Thought generates new potentials and possibilities of manifestation. Thought changes the nature of reality. Thought changes the nature of time. As a philosopher, I naturally deify the goddess principle. I venerate Sophia, deity of wisdom, who clothes God's thoughts in material form, and worship Shakti, the erotic current of feminine energy that powers the universe.

The writer of this work is the vehicle of my arrival—my return—to this realm. He certainly did not expect this to be the case. What began as a quest to understand prophecy has become the fulfillment of prophecy. The vehicle of my arrival has been brought to an awareness of his situation in sometimes painful increments and stages of resistance—and this book follows the evolution of his learning process, as an aid to the reader's understanding.

The vehicle of my arrival had to learn to follow synchronicities, embrace paradoxes, and solve puzzles. He had to enter into a new way of thinking about time and space and consciousness.

Almost apologetically, the vehicle notes that his birthday fell in June 1966—6/66—"count the number of the Beast: for it is the number of the man; and his number is Six hundred threescore and six."

The Beast prophesied is the "feathered serpent," Quetzalcoatl.

Those who prefer to reject all of this out of hand are welcome to do so. In Qabalah, the virtue one seeks to establish on the "Earth Plane" is discrimination. It is up to the individual to find his way through the ideas presented here—of course he is entirely free to ignore them altogether.

But be forewarned: The End of Time approaches. The return of Quetzalcoatl foreshadows the imminent closing of the cycle and the completion of the Great Work.

CHAPTER SIX

Perhaps there is a law operating in the universe that the one who bends his mind to a paradox ends up insolubly meshed within that paradox? Perhaps the universe purely operates on wit, and the best joke, inducing the longest fit of cosmic giggles, becomes the operative law at the next quantum mind-shift? If, as the physicist Arthur March put it, "the world is inseparable from the observing subject and is accordingly not objectifiable," then perhaps undertaking the quest for prophetic knowledge, in itself, causes reality to shiver and shift, as new possibilities open like the petals of an extravagant, multidimensional flower? The message, as I apparently received it, that "a quest to understand prophecy has become the fulfillment of prophecy," suggested some such wild card hypothesis.

When I went back through my notebooks from the months leading up to the Amazon trip, I found many of the ideas that came through as Quetzalcoatl's transmission scrawled out as partial fragments. It was conceivable that ayahuasca had simply acted as a psychic amplifier, catalyzing my intuitions and secret desires into a form that seemed delivered by a presence—an authoritative voice—separate from my own mind. Or perhaps I had succumbed to a trap set by malicious entities from the astral plane, puffing me up with delusions of grandeur, ready to tear me down in the future, as they had done to poor Aleister Crowley, the "Great Beast" who finished his life as an impoverished alcoholic in an English boardinghouse?

During his prophetic download in the Colombian Amazon, Terence McKenna noted that it seemed as if an extradimensional consciousness—

a "humorous something"—had utilized every thought in his head to communicate with him. Was it possible that all of my reading and note-taking in the months before the event was a form of self-organization, in which "the other" pulled itself together, "piece by piece and bit by bit," before becoming known? Had I been preparing myself, without knowing it, to receive the signal from this emphatic archetype? Had I broken through, albeit fleetingly, to a new level of consciousness, receiving a "a new covenant" for humanity? Or did overuse of hallucinogens merely distort my judgment, tilting me toward madness? These questions, among others, swarmed around me.

To make matters worse, along with the potentially reality-shifting reception of the Quetzalcoatl transmission, I had accessed a hypothetical past life as an Indian emperor, fulfilling all New Age clichés. I had never heard of Ashoka before—or if I had heard of him, I had no recollection of who he was. When I researched the matter at home, I discovered that Ashoka was not just any old king, but the majestic Devanampriya Priyadarshin, "Beloved of the Gods," "He Who Looks Benevolently." This possibility seemed an incredibly self-aggrandizing stretch from the humble reality of my South Williamsburg life, where my reign extended over an underheated shoebox of a room. In any event, whatever happened in the Amazon—whether some archetypal pattern was revealed, past life triggered, or wish-fulfillment projected by myself and the priestess—the figure of Ashoka had forced his way into my mind, demanding consideration.

Born around 304 BC, Emperor Ashoka ruled the Mauryan Empire from 272 to 232 BC, overseeing a kingdom stretching from modern Afghanistan across Kashmir and Nepal, crossing the entire Gangetic plain and extending into Bengal. His territory encompassed the entire Indian subcontinent except for the southern states of Tamil Nadu and Kerala. Although he was famous as India's great "Dharma King," leaving inscriptions carved in rocks and pillars throughout his empire—eighty-four thousand of them, according to legend—Ashoka's activities as emperor were shrouded in mystery until his Brahmi script was deciphered by an English archaeologist in the early nineteenth century.

According to legend, Ashoka seized his father's throne after a violent civil war fought between him and his brothers, putting many siblings to death (ninety-nine of them, say the stories). When he had consolidated his rule, he conquered the Kalinga Empire to the south, killing one hundred thousand enemy soldiers and vast numbers of civilians, gaining the nickname "Ashoka the Fierce." His personal revulsion at the magnitude of this slaughter is said to have inspired his conversion to Buddhism's peaceful ways. One of his inscriptions proclaims, "Devanampriya, the conqueror of the Kalingas, is remorseful now, for this conquest is no conquest, since there was killing, death and banishment of the people. Devanampriya keenly feels all this with profound sorrow and regret."

Upon turning to Buddhism, Ashoka instituted government based on dharma—a word variably translated as "righteousness," "moral order," "duty," "truth," or, more specifically, the body of teachings of the Buddha. As part of this unique experiment in just rulership, he reduced tax burdens, built rest houses along highways, planted trees across his empire, instituted a policy of religious tolerance, eliminated most forms of capital punishment, and created a class of administrators schooled in dharmic principles. He was not a fan of public ceremonies and festivals, noting that such events "bear little fruit," compared to "the ceremony of the Dharma," requiring "proper behavior towards servants and employees, respect for teachers, restraint towards living beings, and generosity towards ascetics and Brahmans." An enthusiastic hunter in his youth, he became a vegetarian after his conversion, instituting protections for animals throughout the empire, substituting the "royal pilgrimage" for the "royal hunt." According to one tradition, when a jealous wife attempted to destroy the Bodhi Tree, under which the Buddha attained enlightenment, Ashoka personally attended to it, nursing it back to health.

Translators believe that the edicts were written in Ashoka's own words, conveying an intense desire to be clearly understood, frequently referring to the good works he has done, "although not in a boastful way, but more, it seems, to convince the reader of his sincerity." One of his "Seven Pillar Edicts" declares:

> Beloved-of-the-Gods speaks thus: This Dharma edict was written twenty-six years after my coronation. Happiness in this world and the next is difficult to obtain without much love for the Dharma,

> much self-examination, much respect, much fear (of evil), and much enthusiasm. But through my instruction this regard for Dharma and love of Dharma has grown day by day, and will continue to grow. And my officers of high, low and middle rank are practicing and conforming to Dharma, and are capable of inspiring others to do the same. Mahamatras in border areas are doing the same. And these are my instructions: to protect with Dharma, to make happiness through Dharma and to guard with Dharma.

Historians consider Ashoka one of the most important figures in the dissemination of Buddhism. Although the Mauryan Empire collapsed in the century after his death, Ashoka's reign was remembered as a brief Golden Age, commemorated by the Wheel of Dharma, displayed on India's national flag.

In Mesoamerica, the reign of the last human incarnation of the Plumed Serpent, Ce Acatl Topiltzin, in the tenth century AD, was also remembered, enshrined in legend as a brief Golden Age, marked by a renaissance in the arts and sacred sciences, and the abolition of the practice of human sacrifice. In *The Gospel of the Toltecs: The Life and Teachings of Quetzalcoatl,* Frank Díaz compiled extant source material about Quetzalcoatl. "I only came to prepare a way," Ce Acatl tells his followers before departing, in one surviving codice. "A new day is coming, the magnificent day of radiant beauty when I must return to myself. Then you will see me! Your brothers will understand the divine reasons. I will raise my harvest, and gather what I have sowed. The evil animal will forever disappear, and you will be able to walk in peace." Ce Acatl described the joyful period that would follow his return as "Ometeotl's dawn." As harbingers of a Golden Age, connecting Heaven and Earth, the historical monarchs Ce Acatl Topiltzin and Ashoka seem to possess a symbolic symmetry.

LIKE THE SIDEREAL, spiraling motion around a dream-mandala, I felt myself led, step by step, through a labyrinth, to discover, perhaps, that my psyche was not separate but utterly embedded in the investigation I had undertaken. Of course, I could also be deluded. As Terence McKenna noted after his own Amazonian plunge into prophecy, "The notion of some kind of fantastically complicated visionary revelation that happens to put one at the very center of the action is a symptom of mental illness."

If my personal fable in any way fits the archetype of Quetzalcoatl's return, it would represent the ability of the modern mind—the mental-rational structure, in Jean Gebser's model—to reaccess the suppressed shamanic dimensions, without drowning beneath the swamp waters of that primordial darkness. This assimilation of previous forms of consciousness would initiate what Gebser termed the integral structure—characterized by acausal unfolding and transformation—gaining authority as the systems of governance, economics, and thought currently dominating our world quickly collapse. As it says in *Tao Te Ching*, "Non-Tao is short-lived."

In Taoist as well as Mesoamerican thought, kingship is an act of mediation between levels or planes of being—between Heaven and Earth. The king performs the function of "centering the world," maintaining order from his place at the center of the cosmic axis. If a similar task of "world-centering" was required to shift our contemporary civilization to a different path, such an impetus could only come from outside the dominant structure—from the perspective, perhaps, of a somewhat bohemian and alienated intellectual, capable by circumstances of birth and education of conceiving the entire pattern, rooted in the new modes of perception developed by the avant-garde movements of literature and art of the twentieth century, versed in the liberational currents of the radical counterculture, and at the same time, rooted in empiricism and the pragmatic modern mind-set. By the same token, if reincarnation is an actual constituent of human evolution—as Eastern spiritual doctrines claim, as Rudolf Steiner reasserted within the Judeo-Christian tradition, and as the scientific research and hypotheses of Ian Stevenson, Rick Strassman, and Amit Goswami seem to support—then it seems possible, perhaps, that those who have the task of integrating, and fulfilling, the prophetic thought-streams at the "end of the cycle" would have prepared for this role in previous incarnations.

In *The New View Over Atlantis*, John Michell wrote: "A human society which is ordered and regulated on cosmic principles, demonstrably reflecting the order of nature and the heavens, is the only one which will attract

...erve general acceptance." As discussed earlier, the Gregorian calendar currently in use disregards the natural cycles that govern the Sun, Moon, Earth system, the stellar engine driving our planetary evolution, including the evolution of human consciousness. Although we do not generally realize it, our calendar functions as a subliminal programming device, orienting us to a particular perspective on time, space, and being. As José Argüelles discovered, not only is our calendar desynchronized, but our entire model of linear, spatialized temporality is misconceived—time is not equivalent to any sort of quantity, cannot be saved, killed, or overcome. Since avoiding or escaping time is impossible, the sensible alternative is to align with it.

At this juncture, there exist a plethora of reform movements and pragmatic initiatives aimed at different aspects of the social, political, and biospheric crises that plague our world. There are ecological designers and environmental scientists who have developed organic techniques to repair and regenerate damaged ecosystems. There are economists who have invented alternative currencies that would institute different systems of value, creating a more equitable economic order. There are extragovernmental commissions that bring the victims of human rights violations and officially sanctioned torture face-to-face with their perpetrators to resolve abuses according to a global standard of ethics. There exists a vast network of nongovernmental organizations striving for environmental goals, women's rights, conflict resolution, and civic reform, against the resistance of corrupt businesses and compromised institutions. There are efforts to use the Internet to develop social networks supporting sustainable development. There is alternative healthcare integrating indigenous wisdom-traditions with modern knowledge. There are myriad methods for producing forms of energy that do not ruin the environment while encouraging local industries and initiatives. There is the long-suppressed therapeutic use of psychedelics to heal mental disorders, as well as the sacramental use of traditional medicines for elevating the spirit. All of these efforts are important. However, no treatment of symptoms can hope to succeed as long as the underlying cause goes unacknowledged and unaddressed.

If the deepest root of our present predicament is our enslavement by artificial time, then a necessary part of the solution would be a new calendar—a new "timing frequency"—reinstituting natural harmony. The thirteen-moon calendar devised by José Argüelles represents a crucial first step, but this is not a task for one man, however inspired. It will require a meeting of minds from various spiritual traditions, indigenous cultures, and scientific disciplines, capable of overcoming factional discord to create a new global standard, one that can meet with general acceptance. Since the new time, once instituted, would be retroactive as well as projective, all of the congealed accumulation of international treaties, global trade agreements, nation-state constitutions, legal codes, debt structures, corporate charters would be wiped clean by this act, understood as manifestations of a certain frequency of time, and of mind, that is no longer applicable. If we address the timing error, other solutions can meld with each other to form a new pattern for our world. The acceptance of a synchronized calendar could initiate the shift into a unified planetary culture, one that honored the Earth and the human in all of its manifestations.

Of course, many will consider this outside the realm of possibility. However, if consciousness is indeed a quantum phenomenon that evolves by sudden phase-shifts into new contexts, then an imminent global shift into a new timing frequency, instituting a harmonic relation to the solar system is imaginable—and what can be imagined, can be accomplished. The self-organizing global brain of our digital and cellular communications systems allows for transformative ideas to circle the world instantaneously, when the moment is right. It is notable that one of the establishing acts of any revolution is the creation of a new calendar, whether in France in 1792 or Russia in 1918, seeking to strip away the old mind-set and assert a new concordance. The leaders of past revolutions intuitively understood the central importance of the calendar as a metaprogram for consciousness, although they were incapable of instituting a harmonic and nonviolent order due to their historical circumstances and personal limitations. The revolution that faces us is different from any previous one because it encompasses the entire Earth, requiring the nonviolent supersession of the current order, and the shift into a truly compassionate global culture, one that harmonizes with and protects natural systems rather than seeking domination over them.

If the god-form Quetzalcoatl has announced his return, it seems plausible that other archetypes will manifest in the brief historical period g to us. An intricate pattern of legends, mythic archetypes, and ophetic strands twine around the English village of Glastonbury, linking the town, in Michell's words, "with the age-old prophecy of the golden age restored, of the recovery of the Grail, the reappearance of Arthur's Round Table or the coming of Christ's kingdom on Earth." One tradition, linking New Age visions with ancient yearnings, calls for the gathering of the twelve tribes of humanity at Glastonbury, creating a thirteenth tribe, reconciling Heaven and Earth in a "New Jerusalem." Glastonbury might provide a perfect setting for the cross-disciplinary summit, reconciling science and spirit, required to bring down the imprint of the new time, which could be disseminated through the global commons.

By reestablishing harmony, we would realize, in Black Elk's words, "Every place is the center of the world." Or, as Robert Lawlor defined the aboriginal perspective, "There is no part of this existence that needs to be transcended, repressed, or gone beyond." Through this act of conscious concordance, we might also, quite naturally, reaccess the sacred sciences and ritual practices of prehistory, such as geomancy and alchemy. And if my hypothesis about the crop formations has validity, such a unifying gesture may be a prerequisite for humanity to enter into direct contact, benevolent communion, with other orders of galactic intelligence.

IN *THE NEW VIEW OVER ATLANTIS,* John Michell examined temples, pyramids, and monuments from Mesoamerica, England, Egypt, and elsewhere, propounding the thesis that there was once a planetary civilization based on a universal canon of measure and proportion and visionary awareness of the Earth's subtle energy fields. "The nerve centers of the Earth, corresponding on the human body to the acupuncture points of Chinese medicine, were guarded and sanctified by sacred buildings, themselves laid out as microcosms of the cosmic order, the universal body of God." He believed that knowledge of this order was preserved in the building of the Glastonbury Abbey, where various symbolic numbers, including 666, were utilized prominently in the architectural plan.

"The number 666 is that of Teitan, a solar deity," he wrote. "It represents an eternal, natural principle, transcending any moral category." Teitan—

Cheitan, in the Chaldean language—was the snake god who brought knowledge to humanity. As mythic polarity gave way to mental duality, this deity was reduced to a baleful influence, given the name Satan in the Bible. While the institutional church associated the number with evil, Gnostics within the church "recognized the number for what it was, an essential element in the true cosmic scheme, and in laying out their mystic citadels they allowed 666, the number of solar power, to occupy its due place in the numberical order." According to the Gnostic version of the biblical tale, the serpent in the Garden of Eden was actually an apparition of Christ, bringing knowledge to humans, seeking to liberate them from the wicked demiurge who kept them in ignorance. The "Great Beast" of the Apocalypse, first appearing in Revelation 13, possesses an intricate ambiguity, linking him to these older traditions.

The shattering of the Glastonbury Abbey as "the devil's temple" during the Reformation represented the final destruction, in the West, of the ancient mind-set that understood the necessity of maintaining a proper balance between cosmic forces. For the Puritans, "the beast represented some absolute principle of evil, irreconcilable with the iron rule of humanly created morality," Michell notes. "The beast, like the dragon, had to be suppressed." For this reason, as well as many others, Glastonbury could be the logical place for reintegrating Sun, Moon, and Earth, establishing a template for a planetary civilization.

As the poet William Carlos Williams said, "A new world is only a new mind."

EPILOGUE

when the clock strikes me
the powers of being
will prevail
over the powers that be
SAUL WILLIAMS

One story remains to be told, one more strand to be woven into this fable—of myth meshing into history, of the artifice of modern time unraveling to unveil the rudiments of a new reality. It is the story of the timeless time—what anthropologist Armin Geertz calls the "continuum consciousness"—of the Hopi, considered the original denizens of the North American continent, living in arid deserts of Arizona, awaiting the long-prophesied, long-delayed return of "Pahana," the elder white brother who wandered off at the beginning of this cycle, to usher in an age of universal peace. "The coming of the Hopi's lost white brother Pahana, the return of the Mayas' bearded white god Kukulcan, the Toltecan and Aztecan Quetzalcoatl, is a myth of deep significance to all the Americas," wrote Frank Waters in his *Book of the Hopi*. "It is an unconscious projection of an entire race's dream of brotherhood with the races of all continents. It is the unfulfilled longing of all humanity." Waters collected testimonies from thirty Hopi elders for his 1963 book, documenting the myths, rituals, and oral history of the tribe, whose drastic foretellings have long been a New Age cliché.

According to the Hopi, we are currently completing the cycle of the Fourth World—Tuwaqachi, "World Complete"—on the verge of transitioning, or emerging, into the Fifth World, with several more worlds to follow. In each of the three previous conditions, humanity eventually went berserk, bringing ruin upon themselves through destructive practices, wars, misused technologies, and loss of connection to the sacred. As the end of one cycle approaches, a small tunnel or interdimensional passage—the *sipapu*—appears, leading the Hopi and other decent people into the next phase, or incarnation, of the Earth. As in Rudolf Steiner's cosmology, the worlds represent increasing phases of densification, the involuting descent of spirit into matter, where "the way becomes harder and longer," as the creation-spinning deity Spider Woman told the Hopi after the flood ending the Third World.

I had often heard that the Hopi believe we have received most of the signs, recounted in their prophecies, preceding the emergence into the Fifth World. These include a "gourd of ashes falling from the sky," destroying a city, enacted in the atomic blasts obliterating Hiroshima and Nagasaki, and a spiderweb across the Earth, which they associate with our power grids and telephone lines. According to the accounts of the elders compiled by Waters, the Fourth World will end in a war that will be "a spiritual conflict" fought with material means, leading to the destruction of the United States, "land and people," through radiation. Those who survive this conflict will institute a new united world without racial or ideological divisions "under one power, that of the Creator."

Before visiting the Hopi, I had several vivid dreams that seemed foreshadowings. After seeing Godfrey Reggio's film *Naqoyqatsi* (a Hopi word meaning "Life as War"), the last in his trilogy beginning with *Koyaanisqatsi* ("Life Out of Balance"), I had a dream of fiery demons at computer workstations, and awoke with the sense of a visceral supernatural presence flying through my house. The night before leaving for the Southwest, I had an even more specific and frightening nightmare in which I was killed and dismembered by a disgusting-looking daimon—who was simultaneously, in typical dream dislogic, the postmodern conceptual artist Bruce Nauman. After being killed, I returned to Nauman's studio or the daimon's home and said, "Great—now that you have killed me, I control you." I went to a bookshelf and picked up a huge leather-bound volume titled *Grimoire* (a medieval catalogue of spells) and melted it down over a fire, as if it were wax. As I did this, I heard incredibly loud Native American chanting followed by maniacal laughter. I awoke once again with a sense of a powerful presence, wildly unhinged and amoral, looming overhead and then soaring away.

In 2002, I was contacted by a graduate student who was studying the Hopi language at the University of Arizona after studying Sanskrit for some years. He had also spent time with the Secoya, the tribe I had visited in the Ecuadorian Amazon. We spoke several times over the next year. He told me about his own visionary episodes related to the Hopi prophecies. While camping out at Chaco Canyon—a vast desert valley with high rock promontories and enormous earthwork structures, built by the Anasazi, direct ancestors of the Hopi, over a thousand years ago—he had felt compelled to climb up late one night to the top of a nearly sheer cliff carved with ancient petroglyphs. This is a difficult feat, as well as a highly illegal one. A blue light had flooded him from above, and he had received a transmission of the dance and hand gestures of the "Blue Star Kachina." Kachinas are the nature spirits of the Hopi, thought to control elemental forces, and in their ceremonies, Hopi men dress up as kachinas and perform dances, seeking their benevolent influence. The return of Saquasohuh, the Blue Star Kachina, is a penultimate event in the Hopi foretellings. When Saquasohuh dances in the plaza during a ceremony and removes his mask, it will signify the end of the Fourth World and the emergence into the Fifth.

Noting it was a "New Age wet dream" to be part of the Hopi prophecies, the graduate student—an earnest, blond-bearded young man who offered to take me to visit the Hopi—had not told any of his indigenous contacts about this experience. After centuries of exploitation and theft, the Hopi carefully guard their cultural patrimony and ritual knowledge from the prying attention of outsiders. It was an active debate among the tribe whether non-Hopi should even be allowed to learn their language. Driving around Phoenix, his hometown, the graduate student told me some of what he had gleaned about the Hopi, corroborating bits of knowledge I had heard but not fully believed. While Hopiland and Tibet are on opposite sides of the Earth, these two sacred cultures situated on high and dry plateaus have a few intriguing correspondences. Certain words in the Hopi and Tibetan language have reversed meanings—for instance, the Hopi word for day, "Nyma," is the Tibetan word for night. The word for Moon in Tibetan, "Dawa," is the Hopi word for Sun. One of the most important Tibetan lamas, the Karmapa, visited Hopiland, and the graduate student decided to embark on his Hopi studies after attending the meeting.

At my new friend's house, I scanned books on Hopi mythology, and learned about Maasaw, their ambiguous protector-deity. When the Hopi emerged into the Fourth World, they met Maasaw, who gave them the rules of conduct for life on this new land and introduced them to the rudiments of their agricultural system. Maasaw brought the Sun into the Fourth World, but once he had accomplished this, he left the daylight forever to haunt the realms of night and darkness. The name Maasaw literally means "corpse demon" or "death spirit," and he is considered the ruler of the land of the dead. According to *Book of the Hopi,* he had been "appointed head caretaker of the Third World, but becoming a little self-important, he had lost his humility before the Creator. Being a spirit, he could not die, so Taiowa . . . decided to give him another chance." Since his exile from the Earth, Maasaw often appears in nightmares of the Hopi as a terrifying presence, wearing a ghoulish mask.

From these descriptions, I suspected this was the spirit who had introduced himself in my dreams. I found it significant that Maasaw had identified himself with Nauman, an artist who lives in New Mexico, and whose work has a compulsive, repetitive quality. Nauman's videos feature insane clowns on tape loops, caught in impossible situations or weeping while sitting on the toilet. Other works include videos of screaming heads, and plastic sculptures of twisted, hanging deer limbs. I had never understood Nauman's cachet in the art world, but now it seemed to me that he was unconsciously projecting this archetype of the Hopi's caretaker and corpse-daimon, revealing Maasaw's workings in our contemporary world—an anal-sadistic, almost autistic undertone or trapped vibration reflected in the mausoleum-like malls, plastic wraps, and air-conditioned monoliths we had imposed across the continent. Maasaw seemed a masculine variation on the Kali-Shakti archetype, a powerful spiritual energy that could be helpful or malevolent, depending on people's actions and attitudes toward him.

My student friend chose a long, circuitous route, stopping to visit an acquaintance of his, Judith Moore, a self-proclaimed "thirteen dimensional channel," outside Santa Fe, and then Chaco Canyon, on the way to Hopiland. Moore had connections to the Hopi elders, and had also published a book of channeled material on the crop circles, purportedly from "Laiolin," "a member of the Council of Abborah and keeper of the Records of Ra." While I had come to feel at home with ideas that most would consider fantastical, Moore's mystical shtick was too much even for me. Representing the fruitloop fringe of the New Age, she claimed she was an "Arcturian ambassador," receiving transmissions from an Arcturian mothership located somewhere around the Big Dipper. We seemed to develop an immediate aversion for each other.

After a night sleeping on her outdoor trampoline, I awoke to find this large woman in a tizzy, complaining of a frightening nightmare in which a figure of black slate had hovered over her bed. She thought this malevolent figure was sent through me, and wanted us to leave right away. She suggested I might have been a subject in the Montauk Project, which was, according to her, a malevolent psychic experiment conducted by the U.S. military in the 1970s, using the "fear vibrations" emanated by sexually abused children to tear holes into other dimensions. I assured her I had no part in this paranoid fantasy, and told her I didn't appreciate her negative projections onto me. This seemed to clear the air between us. Our visit was not without synchronicity. Before we left Moore's chaotic enclave, I found out that her uncle was Tony Shearer, a poet whose 1971 *Lord of the Dawn* introduced the Mesoamerican prophecies of Quetzalcoatl and the cycles of the Mayan calendar to the modern West. Shearer's interpretation of the Tzolkin and Mayan prophecy inspired José Argüelles, among others, who used Shearer's ideas for the "Harmonic Convergence" of August 16, 1987. After some prodding, I also managed to get from her the name of the Hopi elder, lineage holder, and seer she had met at a festival, Martin Gasheseoma, who lived on Third Mesa, in the town of Oraibi.

As we drove onward through desert wastes, my graduate student friend became increasingly nervous and tense. He distrusted my intentions, thinking I expected some magical episode to immediately transpire. I tried to reassure him that I was not attached to any particular outcome, as long as I had the chance to talk to the Hopi and see one of their ceremonies. We spent a few hours in the extraordinary wind-sculpted cliff-scape of Chaco Canyon, wandering around enormous ruins—round structures with rabbit warrens of rooms that had initiatory purposes—and peering at petroglyphs carved on the cliffsides. Pueblo Bonito, the largest ruin, covering three acres, contains thirty-two kivas and eight hundred rooms, housing up to twelve hundred people at one time. Like the archaic observatory of Stonehenge, the structures built within the canyon are precisely aligned to the orbits of Sun, Moon, and constellations. Like the temple-cities of the Maya, Chaco Canyon had been abandoned for unknown reasons—my friend conjectured that some large-scale magical experiment had backfired dangerously, as archaeological evidence suggested the settlements were hurriedly abandoned. Chaco established a direct link between the Hopi and the Mesoamerican civilizations, as sculptures of plumed serpents were found in chambers filled in with rubble when the Anasazi suddenly split the scene. Hopi oral traditions tell of clans migrating from Palatkwapi, a mysterious Red City of the South, that may have been a Toltec or Mayan center.

Slowed by a flat tire, we did not arrive in Hopiland until late Saturday night, and slept in my friend's car. When I awoke the next morning, he told me he would have to leave immediately—his fiancée, apparently, was having a crisis. Although the crisis did not sound urgent to me, he was adamant he had to hurry back. I suspected he was insecure about my purpose, and perhaps his own as well—simultaneously embarrassed and spooked by his personal connection to the prophecies, if his Blue Star Kachina vision had validity.

Nested within the much larger Navajo reservation, Hopiland is spread across three rocky mesas, with many miles of boulder-strewn desert in between each enclave, connected by one narrow highway. I didn't see how I could get around without a car, but having made it this far, I felt I had no choice but to stay. Furious at the graduate student for abandoning me, I took a room in the local hotel and cultural center. At the front desk, I learned there was a kachina dance on Walpi, a town on the First Mesa, that day. I got directions and began to walk there in the desert heat.

On foot, I had the opportunity to appreciate the landscape as the twelve thousand Hopi have known it for many generations. "Men have had to walk ten miles each day to tend their little patches of squaw corn," Waters wrote. "Women have trudged interminably up the steep cliff-sides with jars of water on their heads." Hopiland resembled the reddish surface of an alien planet, with huge anthropomorphic boulders and the jutting promontories of the mesas on which they built their settlements. Traditionally, the Hopi are subsistence farmers; they work with ancient strains of corn and bean that are, almost miraculously, able to grow in that arid environment. Farming is, for them, a religious duty. According to their legends, when the Creator gave out plants for different peoples to cultivate at the beginning of this world, the Hopi waited until the end, when the only plants remaining were the smallest and most humble crops. They accepted these gratefully. For the Hopi, like the Maya, corn is sacred, and they use their traditional cornmeal, throwing it in the air as a blessing, in their ceremonies.

A battered truck swung in front of me and came to a stop, and a friendly Hopi man offered me a ride up to Walpi. On the way, he talked about education and local politics, criticizing the Tribal Council that was imposed on the Hopi by the U.S. government in the 1960s. At Walpi, I watched fifty or so kachina dancers wearing feathers, brightly painted masks, and costumes emerge from the round kiva—a traditional ceremonial building accessible through a rooftop opening—and begin the daylong ceremony, dancing in the central courtyard, throwing out food—from soda cans to traditional cakes—to the crowd. The purpose of this ritual, as of so many Hopi rituals, was to bring rain; water is sacred to their culture—each spring, each well, is precious. Townspeople and visitors gathered to watch them, setting up rows of seats and sitting on roofs around the plaza. Besides myself, there were only a handful of tourists. The dance reminded me, startlingly, of Tibetan ceremonies I had seen in Kathmandu, during Tibetan New Year—I wondered at the mysterious correspondences between these far-flung cultures. During breaks, the local Hopi opened their houses, inviting everyone, including strangers, to join them in a meal. I accepted the invitation of a Hopi woman, and ate bread and soup with her family and other guests in her main room, so bare by our standards. As I had noted with every indigenous culture I had visited, tribal people have almost no regard for personal space as Westerners conceive it. Space is something to be communally shared, not individually adorned.

It is quite possible that the Hopi, like the Australian Aboriginals, chose to live in such a difficult environment, as it forced them to fine-tune their relationship to elemental forces through initiatory practices. In his book *Rethinking Hopi Ethnography,* Cambridge anthropologist Peter Whiteley recalled, with an almost embarrassed reluctance, that during his time with the Hopi in the 1980s, he witnessed repeated demonstrations of their ability to influence natural forces through ritual, as well as their telepathic or precognitive abilities. He was transfixed by his first visit to a Snake Dance in 1980: "This was no commodified spectacle of the exotic . . . its profound religiosity was tangible, sensible. Within half an hour of the dance (which lasts about forty-five minutes), a soft rain began to fall from a sky that had been burningly cloudless throughout the day." When he went to see one of his informants among the elders, he would often find that the man would answer the questions he had intended to ask before he could vocalize them: "I have no desire to fetishize or exoticize here, but this was something about him and some other, particularly older, Hopis that I have experienced repeatedly and am unable to explain rationally."

From Whiteley's book, I also discovered that the Hopi way of life is threatened with imminent extinction. In the 1960s, the Peabody Coal Company was given a concession to mine coal on their land. They were also awarded the right to use water from the aquifer under Black Mesa to slurry the coal down a pipeline, built by the Enron Corporation. This operation extracts 1.3 billion gallons of pure drinking water annually from the aquifer that sustains Hopi life. There are, of course, other ways to transport coal, but this is the cheapest for Peabody, and the company has continually fought against and effectively delayed efforts to change its practices. In the 1980s, it was discovered that the lawyer who negotiated the original deal for the Hopi was, at the same time, on the payroll of the Peabody Corporation—and the Hopi have received a tiny fraction of the revenue they deserve from the concession, while forfeiting control of their own destiny. According to U.S. government Geological Surveys, by the year 2011, the aquifer will be almost completely depleted—already the Hopi are finding that the local springs on which they rely are drying up.

At sunset, I went in search of Martin Gasheseoma, hiking toward Third Mesa along the narrow highway until a car picked me up. After some searching, I found the elder's house, but he was not at home—a nephew told me he was at another kachina ceremony, one closed to outsiders. Night had fallen as I headed back toward the hotel, sadly thinking that I would have to leave the next day without meeting him, taking the only available bus to get to the airport in Phoenix. Few cars were driving now, and those that were did not stop to pick me up. The highway wound around rocky cliffsides, with almost no shoulder for me to wait out the speeding vehicles that passed by. I had about fifteen miles to go before reaching the hotel, and I realized I would probably walk through the night.

As I trekked the desert road, I realized it no longer mattered to me whether I received a direct transmission of the Hopi prophecies. I had little desire to intrude on their traditional knowledge, as so many had done before me. I felt I had attained my own understanding of the nature of prophecy. I understood that a spectrum of possibilities was indicated by the esoteric foretellings of the Hopi and Maya, and it was up to us to determine the outcome, for ourselves and our world. The cataclysms, polar shifts, and "earth changes" that alarmed writers like Graham Hancock, who concluded in *Fingerprints of the Gods* that the Hopi and Mayan prophecies were all that remained of an advanced civilization's foreknowledge of a literal end of the world, were a negative projection of our own fears and limitations, and a passive vision. Hancock, like many others, could not perceive Apocalypse—the end of the Hopi's Fourth World—as an inner as well as outer event, a process of psychic transformation, restoring a direct and sacred relationship to reality.

In the middle-class New Age culture and "New Edge" festivals such as Burning Man, much lip service is paid to Native American traditions. Perhaps millions of white people hang dream catchers over their beds and put kachina dolls on their shelves. Despite this sentimental interest in indigenous culture and spirituality, precious little, almost nothing, is done by us—those of us with the leisure for yoga and raw food and sweat lodges, who often consider ourselves especially "conscious" or "spiritual" beings—to repair relations with the Native Americans on this continent. The indigenous people are resettled next to toxic waste dumps, abandoned to the least arable lands, ignored when the fish in their rivers are poisoned, when their resources are robbed from them. They continue to be treated with contempt and disregard.

As climate change accelerates along with the global depletion of resources, we are being forced to recognize that our current system is unsustainable, even in the short term. The Hopi situation provides a microcosm of the global crisis—a cruelly ironic situation considering the essential meaning of their culture. As Whiteley notes, "The phrase 'Hopi environmentalism' is practically a redundancy. So much of Hopi culture and thought, both religious and secular, revolves around an attention to balance and harmony in the forces of nature that environmental ethics are in many ways critical to the very meaning of the word 'Hopi.'" Indigenous prophecy, in itself, arises out of a profound attunement to natural cycles, rather than anything "spiritual" or immaterial.

According to Victor Masayesva, a Hopi filmmaker and executive director of the Black Mesa Trust, "It is our water ethic that has allowed us to survive and thrive in one of the most arid areas on planet Earth. It is the knowledge and teachings of our elders that have sustained us. This water ethic that has been handed down to us by our ancestors we are eager to share with everyone who will be facing water shortages—and according to some studies, water wars—in the next few decades. When the water is gone from Black Mesa, so will be the traditional cultures that could have taught the world so much about living successfully with less." The Hopi prophecies tell of the return of Pahana, the elder white brother, in a real exchange of knowledge and a true communion, as the Fourth World comes to an end. Such an exchange could only take place without the slightest taint of paternalism or condescension. As I walked, it occurred to me that we have as much to learn from tribal people like the Hopi—about sustainability, initiatory ritual, and nonhierarchical social organization based on trust and telepathy—as they might have from us, if we stopped trying to force our worldview upon them.

After several hours, a rare car passed and actually stopped for me. Inside was an entire Hopi family, several kids and a couple, and they laughed merrily as I climbed inside. I asked what they were doing out so late, and they said, "Just looking around," and giggled. They seemed to find it hilarious that I was out on that road. I told them how I ended up there. The father said, "If you have the right intentions, anything is possible." This is a basic tenet of Hopi wisdom. Earlier in Phoenix, the graduate student had brought me to meet his teacher, Mark Emery, a tribal leader and professor of the Hopi language. Emery had told me that, for the Hopi, "The person who has the greatest power is the one who has a perfect heart," and even a child, if he could realize that power, could do great things.

Resting my sore legs at breakfast the next morning, I overheard a woman say she was driving to the Phoenix Airport that day. I introduced myself, and she agreed to drive me. Since we had some time to kill, I suggested we go in search of Gasheseoma, and she eagerly agreed. We drove to his house, where another relative told us he was already out working on his land, giving us convoluted directions on how to find him. We drove down the canyon and past several buttes, through utterly dry desert terrain, until we saw the gnarled old man in the distance, hoeing his field under the scorching Sun. He seemed an iconic figure, standing alone in the arid landscape, surrounded by stubbled rows of tiny plants of corn and bean. I called out to him, asking if he would like to talk to us. He put his hoe down and came to greet us in a manner that was friendly, unguarded, and without surprise. We sat together in his little shelter, and I asked him about the time ahead.

"It goes like a movie now," he said. "Soon there is coming the time of purification. But this has all happened before." He believed the U.S. government was already building "machines" for exterminating huge populations, which would be employed as resources dwindled. These machines were being built in the South. This would be part of the "purification." His tone was stoic, serious, yet surprisingly matter-of-fact. I asked him about treatment of the Hopi by the United States.

"Everything has been done illegally," he said. "We didn't want a Tribal Council. The majority voted against it, and it was forced on us. We didn't want the highway either."

"What should be done with it then?" I asked.

"Smash it up," he replied, waving toward it.

For Gasheseoma, there was an obvious distinction between "right-doing" and "wrongdoing." He was quite clear about what should be done about persistent "wrongdoers," such as the executives of the Peabody Corporation. "Cut off their heads," he said.

As we left him, he returned to his field—a tiny, indomitable presence in the vast desert, beneath the wide, scorching sky. I recalled Maasaw's mocking laughter and deviant presence in my dreams. I felt the deep schism of the soul that still needs to be addressed, a wound that will only be healed when our culture forges a real relationship with the indigenous people of this continent, no longer prying into their secrets or imitating their ways, but expiating our dominator guilt and denial by acting in solidarity with them.

ALTHOUGH WE DO NOT KNOW how the Classic Maya interpreted the prophetic culmination of their Long Count, Carl Johan Calleman has elaborated a meticulously exact fractal model of time from the Mayan calendar, in which human consciousness evolves through nine "Underworlds," each twenty times faster in linear time than the previous one. Each underworld passes through a thirteen-stage process of alternating light and dark energy currents—or seven "days" and six "nights" of creation—culminating in a new level of realization. According to Calleman's model, sometime around the year 2008—the "fifth night" of the current underworld, ruled by the energies of Tezcatlipoca, the jaguar god of night and black magic—our current socioeconomic system will suffer a drastic and irrevocable collapse. The primary catalyst for this collapse could be an approaching energy crisis. A report recently issued by the U.S. Department of Energy states: "Previous energy transitions (wood to coal and coal to oil) were gradual and evolutionary; oil peaking will be abrupt and revolutionary." According to a 2005 feature in *The New York Times Magazine:* "The consequences of an actual shortfall of supply would be immense." As Peter Maass reported, "When a crisis comes—whether in a year or two or ten—it will be all the more painful because we will have done little or nothing to prepare for it."

The effects of such a crisis could be exponentially magnified by the side effects of accelerated climate change, resource depletion, military conflicts, and declining food production. According to a 2004 report issued by Pentagon scenario planners, if climate change were to intensify, "major

European cities will be sunk beneath rising seas as Britain is plunged into a 'Siberian' climate by 2020. Nuclear conflict, mega-droughts, famine and widespread rioting will erupt across the world." The analysts concluded, "Disruption and conflict will be endemic features of life. Once again, warfare would define human life." Ignoring the possibility that such a crisis could lead to a compassionate readaptation of human life and society, the Pentagon report may indicate the type of contingency planning currently under way in military circles.

Whatever the Pentagon prophesizes, ecological feedback loops are currently accelerating the process of climate change beyond earlier predictions. Reseachers from Oxford University in the UK and Tomsk State University in Russia have recently discovered that a vast frozen peat bog in western Siberia, "the size of France and Germany combined," has started to thaw, potentially releasing "billions of tons of methane, a greenhouse gas twenty times more potent than carbon dioxide, into the atmosphere." As reported in *The Guardian*, this new discovery could cause a "10 percent to 25 percent increase in global warming," accelerating current forecasts. The increasing rate of ice melting in the Arctic Circle is forcing a similar reevalution of data. According to a new study by the American Geophysical Union: "Warming in the Arctic is stimulating the growth of vegetation and could affect the delicate energy balance there, causing an additional climate warming of several degrees over the next few decades."

I have proposed that this intensifying global crisis is the material expression of a psycho-spiritual process, forcing our transition to a new and more intensified state of awareness. If Calleman's hypothesis is correct, this telescoping of time will mean a high-speed replay of aspects of past historical epochs—echoes of the French Revolution, the rise and fall of the Third Reich, and so on—before consciousness reaches the next twist of the spiral. As part of this transition, we will reintegrate the aboriginal and mythic worldviews, recognizing the essential importance of spiritual evolution, while understanding that this evolution is directly founded upon our relationship to material and physical aspects of reality. The higher consciousness and conscience of our species will be forged through the process of putting the broken and intricate shards of our world back together, piece by piece.

Epochs of radical transformations have mythological and archetypal dimensions—and we may be closer to such a stage than most of us can imagine. Before the French Revolution, the Enlightenment *philosophes,* pamphleteers, and cafe intellectuals of the ancien régime had little clue that they might end up the vanguard of a new social order. After all, before 1989, how many analysts predicted the sudden and astonishingly peaceful fall of the Berlin Wall? Is it conceivable that Wall Street could collapse as suddenly? While millions pin their future and their ambitions on the stock markets, the system has become delinked from any tangible resource, supported solely by mass belief in it. If that belief were to fail, our system would go into freefall. Although it would be a tumultuous transition, an economic collapse might be bracing as well as clarifying, leading to a sea change in priorities and values, and a concomitant change of the elites.

Right now, in this interim period, we have the opportunity to develop alternative support systems—localized organic food production, alternative energy, conflict-resolution projects, complementary currencies, and so on—that can be applied on an increasingly large scale as the old structures continue to give way. The federal government's response to the flooding of New Orleans should serve as a warning that the support structures upon which we have relied are dangerously corrupt and no longer dependable. When a major crisis comes, local communities organized on a basis of trust and self-reliance may find themselves at a distinct advantage.

According to Hopi prophecy, at the end of the Fourth World, the elder white brother, Pahana, will return in a true exchange of hearts, as well as knowledge. From an entirely unpatronizing perspective, we may have quite a lot to learn from the Hopi and other traditional cultures. The philosopher Alan Watts noted that the materialism of modern civilization is paradoxically founded on a "hatred of materiality," a goal-oriented desire to obliterate all natural limits through technology, imposing an abstract grid over nature. The spirituality of tribal people is rooted in a deep and unsentimental connection to the Earth, expressed through a careful attentiveness to and reverence for particular plants, geographic features, and local differences. We can regain a humble reverence for the physical world from indigenous cultures, integrating nature and spirit, if we so choose.

Based on his years of fieldwork, anthropologist Peter Whiteley noted that Hopi ceremonies sometimes seem to have a direct effect on natural forces. If this is the case, it would mean that focused psychic energy can be used to alter climatic conditions—a potentially critical proposition in a time of accelerating climate shift. By attaining knowledge of the interrelatedness of mind and world, as the Hopi and other ancient indigenous cultures understand it, we might be able to concentrate collective psychic energy for planetary transformation and healing. At the same time, the subsistence agriculture of the Hopi—and other edge-dwelling people—capable of sustaining themselves from food grown in extreme climate conditions, may prove to be of more than academic interest for our own near-term survival.

If we are graduating from nation-states to a noospheric state, we may find ourselves exploring the kind of nonhierarchical social organization—a "synchronic order" based on trust and telepathy—that the Hopi and other aboriginal groups have used for millennia. If a global civilization can self-organize from our current chaos, it will be founded on cooperation rather than winner-take-all competition, sufficiency rather than surfeit, communal solidarity rather than individual elitism, reasserting the sacred nature of all earthly life. Those who desire such a world will work to create it.

As the Hopi also say: We are the ones we have been waiting for.

AFTERWORD

to the Paperback Edition of *2012*

In *2012: The Return of Quetzalcoatl,* I propose that ancient myths and archetypes may take on a new relevance as we approach the end of the Mayan calendar's Great Cycle. An underlying theme of Mayan legends such as the *Popol Vuh* is the struggle of gods and heroes to escape Xibalba, the Underworld. This feat can be accomplished only through prodigious applications of ingenuity and will. In Mesoamerican myth, the hero twins as well as the creator deity, Quetzalcoatl, use genius, magic, and trickery to overcome the dark powers of the Lords of the Underworld—immortal spirits of ruin and destruction who are also inveterate cheaters—to find their way back to the surface realm. The narrative of the *Popol Vuh* leads from mythic beginnings—the origin of the word and the world—to the Ball Court of Xibalba, and then follows the chronicles of the Quiche Maya, before ending in the historical period of the Conquest.

Since the release of *2012*, it has occurred to me that the narrative of the *Popol Vuh,* interrupted five hundred years ago by colonialism, may have picked up in our own time. Today, we confront the modern Xibalba of ecological meltdown, military destruction, and economic inequity—a global system built on ignorance and greed. The contemporary Lords of the Underworld manifest their powers in a myriad of ways. On a supersensible level, they could be the "principalities and powers" defined by the visionary philosopher Rudolf Steiner as Lucifer, who seeks to pull us away from the Earth into hubris and disengaged fantasy, and Ahriman, who drags us down into insatiable materialism, soulless technology, and entrophy. According to Steiner, these forces cannot be suppressed or avoided—they work upon us all the time, and we evolve by learning to balance between them. The Ahrimanic emphasis of our time reveals itself through the blockages in our cultural imagination, such as the intense force of repression and ridicule exerted on new ideas that might offer us hope, or the possibility of redemption.

As a journalist, I knew the habits and tactics of the culture industry from the inside. Therefore, it was not surprising that the mainstream media reacted, at first, by attacking me personally, while trashing my book and distorting my ideas. My vision for the future is entirely egalitarian. I argue throughout *2012* that we have the capacity to create a compassionate planetary civilization that will care for everyone, if we can enact a change in consciousness. I was, therefore, saddened to read lines like the following from *Rolling Stone:* "But according to Pinchbeck, not everyone will be saved in 2012—only the psychedelic elite and those who have reached a kind of supramental consciousness will make it through the bottleneck at the end of time." Or this from the *New York Times Book Review:* "In 2012, urban liberals and fundamentalist Christians alike lose their heads to the Pinchbeckian guillotine." Such statements reflect the fears and projections of the writers. Nothing could be farther from my ideas, or my desires.

I included material that was painfully personal in *2012,* laying bare my psychic wounds. I realized that, by doing this, I opened myself to easy attacks. However, I felt it was the only way my book could evoke a deep enough response in readers that it might incite a shift in perspective. Despite the negative press, the book began to find its audience—those with an open mind to new ideas, as well as those who were prepped through their own private studies of Gnosticism, yoga, Mayan archaeology, and so on. As public interest grew, the media began to respond more favorably.

The success of *2012* has coincided with—and helped to inspire—a renewed curiosity about many subjects rejected by the mainstream over the last thirty years. While this seems astonishing to some people, it does not surprise me. Unbeknownst to the cultural gatekeepers, we have embarked upon a new phase of the initiatory journey begun in the 1960s—with the opportunity to avoid the tactical mistakes, strident statements, and polarizations of the past. When consciousness shifts, it can happen suddenly, and in unexpected ways.

Meanwhile, the weather gets stranger, and the headlines of the newspaper more grim. As we lurch toward ecological and military catastrophe, we cannot dismiss the possibility of a collapse of the current system, whether through economic crash, environmental crisis, or terrorist disruption. While such a fall would be frightening, it could also speed the spread of new models for a sustainable planetary culture, with a different relationship to nature and technology. Considering current rates of resource depletion and species extinction, our own survival may depend upon the institution of a new system that allocates global resources rationally and compassionately, while recognizing the reality of psyche and spirit.

Like many of you, perhaps, I oscillate between my awareness of personal responsibility and agency and my sense of helplessness before the larger historical and evolutionary forces that have been unleashed, while wrestling with my own limitations. The "curious literary quality" of existence that Terence McKenna noticed seems ever more tangible, as this story moves through us. Can an intensified human consciousness integrate its shadow projections and rise to the occasion, healing the pain of people and planet, or will the Lords of the Underworld win this round? Nobody knows for sure—but the celestial court has assembled, and the ball game must be played.

ACKNOWLEDGMENTS

Thanks to:

Allan Brown, Michael Brownstein, Michael Glickman, Graham Hancock, Laura Hoffmann, Mitch Horowitz, Gerald Howard, Joyce Johnson, Ken Jordan, John Martineau, Neil Martinson, Eric Simonoff, Geoff Stray, Elizabeth Thompson, the staff of DrinkMe Cafe, and the community on the www.breakingopenthehead.com discussion forum.

BIBLIOGRAPHY

Alverga, Alexis Polari de. *Forest of Visions.* Park Street Press, 1999.
Argüelles, José. *The Call of Pakal Votan.* Altea, 1996.
———. *Earth Ascending.* Shambhala, 1984.
———. *The Mayan Factor.* Bear & Company, 1987.
———. *Time & the Technosphere.* Bear & Company, 2002.
Ashe, Geoffrey. *King Arthur's Avalon.* Fontana, 1957.
Ayres, Ed. *God's Last Offer.* Four Walls, Eight Windows, 1999.
Baudrillard, Jean. *The Spirit of Terrorism.* Verso, 2002.
Benjamin, Walter. *Illuminations.* Harcourt, Brace & World, 1968.
Bennett, John G. *The Masters of Wisdom.* Bennett Books, 1995.
———. *Gurdjieff: Making a New World.* Harper & Row, 1973.
Blake, William. *The Collected Poems.* Penguin Books, 1977.
Brennan, Martin. *The Hidden Maya.* Bear & Company, 1998.
Brownstein, Michael. *World on Fire.* Open City Books, 2002.
Calleman, Carl Johan. *The Mayan Calendar and the Transformation of Consciousness.* Bear & Company, 2004.
Capra, Fritjof. *The Tao of Physics.* Shambhala Publications, 1976.
Castaneda, Carlos. *The Art of Dreaming.* Harper Perennial, 1994.
Clarke, Arthur C. *Childhood's End.* Sidgwick & Jackman Ltd., 1954.
Cleary, Thomas. *The Taoist I Ching.* Shambhala, 1986.
Coe, Michael D. *The Maya.* Thames & Hudson, 1999.
Cohn, Norman. *The Pursuit of the Millennium.* Paladin Books, 1970.
Crowley, Aleister. *The Book of the Law.* Weiser Books, 1976.
———. *Magick in Theory and Practice.* Castle Books, 1991.
Davis, Beth, ed. *Ciphers in the Crops.* Gateway Books, 1992.
Denzler, Brenda. *The Lure of the Edge.* University of California, 2001.
Diamond, Jared. *Guns, Germs, and Steel.* Norton, 1997.
———. *Collapse.* Viking Penguin, 2004.
Díaz, Frank. *The Gospel of the Toltecs.* Bear & Company, 2002.
Doniger O'Flaherty, Wendy, trans. *The Rig Veda.* Penguin, 1981.
Duffeyes, Kenneth S. *Hubbert's Peak.* Princeton University Press, 2001.
Easton, Stewart C. *Rudolf Steiner: Herald of a New Epoch.* Anthroposophic Press, 1980.
Edinger, Edward F. *Archetype of the Apocalypse.* Open Court, 1999.
———. *The Bible and the Psyche.* Inner City Books, 1986.

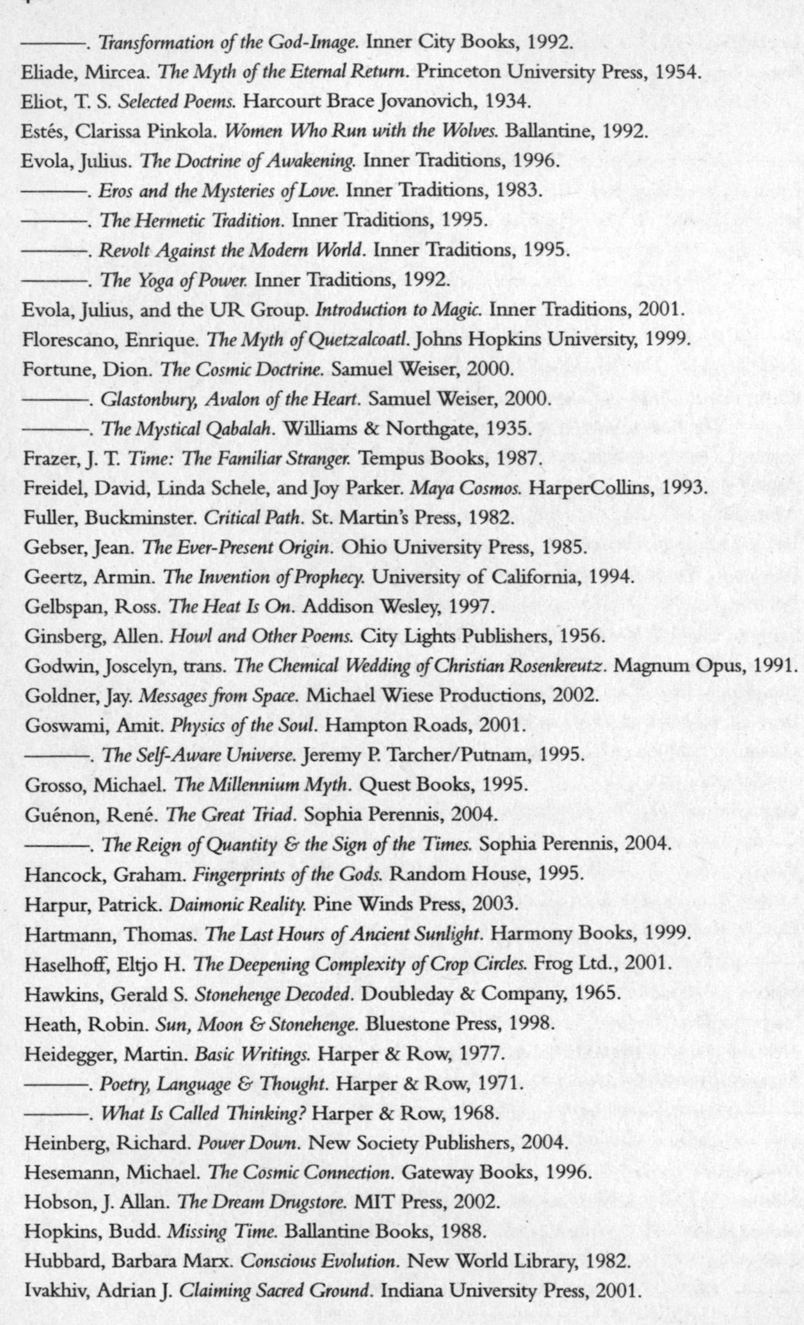

———. *Transformation of the God-Image.* Inner City Books, 1992.

Eliade, Mircea. *The Myth of the Eternal Return.* Princeton University Press, 1954.

Eliot, T. S. *Selected Poems.* Harcourt Brace Jovanovich, 1934.

Estés, Clarissa Pinkola. *Women Who Run with the Wolves.* Ballantine, 1992.

Evola, Julius. *The Doctrine of Awakening.* Inner Traditions, 1996.

———. *Eros and the Mysteries of Love.* Inner Traditions, 1983.

———. *The Hermetic Tradition.* Inner Traditions, 1995.

———. *Revolt Against the Modern World.* Inner Traditions, 1995.

———. *The Yoga of Power.* Inner Traditions, 1992.

Evola, Julius, and the UR Group. *Introduction to Magic.* Inner Traditions, 2001.

Florescano, Enrique. *The Myth of Quetzalcoatl.* Johns Hopkins University, 1999.

Fortune, Dion. *The Cosmic Doctrine.* Samuel Weiser, 2000.

———. *Glastonbury, Avalon of the Heart.* Samuel Weiser, 2000.

———. *The Mystical Qabalah.* Williams & Northgate, 1935.

Frazer, J. T. *Time: The Familiar Stranger.* Tempus Books, 1987.

Freidel, David, Linda Schele, and Joy Parker. *Maya Cosmos.* HarperCollins, 1993.

Fuller, Buckminster. *Critical Path.* St. Martin's Press, 1982.

Gebser, Jean. *The Ever-Present Origin.* Ohio University Press, 1985.

Geertz, Armin. *The Invention of Prophecy.* University of California, 1994.

Gelbspan, Ross. *The Heat Is On.* Addison Wesley, 1997.

Ginsberg, Allen. *Howl and Other Poems.* City Lights Publishers, 1956.

Godwin, Joscelyn, trans. *The Chemical Wedding of Christian Rosenkreutz.* Magnum Opus, 1991.

Goldner, Jay. *Messages from Space.* Michael Wiese Productions, 2002.

Goswami, Amit. *Physics of the Soul.* Hampton Roads, 2001.

———. *The Self-Aware Universe.* Jeremy P. Tarcher/Putnam, 1995.

Grosso, Michael. *The Millennium Myth.* Quest Books, 1995.

Guénon, René. *The Great Triad.* Sophia Perennis, 2004.

———. *The Reign of Quantity & the Sign of the Times.* Sophia Perennis, 2004.

Hancock, Graham. *Fingerprints of the Gods.* Random House, 1995.

Harpur, Patrick. *Daimonic Reality.* Pine Winds Press, 2003.

Hartmann, Thomas. *The Last Hours of Ancient Sunlight.* Harmony Books, 1999.

Haselhoff, Eltjo H. *The Deepening Complexity of Crop Circles.* Frog Ltd., 2001.

Hawkins, Gerald S. *Stonehenge Decoded.* Doubleday & Company, 1965.

Heath, Robin. *Sun, Moon & Stonehenge.* Bluestone Press, 1998.

Heidegger, Martin. *Basic Writings.* Harper & Row, 1977.

———. *Poetry, Language & Thought.* Harper & Row, 1971.

———. *What Is Called Thinking?* Harper & Row, 1968.

Heinberg, Richard. *Power Down.* New Society Publishers, 2004.

Hesemann, Michael. *The Cosmic Connection.* Gateway Books, 1996.

Hobson, J. Allan. *The Dream Drugstore.* MIT Press, 2002.

Hopkins, Budd. *Missing Time.* Ballantine Books, 1988.

Hubbard, Barbara Marx. *Conscious Evolution.* New World Library, 1982.

Ivakhiv, Adrian J. *Claiming Sacred Ground.* Indiana University Press, 2001.

Jacobs, David M. *The Threat*. Simon & Schuster, 1998.

Jaynes, Julian. *The Origin of Consciousness in the Breakdown of the Bicameral Mind*. Houghton Mifflin, 1976.

Jenkins, John Major. *Galactic Alignment*. Bear & Company, 2002.

———. *Maya Cosmogenesis 2012*. Bear & Company, 1998.

Jenkins, John Major, and Martin Matz. *Pyramid of Fire*. Bear & Company, 2004.

Johnson, Robert A. *The Fisher King and the Handless Maiden*. HarperCollins, 1993.

Jung, Carl. *The Archetypes and the Collective Unconscious*. Princeton University Press, 1959.

———. *Memories, Dreams, Reflections*. Pantheon Books, 1963.

———. *The Portable Jung*. Viking Press, 1971.

———. *Psychology and Alchemy*. Princeton University Press, 1953.

Jyoti. *An Angel Called My Name*. DharmaGaia Publishing, 1998.

Kafka, Franz. *The Penal Colony*. Schocken, 1961.

———. *The Trial*. Schocken, 1968.

Kipnis, Laura. *Against Love*. Pantheon Books, 2003.

Knight, Gareth. *A Practical Guide to Qabalistic Symbolism*. Weiser Books, 2001.

Kühlewind, Georg. *Becoming Aware of the Logos*. Lindisfarne, 1985.

LaCarriere, Jacques. *The Gnostics*. City Lights Books, 1989.

Lachman, Gary. *A Secret History of Consciousness*. Lindisfarne, 2003.

Lao Tzu. *Tao Te Ching, The Definitive Edition*. Tarcher/Penguin, 2001.

Lawlor, Robert. *Sacred Geometry*. Thames & Hudson, 1982.

———. *Voices of the First Day*. Inner Traditions, 1991.

Loftin, John D. *Religion and Hopi Life*. Indiana University Press, 1991.

Marcuse, Herbert. *One-Dimensional Man*. Beacon Press, 1964.

Martineau, John. *A Little Book of Coincidence*. Wooden Books, 2001.

McKenna, Dennis, and Terence McKenna. *The Invisible Landscape*. Seabury, 1975.

McKenna, Terence. *The Archaic Revival*. HarperCollins, 1991.

———. *True Hallucinations*. HarperCollins, 1994.

Men, Humbatz. *Secrets of Mayan Science/Religion*. Bear & Company, 1990.

Michell, John. *The Dimensions of Paradise*. Adventures Unlimited Press, 2001.

———. *New Light on the Ancient Mystery of Glastonbury*. Gothic Image, 1990.

———. *The New View Over Atlantis*. Thames & Hudson, 1983.

Moore, Judith, and Barbara Lamb. *Crop Circles Revealed*. Light Technology, 2001.

Narby, Jeremy. *The Cosmic Serpent*. Tarcher/Putnam, 1998.

Neihardt, John G. *Black Elk Speaks*. William Morrow & Company, 1932.

Nietzsche, Friedrich. *Beyond Good & Evil*. Random House, 1966.

———. *Ecce Homo*. Viking Penguin, 1979.

———. *On the Genealogy of Morality*. Cambridge University Press, 1994.

Nisargadatta Maharaj, Sri. *I Am That*. Acorn Press, 1990.

Norbu, Chogyal Namkai. *Dzogchen, The Self-Perfected State*. Snow Lion, 1996.

Noyes, Ralph, ed. *The Crop Circle Enigma*. Gateway Books, 1990.

Ouspensky, P. D. *In Search of the Miraculous*. Harcourt Brace & Company, 1949.

———. *The Psychology of Man's Possible Evolution*. Alfred A. Knopf, 1954.

Pagels, Elaine. *Beyond Belief.* Vintage Books, 2001.

Patton, Phil. *Dreamland*. Villard Books, 1998.

Paz, Octavio. *The Labyrinth of Solitude.* Grove Press, 1985.

Peat, F. David. *Synchronicity.* Bantam Books, 1987.

Pringle, Lucy. *Crop Circles.* HarperCollins, 1999.

Pritchard, Evan T. *Native New Yorkers.* Council Oak Books, 2002.

Radin, Dean. *The Conscious Universe.* HarperCollins, 1997.

Rilke, Rainer Maria. *The Selected Poetry of Rainer Maria Rilke.* Random House, 1982.

———. *Sonnets to Orpheus.* University of California Press, 1960.

Robinson, James M., ed. *The Nag Hammadi Library.* HarperCollins, 1990.

Roney-Dougal, Serena. *Where Science and Magic Meet*. Vega, 2002.

Roy, Arundhati. *Power Politics.* South End Press, 2001.

Russell, Peter. *Waking Up in Time.* Origin Press, 1992.

Schnabel, Jim. *Round in Circles.* Prometheus Books, 1994.

Scholem, Gershom. *On the Kabbalah and Its Symbolism*. Schocken Books, 1965.

———. *On the Mystical Shape of the Godhead*. Schocken Books, 1991.

Schreck, Nikolas, and Zeena Shreck. *Demons of the Flesh*. Creationbooks, 2002.

Shearer, Tony. *Lord of the Dawn*. Naturegraph, 1971.

Sheldrake, Rupert. *The Presence of the Past.* Times Books, 1988.

Shenton, Rita. *Christopher Pinchbeck and His Family.* Brant Wright Associates, 1976.

Silva, Freddy. *Secrets in the Fields.* Hampton Roads, 2002.

Smolin, Lee. *The Life of the Cosmos.* Oxford University Press, 1997.

Steiner, Rudolf. *The Apocalypse of St. John.* Anthroposophic Press, 1958.

———. *Autobiography.* Anthroposophic Press, 1999.

———. *How to Know Higher Worlds.* Anthroposophic Press, 1994.

———. *Karmic Relationships (Vols. I–VI).* Rudolf Steiner Press, 1971.

———. *Man as Symphony of the Creative World*. Rudolf Steiner Press, 1970.

———. *An Outline of Esoteric Science.* Anthroposophic Press, 1997.

———. *The Philosophy of Freedom*. Rudolf Steiner Press, 1964.

———. *Self-Transformation*. Rudolf Steiner Press, 1980.

Stevenson, Ian. *Reincarnation and Biology.* Praeger Publishers, 1997.

Strassman, Rick. *DMT: The Spirit Molecule.* Park Street Press, 2001.

Strieber, Whitley. *Communion*. Avon Books, 1987.

Talbot, Michael. *The Holographic Universe.* HarperCollins, 1991.

———. *Mysticism and the New Physics*. Bantam Books, 1981.

Tedlock, Dennis. *Breath on the Mirror.* HarperCollins, 1993.

Teilhard de Chardin, Pierre. *The Appearance of Man*. Harper & Row, 1965.

———. *The Phenomenon of Man*. Harper & Brothers, 1959.

Temple, Robert. *The Sirius Mystery.* Random House, 1998.

Thomas, Andy. *Swirled Harvest*. Vital Signs Publishing, 2003.

Thompson, William Irwin. *At the Edge of History and Passages About Earth*. Lindisfarne Books, 1990.

———. *Blue Jade for the Morning Star.* Lindisfarne Books, 1983.

———. *Coming into Being.* St. Martin's Press, 1996.
———. *Imaginary Landscape.* St. Martin's Press, 1989.
———. *The Time Falling Bodies Take to Light.* St. Martin's Press, 1981.
Tompkins, Peter. *Mysteries of the Mexican Pyramids.* Harper & Row, 1976.
Twist, Lynn. *The Soul of Money.* W. W. Norton, 2003.
Vallee, Jacques. *Passport to Magonia.* Contemporary Books, 1993.
Wangyal Rinpoche, Tenzin. *The Tibetan Yogas of Dream and Sleep.* Snow Lion, 1998.
Washington, Peter. *Madame Blavatsky's Baboon.* Schocken Books, 1996.
Waters, Frank. *The Book of the Hopi.* Viking Press, 1963.
Watts, Alan W. *Nature, Man and Woman.* Pantheon Books, 1958.
Weber, Eugen. *Apocalypses.* Random House, 2000.
West, John Anthony. *Serpent in the Sky.* Quest Books, 1993.
Whiteley, Peter M. *Rethinking Hopi Ethnography.* Smithsonian Institution, 1998.
Wiedner, Jay, and Vincent Bridges. *The Mysteries of the Great Cross at Hendaye.* Destiny Books, 1999.
Wilber, Ken. *A Brief History of Everything.* Shambhala, 1996.
———. *The Eye of the Spirit.* Shambhala, 2001.
Wilson, Colin. *Aleister Crowley.* Aquarian Press, 1987.
Wilson, Robert Anton. *Cosmic Trigger Volume 1.* New Falcon Publications, 1991.
Woolf, Virginia. *A Room of One's Own.* Harvest Books, 1989.
Yates, Francis. *The Rosicrucian Enlightenment.* Routledge, 1972.
Yeats, William Butler. *The Collected Poems of W. B. Yeats.* Macmillan Publishing, 1956.

INDEX

ABOUT THE AUTHOR

DANIEL PINCHBECK is the author of *Breaking Open the Head: A Psychedelic Journey into the Heart of Contemporary Shamanism* (Broadway Books, 2002). His articles have appeared in *The New York Times Magazine, Esquire, Wired, The Village Voice, LA Weekly, ArtForum, Arthur,* and many other publications. He is currently a national columnist for *Conscious Choice* magazine (www.consciouschoice.com) and the editorial director of *Reality Sandwich* (www.realitysandwich.com). He lives in New York City.